LASERS AND THEIR APPLICATIONS IN PHYSICAL RESEARCH

LAZERY I IKH ISPOL'ZOVANIE V FIZICHESKIKH ISSLEDOVANIYAKH

ЛАЗЕРЫ И ИХ ИСПОЛЬЗОВАНИЕ В ФИЗИЧЕСКИХ ИССЛЕДОВАНИЯХ

The Lebedev Physics Institute Series

Editors: Academicians D. V. Skobel'tsyn and N. G. Basov

P. N. Lebedev Physics Institute, Academy of Sciences of the USSR

Recent Volumes in this Series

Proceedings (Trudy) of the P. N. Lebedev Physics Institute

Volume 91

Lasers and Their Applications in Physical Research

Edited by

N. G. Basov

P.N. Lebedev Physics Institute
Academy of Sciences of the USSR
Moscow, USSR

Translated from Russian by
Donald H. McNeill

**SPRINGER SCIENCE+
BUSINESS MEDIA, LLC**

Library of Congress Cataloging in Publication Data

Main entry under title:

Lasers and their applications in physical research.

(Proceedings (Trudy) of the P. N. Lebedev Physics Institute; v. 91)
Translation of Lazery i ikh ispol'zovanie v fizicheskikh issledovaniiakh.
Includes bibliographies.
1. Lasers – Addresses, essays, lectures. 2. Quantum electronics – Addresses, essays, lectures. I. Basov, Nikolaĭ Gennadievich, 1922-   II. Series: Akademiia nauk SSSR. Fizicheskiĭ institut. Proceedings; v. 91.
QC1.A4114 vol. 91 [QC689]   530'.08s [535.5'8]   78-13582
ISBN 978-1-4757-0012-1   ISBN 978-1-4757-0010-7 (eBook)
DOI 10.1007/978-1-4757-0010-7

The original Russian text was published by Nauka Press in Moscow in 1976 for the Academy of Sciences of the USSR as Volume 91 of the Proceedings of the P. N. Lebedev Physics Institute. This translation is published under an agreement with the Copyright Agency of the USSR (VAAP).

PREFACE

In this volume an approach is developed for describing the dynamic processes in lasers based on a spectral representation of the polarization of the material and the radiation field. Reviews are given of results on heterojunction lasers based on an entire series of new semiconductor solution systems and of results on the problem of controlling the spectral composition, directionality, and polarization of the output from heterolasers. Different materials are studied for use as active media in high-power pulsed Raman lasers. The state of theoretical and experimental work on laser ranging of the moon is discussed. It is shown that the basic characteristics of the earth—moon system can be determined from one to three orders of magnitude more accurately by using laser ranging measurements than by using other optical methods.

This anthology is intended for physicists doing research on and working with lasers.

CONTENTS

CONTENTS

FLUCTUATING INTENSITY REGIMES IN
LASERS AND MASERS

V. A. Dement'ev, T. N. Zubarev,
and A. N. Oraevskii

The conditions for spiking in lasers (and masers) are studied. A method is developed for analyzing the equations describing laser processes which is a variant of the Fourier method in combination with the small-parameter method applied to this particular problem. This method makes it possible to consider in a unified way the stationary states of lasers, their stability, and transition processes. The following laser models are analyzed: standing wave and traveling wave lasers, lasers with a point active medium, with an inhomogeneous luminescence line, and with a Q-dispersive cavity. The effect of pumping instability and the cavity parameters on the operating regime of a laser is studied. Stochastic methods are used to study random lasing and a laser with noise pumping. These cases have made it possible to formulate criteria for the different spiking regimes in various types of lasers and masers. An attempt is made to compare some of the conclusions of this theory with available experimental data.

INTRODUCTION

Research on the dynamics of lasers and masers has been going on for more than ten years. During this period much experimental and theoretical work has been done on fluctuating intensity regimes (spiking) in molecular masers and various types of lasers [1-9]. It is found experimentally that under certain conditions a spikeless lasing regime is replaced by chaotic or regular spiking. In some cases spiking is regarded as a favorable factor while in others it is undesirable. The spikes have different characteristics. There are the so-called free lasing spikes (usually lasting milliseconds), the envelope of which is modulated at a low frequency (of the order of a kilohertz); self Q-switching spikes (lasting nanoseconds in lasers), which are a sequence of giant pulses with a repetition rate of the order of the effective lifetime of the upper working level; and ultrashort pulses (lasting picoseconds in lasers) with a repetition rate equal to the transit time of a photon through the cavity. The conditions for disruption of spikeless operation of a laser depend very subtly on a whole group of parameters. Detailed experimental studies have been made of the effect of such factors as the mode composition of cavities [10-12], the amount of pumping [4, 10, 13-16], and the stability of the laser parameters [17]. Definite progress has been made in the theoretical description of laser operation. Undamped oscillations in the density of photons in inverted systems have been obtained in [3, 18-21, 161]. Nonetheless, up to now (1) there is no single opinion on the nature of spiking; (2) no approach has been proposed that can uniquely yield all the experimentally observed regimes; and (3) the theoretical models are often very far from experiment, so it is difficult to compare the theoretical results with experiment. In this article we summarize some of our results on formulating a theory which satisfies these requirements.

Up to now the theory of lasers has involved three approaches [22]. The first approach consists of work on the quantum theory of the laser [23-26] in which the active medium and the electromagnetic field are considered from the standpoint of quantum theory. This is the microscopic level of investigation and, as a result, the quantum statistics of laser radiation has come to be understood and profound analogies have been discovered between laser instabilities and physically well-known phase transitions and critical phenomena near a state of thermodynamic equilibrium. The second approach combines work on the macroscopic level [22] and information theory to describe processes in lasers. The third approach involves a semimicroscopic treatment with a quantum-mechanical analysis of the active medium and a classical examination of the field [3, 27]. The quasiclassical equations are sufficiently general for studying the dynamics of lasers and are the basis of this paper as well as most work on the theory of lasers. This is why a number of problems in laser kinetics lie beyond the scope of this article. For example, we do not discuss the operating regimes of lasers with a transverse inhomogeneity in the field [28, 29], multicomponent media [30-32], multichannel lasing [33], the effect of the dependence of losses and the refractive index of materials on the field strength [3, 34], the effect of the relationship between the pumping and lasing channels [35] on the operation of lasers, and so on.

The active medium of a laser is essentially an ensemble of quantum objects with two working levels† which interact with one another through an electromagnetic field. In the absence of energy sinks and sources the probability of a two-level molecule situated in an electromagnetic field being in one of the levels oscillates at a frequency equal to the ratio of the product of the moment of the transition (dipole or magnetic) times the field strength to Planck's constant [36]. In the microwave range the working transition is usually a magnetic dipole transition and the frequency of the oscillations in the inverted population has an obvious physical meaning: It is the nutation frequency of the magnetic moment of the molecule about the direction of the magnetic field. In the optical range the working levels are part of an electric dipole transition which has no simple classical analog, but this frequency is still called the optical nutation frequency.

The operation of a laser cannot be understood without including the processes of dissipation and pumping. Dissipative processes in the laser are described by three phenomenological relaxation constants. The lifetime of the upper working level of the molecule, or the longitudinal relaxation time of the material, determines the rate of change of the population inversion in the material. The width of the luminescence line, which is inversely proportional to the transverse relaxation time of the material,‡ defines the duration of a coherent wavetrain in spontaneous emission of the molecule. The spectral width of a mode, which is inversely proportional to the lifetime of a photon in the cavity, characterizes the rate of damping of the field in the cavity. The effect of pumping is described by the change from the stationary value of the inversion in the absence of a field. When, due to pumping, more energy enters the system than is lost, the laser is self-excited. After the pump is turned on, the threshold population inversion is established after a time of the order of the lifetime of the laser level. The spectra of the modes in the laser output are formed after a further delay equal to the photon lifetime in the cavity. If the product of the transit time of the photons across the cavity and the optical nutation frequency is of the order of unity, then the electromagnetic wave strongly modulates the absorption coefficient or refractive index of the material, the modes may be coupled in phase (as when they are synchronized by an external force), and the laser emits ultrashort pulses.

---

† In the following we shall speak of a two-level molecule for brevity.

‡ The terms "longitudinal" and "transverse" relaxation times of the material, as well as the term "nutation frequency," have been taken from microwave terminology.

The envelope of the ultrashort pulses is formed due to the development of deviations in the partial intensities of the modes from their stationary values. These intensity deviations undergo self-consistent relaxation (at customary pumping intensities) oscillations together with the deviations in the inverted population and polarization of the material. Oscillations at the relaxation frequency arise in the same way as in the predator–prey (Volterra's) problem. The relaxation oscillations in the output intensity may be damped or may grow. In the latter case the stationary envelope of the ultrashort pulses is a sequence of giant pulses which repeat at a frequency equal to the lifetime of the laser level since the stationary regime is determined by gross energy factors. Thus, relaxation and nutation oscillations in the inversion are potential sources of instabilities in the spiking regime of a laser as well. In its most developed form the spike structure of the laser output is a sequence of ultrashort pulses with an envelope made up of giant pulses. In the opposite limiting case the laser operates without spiking. To realize these cases a number of sufficient conditions must be satisfied which differ according to the type of laser, and the resulting lasing regime may be a very poor copy of the picture sketched above.

Laser instabilities depend on a whole series of factors of which the following are most important:

(a) The ratio of the optical nutation frequency to the relaxation constants of the material and field. If this frequency is much greater than the relaxation constants, then the inverted population mainly undergoes nutation oscillations, while in the opposite limiting case the inversion oscillates at the frequency of the relaxation oscillations.

(b) The relations among the relaxation constants of the material. If the linewidth is much greater (less) than the spectral width of a mode, then the polarization (field) lags behind the field (polarization). The lifetime of the upper working level of a laser, except in the case of molecular and spin lasers, is much greater than the other characteristic times, and this often makes it possible to divide spike formation into pumping and emission stages.

(c) The duty cycle of the cavity. The lasing kinetics depend to a great extent on the ratio of the dimensions of the active medium to the wavelength. If the dimensions of the medium are much less than the wavelength, then its dimensions can be neglected and the laser can be treated as having a point active medium.

(d) The degree of spatial inhomogeneity in the field, which is characterized by the relations among the intensities of waves traveling in opposite directions.

(e) Spatial dispersion in the Q factor of the resonator cavity. This explains the different losses of modes with different indices.

(f) The nature of line broadening. The luminescence line may be homogeneously and inhomogeneously broadened. A homogeneous line has a Lorentzian shape while an inhomogeneous line may be described by a Gaussian function or by discrete distributions.

We now present a classification of the instabilities and give a brief review of the contents of this article. First of all, the instabilities may be divided into two large classes. The appearance of instabilities of the first class is determined by the energy balance in the laser or maser, and the condition for instability may have the form of a condition for self-excitation of the laser including saturation. This kind of instability is aperiodic or oscillatory with a frequency equal to the difference between the modal frequencies of the cavity and a growth rate of the order of the width of the mode (in lasers). An instability of this type describes the excitation of the spectrum of axial modes in the kinetic theory of lasers (oscillatory instability) or the destruction of a packet of modes (mode capture or trapping instability) in three- and single-mode lasers (cf. Section 5, paragraph 5.2).

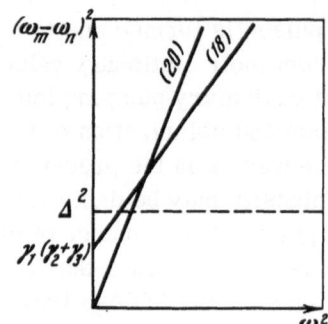

Fig. 1. The dependence of the frequency of nutation (18) and relaxation (20) oscillations in the inversion on the pumping $\gamma_3 > \gamma_1 + \gamma_2$. If $\gamma_3 < \gamma_1 + \gamma_2$ the branches (18) and (20) do not intersect. The dashed line corresponds to a frequency Δ for oscillations in the field, inversely proportional to the transit time of a photon through the cavity.

As an instability of the second class develops, the inversion oscillates at the relaxation or nutation frequency while the growth rate is determined by various factors and is of the order of or less than the Einstein coefficient for spontaneous emission. Various models have been studied.

A. A laser or maser with homogeneous parameters, that is, one using traveling waves with a homogeneous luminescence line and a dispersion-free cavity completely filled with the active medium (Sections 2-4).

1. Single-mode laser or maser (Section 3). The frequencies of the relaxation and nutation oscillations in the inversion depend in different ways on the laser parameters. It may happen that for certain parameter values these frequencies are of comparable magnitude (cf. Fig. 1). Then one speaks of an intersection of the relaxation and nutation branches of the oscillations. The strong interaction between the different forms of oscillation which may occur when the branches intersect is the reason for instability in the spikeless regime in masers, semiconductor lasers, and molecular lasers if the spectral width of a mode is of the order of or much greater than the width of the luminescence line. The mechanism of the instability is as follows. The deviations in the inverted population and other dynamic variables from their values in the spikeless regime primarily excite relaxation (nutation) oscillations in the inversion at small (large) pumping levels in the sense of factor (a). Since the frequencies of the relaxation and nutation oscillations are close, the primary excited oscillation is amplified by interacting with the weakly excited oscillation. The developing instability brings the laser into a self-Q-switched spiking regime at the optical nutation frequency at high pumping levels. Weak and strong regimes of spike excitation are possible.

2. A multimode traveling wave laser or maser (Section 4). The instability in the spikeless regime is a consequence of either the Cerenkov effect or an autoparametric resonance. The Cerenkov instability occurs when the phase velocity of an electromagnetic wave is greater than the propagation velocity of light in a medium (cf. Figs. 2 and 7). With an autoparametric resonance, heating between modes at a frequency equal to twice the optical nutation frequency pumps oscillations at the nutation frequency (cf. Figs. 4 and 7). Whereas with the Cerenkov instability a laser can "weakly" go into a spiking regime at the nutation frequency, the auto-

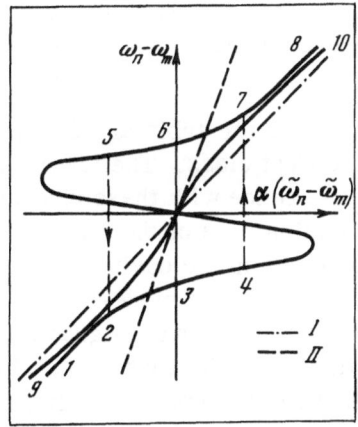

Fig. 2. The dependence of $\omega_n - \omega_m$ on $\tilde{\omega}_n - \tilde{\omega}_m$ for various values of pumping with $\gamma_1 \ll \gamma_3$: 1-8) $\omega_1^2 > 2\gamma_1^2$; 9, 10) $\omega_1^2 < 2\gamma_1^2$. (I) the asymptote $\omega_n - \omega_m = \alpha(\tilde{\omega}_n - \tilde{\omega}_m)$; (2-4-7-5) hysteresis which occurs on tuning the cavity; (II) the straight line $\omega_n - \omega_m = \tilde{\omega}_n - \tilde{\omega}_m$.

parametric resonance yields only a nonlinear instability, that is, a "hard" spike excitation regime at the nutation frequency. The growth rate is inversely proportional to the effective photon lifetime in the cavity with a proportionality coefficient of the order of the square of the ratio of the nutation frequency to the linewidth.

B. Lasers and masers with inhomogeneous parameters (Section 5).

I. Point model (Paragraph 5.1).

1. A single-mode laser with a point active medium behaves as a single-mode traveling wave laser.

2. A two-mode laser with a linewidth much greater than the width of a mode. Mode capture is unstable when the difference between the mode frequencies is of the order of the widths of the modes (cf. Fig. 8) and the stability boundary is "hazardous," that is, for smaller differences between the mode frequencies a nonlinear instability may occur and the laser may undergo a strong transition into a self-Q-switched spiking regime.

3. Three-mode and single-mode laser with a linewidth much greater than the spectral width of a mode. Mode capture is stable for arbitrary differences in the mode frequencies. Apparently only a strong transition to a self-Q-switched spiking regime with mode locking is possible (cf. Fig. 8).

II. Spatial inhomogeneity of the field in the cavity (Paragraph 5.2).

1. Single-mode laser. In the neighborhood of an antinode (node) of a standing wave the field strength is greater (less) than the field strength in a traveling wave laser. Thus, field oscillations at the relaxation frequency are damped more strongly (weakly) near an antinode (node) of a standing wave than in a traveling wave laser. Near the lasing threshold these effects compensate one another on the average. As the pumping level is increased the standing wave begins to burn out the inversion near the nodes by means of multiphoton processes, the hole in the spatial distribution of the population inversion becomes larger, and the damping rate of the spikes becomes greater than in a traveling wave laser.

2. Two-mode laser. Mode capture sets in after a time of the order of the damping decrement of the spikes in a single-mode laser. The spectrum of the modes which are set to oscillating due to spatial burnout of the inversion is formed with a growth rate proportional to the spectral width of the mode, and in a two-mode laser the coefficient of proportionality is of the order of the square of the ratio of the frequency difference between the modes to the spectral width of the modes. Thus, the second process is more effective than the first when the frequency difference between the modes is greater than the frequency of the relaxation oscilla-

tions, and mode capture is unstable at such frequency differences (cf. Fig. 10). The stability boundary is hazardous and the spikes at the relaxation frequency either are damped or grow. In the latter case a self-Q-switched spiking regime results.

3. Three- and multimode laser. Mode capture is unstable for arbitrary frequency differences between the modes because of an instability of the first class. The laser operating regime is described by a complex integral manifold in phase space with these characteristic time scales (in order of appearance): the transverse relaxation time of the material, the transit time of photons across the cavity, the lifetime of photons in the cavity, and the longitudinal relaxation time of the material. In this regime a laser produces mode-locked self-Q-switched spikes. The partial intensities of the modes are established with a growth rate of the order of the spectral width of the mode because of an instability of the first type. The envelope (a giant pulse) is formed in a time of the order of the lifetime of the upper working level as an instability develops at the relaxation frequency. The phases of the modes are synchronized with a growth rate of the same form as in a multimode traveling wave laser, that is, mode synchronization occurs due to optical nutation. It follows from obvious physical considerations that, beginning with a sufficiently large number of synchronized modes, the development of ultrashort pulses should take place in the same way in traveling wave and standing wave lasers; thus, as far as the physics of the phenomenon of synchronization (locking) is concerned, we can repeat all that was said above in Paragraph A.2. From this standpoint the effect of a bleachable filter in mode locking is to isolate the strongest fluctuating emission outburst of the laser which then initiates these dynamic synchronization mechanisms.

III. Spatial dispersion in the Q-factor (Paragraph 5.2).

The spectrum of the laser modes is established during competition among the position dependences of the gain and absorption coefficients. Burning a gap in the spatial distribution of the inversion widens the lasing spectrum, and increased loss of modes as the mode frequencies are shifted further away from the center of the luminescence line narrows the spectrum of the working modes. The first effect exceeds the second for frequency differences between the modes of the order of the square root of the difference in the squares of their widths, and an oscillatory instability at the relaxation frequency appears. The development of the instability brings the laser into a stationary spiking regime of the same type as in Paragraph B.II.3.

IV. Inhomogeneous luminescence line (Paragraph 5.3).

1. Single-mode laser. The field burns a hole of width equal to the homogeneous line broadening in the inhomogeneous luminescence line. If the relaxation frequency for the luminescence centers involved in lasing is equal to the width of this hole, then an oscillatory lasing instability develops at the relaxation. Bifurcations in a spikeless lasing regime at the stability boundary are studied for a laser with a mode width much greater than the line width. The stability boundary may be hazardous or safe and the stationary spiking regime may consist of spikes with a width and period of the order of the transverse and longitudinal relaxation times of the material, respectively.

2. Multimode laser. A new feature is observed in the amplitude characteristic of a two-mode laser in which the transverse relaxation time of the material is much less than the longitudinal relaxation time while the latter is much less than the photon lifetime in the cavity. It is known that if the frequency difference between the modes is less than the homogeneous broadening of the line, then the partial intensities of the modes are reduced by roughly a factor of 2 because of overlapping of the holes burned by the modes in the luminescence line (the so-called Lamb dip). If the frequency difference between the modes is reduced further, to the point that the product of the frequency difference and the longitudinal lifetime of the material is of order unity, then the population inversion follows the field and oscillates with a large

amplitude. The resulting combination frequencies lie outside the amplifying bandwidth of the cavity and are absorbed. Because of this the partial intensities of the modes are further reduced by a factor of 1.5. Thus, against the background of the Lamb dip there is a narrower dip with a width inversely proportional to the longitudinal lifetime of the material.

C. Nonautonomous systems (Section 6).

1. Single-mode laser (Paragraphs 6.1 and 6.3). A parametric instability with modulation of the cavity losses develops when the ratio of the frequency of the external force to the relaxation frequency is a rational number. The amplitude−frequency characteristics of the harmonic (i.e., at the frequency of the external influence) and subharmonic (with a frequency equal to half the lowest modulation frequency) intensity oscillations are determined. When the loss modulation amplitude is greater than the ratio of the damping increment of the relaxation spikes in an autonomous single-mode laser to the spectral width of the mode, then the resonance becomes nonlinear, the maximum of the amplitude−frequency characteristic is shifted toward lower frequencies, and hysteresis phenomena may occur. Modulating the pumping yields these effects at powers such that the optical nutation frequency is much less than the linewidth. In the inverse limit the amplitude−frequency characteristic has a resonance at the optical nutation frequency when the pump is modulated. When the losses are modulated this effect does not occur since at higher pumping levels the nutation frequency is the characteristic frequency of the fluctuations in the inversion; thus, if an external force at this frequency acts on the field (as during modulation of the losses) then, as with a roll dampener on a ship, the inverted population oscillates with a large amplitude while the field hardly oscillates.

2. Multimode laser (Paragraph 6.2). Modulating the losses yields the same type of phenomena as in a single-mode laser. When the refractive index of the material is modulated, we obtain a phase modulation of the output with a maximum in the partial amplitude−frequency characteristic near frequencies of the order of the slow relaxation frequency in a multimode laser. (This frequency is roughly equal to the relaxation frequency of a single-mode laser divided by the square root of the number of excited modes.)

D. Stochastic effects (Section 7).

1. Single-mode laser (Paragraphs 7.1, 7.2, and 7.3). If the external influence modulates the laser parameters over a wide spectrum of frequencies and to such a depth that the system can change its stationary states within a resonance and go from one resonance to another, then the output becomes highly irregular and close to random. For pumping with white noise the noise intensity required for randomization within a resonance is directly proportional to the square of the ratio of the lifetime of a photon in the cavity to the effective lifetime of the upper laser level. The stochastic instability is the most hazardous kind of instability for the operation of a laser. The analysis of the conditions for the stochastic instability relies on the kinetic equations for the radiant energy density. A study of the stationary solutions of these equations shows that the laser output has an almost-periodic turbulence spectrum, i.e., a linear spectrum. If the spectral width of a mode is much less than the linewidth, then a stationary random regime with a continuous spectrum corresponding to developed turbulence in the laser output is possible. The laser emits random spikes with a frequency and width of the order of the relaxation frequency while in the case of almost-periodic turbulence the spikes are grouped in packets whose repetition rate equals the difference between the frequencies in the spectrum. An examination of the stability of the spikeless regime and an analysis of the transition processes shows that the turbulence may develop strongly or weakly. If the cavity frequency equals the frequency of the line, the spikeless regime is stable but strong excitation of a ι developed stationary turbulence is possible. As the cavity frequency is displaced from the frequency of the line, the spectrum of the stationary turbulence is depleted and the number of stationary states is reduced. However, the role of the nonstationary motions is enhanced at

the same time that the threshold for excitation of turbulent emission is reduced. If the shift in the cavity frequency from the line frequency exceeds the product of the nutation frequency and the square root of the maximum possible number of spectral components in the output of a laser whose mode frequency equals the line frequency (Fig. 2), then the spikeless regime is unstable and the resulting turbulence is nonstationary.

2. Multimode laser (Paragraphs 7.3 and 7.4). If the mode spectrum has a single characteristic scale (that is, only axial modes operate), then random relaxation spikes occur in the laser output when the product of the slow relaxation frequency and the lifetime of the upper laser level exceeds unity. If a laser generates a group of transverse modes near the frequency of an axial mode such that the emission is coherent within that group while the emission from groups belonging to different axial modes is incoherent, then the laser may emit partially ordered trains of spikes. Strong excitation of the spiking regime occurs when the slow relaxation frequency is of the order of the distance between the frequencies of the transverse modes. A random self-Q-switched spiking regime appears. Because each group of transverse modes oscillates independently, a giant pulse splits randomly into smaller pulses, whose number is proportional to the number of axial modes generated.

As we proceed to prove these statements, we note that the theory of instabilities may be constructed deductively. As a rule, in order to explain the nature of the instabilities, a different approach is preferred. At first the instabilities in the simplest model of a laser are analyzed. This model is then refined as more new factors are included. The figures also serve this purpose and most of them are qualitative.

1. Derivation and Preliminary Analysis

of the Equations

To model the processes taking place in a laser or maser we shall consider an active medium made up of two-level molecules located in a resonator cavity. The interaction of the electromagnetic field with the matter obeys the following system of equations [1, 2]:

$$\dot{v} = \gamma_1 (1 - v) + 2i\Omega e (\rho^* - \rho), \tag{1}$$

$$\dot{\rho} = -(i\omega_0 + \gamma_2) \rho - i\Omega ev, \tag{2}$$

$$\ddot{e} + 2\gamma_3 \dot{e} - c^2 \nabla^2 e + 2 \frac{\Omega}{\omega_0} (\ddot{\rho} + \ddot{\rho}^*) = 0, \tag{3}$$

where $e = (2\pi n_0 \hbar \omega_0)^{-1/2} E$ (E is the electric field strength†); $\rho = \rho_{12}$ and $v = \rho_{22} - \rho_{11}$, where ρ_{ij} (i, j = 1, 2) is the density matrix of the two-level molecule in the energy representations; that is, the real and imaginary parts of ρ describe the polarization and displacement current of the medium, respectively, and v gives the inverted population of the medium; ω_0 is the frequency of the transition between the working levels of the molecule; $\Omega = (2\pi n_0 \omega_0/\hbar)^{1/2} D$ (D is the absolute value of the dipole moment of the transition and n_0 is the density of inverted molecules in the absence of a field in the stationary state); γ_1^{-1} is the effective lifetime of the upper working level of the molecule or the longitudinal relaxation time of the medium; γ_2 is the luminescence linewidth, inversely proportional to the transverse relaxation time of the medium; γ_3^{-1} is the photon lifetime in the cavity and is inversely proportional to the spectral width of the mode. Table 1 shows the relaxation constants for various types of lasers and masers.

---

† For a maser the electric field and transition dipole moment in Eqs. (1)-(3) are replaced by the magnetic field strength and the transition magnetic dipole respectively. To be specific, in the following we shall speak of an electric dipole transition.

TABLE 1

Type of laser (maser)	γ_1, sec^{-1}	γ_2, sec^{-1}	δ, sec^{-1}	γ_3, sec^{-1}	References
Ruby laser, 300°K	10^3	10^{11}	—	10^7—10^9	[2, 37]
Same, 4—77°K	10^3	10^7—10^9	$2 \cdot 10^{10}$	10^7—10^9	[38]
Dysprosium laser CaF$_2$:Dy^{2+}, 27°K	10^2—10^5	10^9	—	10^8	[39, 40]
Ruby maser	10^2	10^8	—	10^6—10^8	[2, 41]
Semiconductor laser	10^9	10^{12}	—	10^9—10^{12}	[42]
Molecular maser (laser)	10^4	10^4	—	10^6	[1]
He—Ne laser	10^8	10^8—10^9	10^{10}	10^7	[43]
Atmospheric pressure CO$_2$ laser	10^3	10^{10}	10^{10}	10^8	[44—46]
Dye lasers	10^8—10^9	10^{12}	—	10^8—10^9	[47]

Note: δ is the inhomogeneous line broadening (see Section 5).

Since we are usually interested in the electric field in the cavity, it is natural to eliminate the elements of the density matrix from Eqs. (1)-(3). This may be done in a general way, but the weak nonlinearity of the equations and the existence of dispersion make it possible to limit ourselves to the approximation

$$\ddot{\rho} + \ddot{\rho}^* \simeq - \omega_0^2 (\rho + \rho^*) \tag{4}$$

upon substituting Eq. (2) into Eq. (3). After this we find the following equation for the field from Eqs. (1)-(3):

$$\frac{AB}{4\omega_0^2} e = - \Omega^2 \left\{ 1 + \int e\hat{e} \exp\left[- \gamma_1 (t - \tau) \right] d\tau \right\} e, \tag{5}$$

which is an integrodifferential equation with partial derivatives and a retarding kernel $\exp[-\gamma_1(t - \tau)]$ and

$$
\begin{aligned}
A &= - c^2 \nabla^2 + \frac{\partial^2}{\partial t^2} + 2\gamma_3 \frac{\partial}{\partial t}, \\
B &= \omega_0^2 + \frac{\partial^2}{\partial t^2} + 2\gamma_2 \frac{\partial}{\partial t}, \\
\hat{e} &= \frac{A}{\omega_0^2} \left(\frac{\partial}{\partial t} + \gamma_2 \right) e.
\end{aligned}
\tag{6}
$$

Recurrence is a feature of the autooscillatory motions in which all the dynamic characteristics of the system are roughly repeated with a quasi-period depending on the time and accuracy of the repetition [48]. The most general class of recurrent motions which have been studied analytically up to now are the almost-periodic oscillations for which the quasi-period is independent of time but depends on the accuracy of the repetition. If the period is also independent of the accuracy of repetition, then we obtain a periodic oscillation which is a special case of an almost-periodic oscillation. In general a periodic motion is a superposition of a countable number of sinusoidal oscillations with frequencies that are multiples (of a single fundamental), while an almost-periodic motion decomposes into a Fourier series with incommensurable frequencies. The simplest example of an almost-periodic function is, therefore, the beating of two harmonic oscillations with close frequencies. In [49] the basic properties of periodic functions are generalized to almost-periodic functions. In particular, it is proved that arithmetic operations, differentiation, and integration (in the last case the Fourier expansion does not have to contain a free term) are defined in the class of almost-periodic functions.

These simple properties of almost-periodic functions are the basis of their extensive use in various problems in the theory of oscillations. Thus, we seek the stationary states of a laser as solutions of Eqs. (5) in the form of trigonometric series with real frequencies:

$$e = \sum_p e_p \exp(-i\omega_p t), \qquad \omega_{-p} = \overset{*}{\omega_p}, \qquad e_{-p} = \overset{*}{e_p}, \tag{7}$$

where the components e_p are expanded in a series over the eigenfunctions (modes) of the cavity Φ_λ:

$$e_p = \sum_\lambda e_p^\lambda \Phi_\lambda, \qquad -c^2 \nabla^2 \Phi_\lambda = \widetilde{\omega}_\lambda^2 \Phi_\lambda. \tag{8}$$

In general the electromagnetic field has an infinite number of degrees of freedom and is described by partial differential equations. From a mathematical standpoint the stability of the solutions of partial differential equations is a complicated and still mostly unsolved problem. The major techniques and assumptions of stability theory were formulated by Lyapunov for systems with a finite number of degrees of freedom and obeying ordinary differential equations [50]. In recent years it has been shown that they can be generalized to equations with an infinite number of degrees of freedom if the variation equations for the motions whose stability is being studied have a bounded spectrum, that is, the solutions of the variation equations that have the exponential form $\exp(\gamma t)$ are such that $|\gamma| < C$, where C is a constant [51]. The wave equation (3) contains the unbounded operator $c^2 \nabla^2$ whose spectrum $\widetilde{\omega}_\lambda^2$ [see Eq. (8)] extends to infinity. Nevertheless, it is clear that an instability cannot develop on modes whose frequencies differ from the center of the emission line of the material by much more than the width of the line. Intensity fluctuations in such modes are damped rapidly. Because of this rapid process the system lies in the neighborhood of an integral manifold (a manifold of the trajectories) of finite dimensionality. Therefore, the classical methods of studying stability [50] are applicable to the dynamics of lasers for dispersive media, that is, media with a finite linewidth, and in analyzing the stability of the stationary states of a laser we shall seek solutions of the equations in the form of Eqs. (7) and (8) with complex frequencies ω_p.

We shall study transition processes in lasers using the trigonometric expansions (7) with complex frequencies ω_p which vary slowly in time. This variant of the averaging method is known in the literature as the quasiclassical approximation or the WKB method [36; 52]. Thus, the stationary states of a laser and their stability and transitions are analyzed in a unified way with the aid of solutions of the form (7) and (8) with real and complex, constant and slowly time-varying frequencies ω_p. Expansions of this form are used to solve nonlinear problems in dynamic and stochastic theories [53]. This method has many variants [52-54], including Poincaré's small parameter method, the averaging method, the harmonic balance principle, the stroboscopic method, the WKB method, and so on. In the form given above it is well suited to solving various nonlinear problems in the theory of lasers and makes it possible to explain the instabilities as well as to study the effects of the mode composition of the cavity, the inhomogeneity of the laser parameters, and randomness on the operation of the laser.

We shall derive some general, mainly qualitative results which are valid for the entire class of stationary laser states. A stationary state of the field is characterized by a spectrum of output (generated) frequencies ω_p and a spectrum of oscillating (generating) modes $\widetilde{\omega}_\lambda$. In general these are different concepts. In fact, if the frequency difference between the modes is less than the spectral width of a mode γ_3, then such modes may be captured and form a packet of modes which oscillate at a single frequency while different packets, generally speaking, overlap one another. In the space and time Fourier expansion (7) and (8) the quantity $\sum_\lambda |e_p^\lambda|^2$ characterizes the partial intensity of a line in the frequency spectrum while $\sum_p |e_p^\lambda|^2$ gives the partial intensity of a mode in the spectrum of stationary state modes. To characterize the spectra we introduce the moments of the distribution of partial intensities. The first moment determines the location of the center of the spectrum and the second gives the width of the spectrum. It appears that in a stationary laser state there is a simple relationship between the center of the frequency spectrum $\bar{\omega} = \sum_{p,\lambda} \omega_p |e_p^\lambda|^2 [\sum_{p,\lambda} |e_p^\lambda|^2]^{-1}$ and the center of the mode spectrum $\tilde{\omega} = \sum_{p,\lambda} \tilde{\omega}_\lambda |e_p^\lambda|^2 [\sum_{p,\lambda} |e_p^\lambda|^2]^{-1}$. To derive this relationship we substitute Eq. (7) in Eq. (5). We obtain the following equation† :

$$\frac{A_p B_p}{4\omega_0^2} e_p = -\Omega^2 \left\{ e_p + \sum_{\omega_k + \omega_l + \omega_m = \omega_p} \frac{e_k \hat{e}_l e_m}{\gamma_1 - i(\omega_l + \omega_m)} \right\}, \tag{9}$$

where $A_p, B_p,$ and $\hat{e}_p$ are given by Eq. (6) with $i\omega_p$ substituted in place of $\partial/\partial t$. We multiply Eq. (9) by e_{-p}, sum it over $p \geq 0$, integrate it over space, and separate the imaginary part. The fourfold sum (over $p \geq 0$, k, l, m) in the latter equality contains every term together with its complex conjugate because the width of the frequency spectrum of the field is much less than $\bar{\omega}$ ($|\omega_m - \omega_n| \ll \bar{\omega}$) and $\gamma_1 \ll \bar{\omega}$, so it goes to zero. We thus obtain the desired dispersion relation

$$\bar{\omega} = \alpha\tilde{\omega} + \beta\omega_0, \tag{10}$$

where $\alpha = \gamma_2/(\gamma_2 + \gamma_3)$ and $\beta = 1 - \alpha$. This expression means that the average nonlinear frequency pulling and pushing balance one another, and only so-called linear frequency pulling remains. In particular, for a single-mode system with a cavity eigenfrequency $\hat{\omega}_n$ Eq. (10) becomes the well-known formula [27, 55]

$$\omega_n = \alpha\tilde{\omega}_n + \beta\omega_0.$$

The nonlinear terms in Eq. (9) may be resolved into "diagonal" ($\omega_l + \omega_m = 0$) and "nondiagonal" ($\omega_l + \omega_m \neq 0$) terms. The "nondiagonal" terms are due to oscillations in the population inversion with time, and their ratio to the "diagonal" terms is of order $\gamma_1 |\gamma_1 - i(\omega_l + \omega_m)|^{-1}$. We shall call the approximation in which the "nondiagonal" terms are rejected the "diagonal" approximation. In the "diagonal" approximation, in particular, all the results for equilibrium points of the balance or kinetic equations are obtained. This approximation yields a crude picture of the operation of a laser whose frequency spectrum contains no anomalously close frequencies [56], i.e.,

$$|\omega_l - \omega_m| \sim \gamma_1. \tag{11}$$

If close frequencies (11) are contained in the frequency spectrum, then in this frequency interval the population inversion follows the field quasistatically and the retarding kernel $\exp[-\gamma_1(t - \tau)]$ of Eq. (5) becomes of order unity; thus, the nonlinear terms have an especially strong effect on the behavior of the system.

A more subtle effect of the oscillations in the inverted population that is described by the "nondiagonal" terms in Eq. (9) is observed even in studies of the stability of a single frequency regime in Eq. (7) with $e_p = 0$ when $p \neq \pm n$. This kind of regime is also called harmonic [3] or monochromatic [57]. In accordance with the method described above, to study the stability of

---

† If we do not make assumption (4), then the factor ω_0^{-2} in the terms $(A_p B_p)/4\omega_0^2$ and $\hat{e}_p$ goes into ω_p^{-2}. In addition, to avoid misunderstanding, we note that the sum in Eq. (9) is over both positive and negative frequencies.

the stationary lasing regimes it is necessary to substitute Eq. (7) with complex frequencies ω_p into Eq. (5). Formally we obtain the same equation (9) for e_p and ω_p as in the analysis of the stationary regimes. Then we set $e = e_{st} + \delta e$, where e_{st} describes the single-frequency lasing regime and δe is a perturbation with complex frequencies. The solutions of Eq. (9), after linearization in δe, are a superposition of three types of solution:

incoherent perturbations

$$\delta e = e_m \exp\left(-i\omega_m t\right) + \text{c.c.}, \quad \mathrm{Re}\,\omega_m \neq \omega_n \tag{12}$$

and coherent perturbations

$$\delta e = e_m \exp\left(-i\omega_m t\right) + \text{c.c.}, \quad \mathrm{Re}\,\omega_m = \omega_n, \tag{13}$$

$$\delta e = e_m \exp\left(-i\omega_m t\right) + e_l \exp\left(-i\omega_l t\right) + \text{c.c.}, \quad \omega_m + \omega_l^* = 2\omega_n. \tag{14}$$

Coherent perturbations with more than two components do not exist in the linear approximation since the interaction of the field with the material is treated in a dipole approximation. [The nonlinear terms in Eq. (5) are proportional to $e^2 \hat{e}$.] The coherent perturbations (13) and (14) arise in lasers with sufficiently homogeneous characteristics. For example, in the case of the coherent perturbations (14) the electromagnetic waves e_m and e_l yield a nonlinear polarization at the combination frequency $\omega_m + \omega_l^* - \omega_n$. The polarization is a source of a field at this frequency which coincides with the frequency of the third component of the perturbation e_n. The phases of the polarization and field at frequency ω_n are related in the same way as the phases of the external force and the stimulated oscillations. The phase of the polarization is determined by the phases of the combining fields. Thus, a coherent interaction of the electromagnetic waves develops and has the form

$$e_m \, |e_n|^2 \exp\left[i\left(2\varphi_n - \varphi_l - \varphi_m\right)\right]. \tag{15}$$

Similarly, the coherent perturbations (13) are related to coherent interactions of the waves of the type

$$e_m \, |e_n|^2 \exp\left[i2\left(\varphi_n - \varphi_m\right)\right]. \tag{16}$$

Such interactions are not always possible. If the relative phases of the components undergo strong fluctuations over an interval much greater than the duration of a spike and much less than the observation time, then the processes in the laser must be considered stochastically. Phase averaging implies that the interactions (15) and (16) go to zero. A strong dispersion $\gamma_3(\omega_p)$ leads to the same result. Then, if the damping of the field is very great at frequency ω_m or ω_l, the cavity filters out the field components at the combination frequencies. This is the so-called filter hypothesis which is introduced as a basis of the harmonic balance method, for example, in the theory of automatic control [58].

The development of instability with respect to coherent perturbations brings a laser into a synchronous stationary regime in which the phases of the e_p^λ components in Eqs. (7) and (8) are related to one another by certain formulas. An arbitrary stationary state depends in any case on two arbitrary constants. One of these constants gives the phase of the high-frequency carrier oscillations of the signal and the other constant determines the phase of the envelope. The first phase is arbitrary in view of the autonomy of the system, while the second is arbitrary if we neglect effects of the order of the ratio of the linewidth to the working transition frequency. Thus, the phases of the two components of the spectrum are arbitrary. Let us specify some third component. If the frequencies of the three components are chosen to be equidistant $(\omega_k + \omega_m = 2\omega_l)$, then the phase of the third component is related to the phase of the first two components through the nonlinear polarization of the material. These two components contribute to the polarization at a combination frequency equal to the frequency of the third component, and the oscillations are synchronized as when an external force acts on an oscillating system [see Eq. (15)]. Continuing in this way we can isolate a series of equidistant frequencies ω_p in the spectrum that will have the phases of all its components fixed once the phases of the two initial components are specified. If there are no other components in the spectrum, then the lasing is synchronous. Otherwise, we may speak of partial synchronization. Mathematically the conditions for synchronization of the components are given, as follows from this discussion (also see the sum in Eq. (9)], by equations of the form $\omega_k - \omega_l = \omega_m - \omega_p$. The growth of perturbations (13) leads to simple synchronization (locking) or mode capture. Development of perturbations (14) yields a multiple synchronization effect [59]. In a laser the axial cavity modes, for example, may synchronize in this way, and the laser will go into a spiking regime with a spike repetition rate $\sim \gamma_3\Delta$, where $\gamma_3\Delta$ is the frequency separation between adjacent output modes.

In the spectral approach to laser dynamics being applied in this article any modulated output regime is a superposition of oscillations at different frequencies. We now show what set of frequencies may exist in a single-mode laser. To do this we write the eigenvalue equation for the coherent amplitudes $(\varphi_l + \varphi_m - 2\varphi_n = 0)$ of the perturbations (14) relative to the single-frequency regime of a single-mode traveling wave laser:

$$i(\omega_m - \omega_n)\{(\omega_m - \omega_n)^2 - [\gamma_1(\gamma_2 + \gamma_3) + \omega^2]\} + \{(\gamma_1 + \gamma_2 + \gamma_3)(\omega_m - \omega_n)^2 - 2\gamma_3\omega^2\} = 0, \qquad (17)$$

where $\omega = 2D\hbar^{-1}\left(\sum_{n>0} |E_n|^2\right)^{\frac{1}{2}}$ is the optical nutation frequency, i.e., the transition frequency of the molecules from one level to the other for a laser with two working levels with $\gamma_1 = \gamma_2 = \gamma_3 = 0$ [36]. In the single-frequency regime, we obtain, on calculating the amplitude of the field oscillations in terms of the initial laser parameters, the following expression for ω:
$\mathring{\omega} = [\gamma_1\gamma_3^{-1}(\Omega^2 - \gamma_2\gamma_3)]^{\frac{1}{2}}$. The reactive term [first curly bracket in Eq. (17)] goes to zero at the frequency

$$|\omega_m - \omega_n| = [\gamma_1(\gamma_2 + \gamma_3) + \omega^2]^{\frac{1}{2}}, \qquad (18)$$

which approaches the optical nutation frequency ω at high pumping levels. If the relaxation constants are nonzero, then the reactive term dominates at high pumping levels, when

$$\omega \gg (\gamma_2 + \gamma_3). \qquad (19)$$

At relatively small pumping levels the active term [second curly brackets in Eq. (17)] dominates. It is zero if

$$|\omega_m - \omega_n| = \left(\frac{\gamma_2 + \gamma_3}{\gamma_1 + \gamma_2 + \gamma_3}\right)^{\frac{1}{2}} \omega_1 \qquad (20)$$

and goes to zero when $\gamma_1, \gamma_2, \gamma_3 \to 0$. Here $\omega_1 = \sqrt{2\beta}\omega$ is the relaxation frequency of a single-mode laser or maser. The frequency (20) has been evaluated in [60] for a laser (in which $\gamma_2 \gg \gamma_3 \gg \gamma_1$). The existence of a set of frequencies in the spectrum with a characteristic frequency difference between them results in amplitude modulation of the laser output at a frequency equal to this difference. The operation of a laser in a spiking regime with high-amplitude spikes at typical pump powers may then be divided into two stages: a pumping stage of duration ν^{-1} of the order of the lifetime of the upper laser level γ_1^{-1} and an emission stage of duration τ of the order of the photon lifetime in the cavity γ_3^{-1}. When the amplitude of the spikes, e_{max}, is reduced, their duration and the repetition rate increase, so the average output power of the laser does not change. The width and repetition period of the spikes are comparable in order of magnitude for small deviations in the field near the single-frequency operating regime:

$$\nu \sim \gamma_1 \, |\, e_{max}\,|^{-1}, \quad \tau^{-1} \sim \gamma_3 \, |\, e_{max}\,|;$$

$$\nu \sim \tau^{-1} \sim \omega_1 \quad \text{for} \quad |\, e_{max}\,|^2 \sim \gamma_1/\gamma_3. \tag{21}$$

Therefore, relaxation or nutation oscillations are a potential source of amplitude modulation (spiking) in a laser. The relaxation and nutation branches of the oscillations are illustrated in Fig. 1. At ordinary pumping levels the relaxation branch is predominantly excited and the field oscillates in the relaxation regime at frequency ω_1, but the nutation branch has an effect on the stability of the oscillations if the frequency of the nutation oscillations is of the order of the relaxation oscillation frequency. That is, the nutation and relaxation branches intersect or, in general, are close to one another (Fig. 1). For a laser (maser, molecular maser or laser) with ordinary pumping, when $\Omega^2 \sim \gamma_2\gamma_3$, we have from Table 1 that

$$\omega_1 \ll \omega \quad (\omega_1 \sim \omega). \tag{22}$$

However, it should be emphasized that the existence of relaxation and nutation branches is only a potential source (necessary condition) of self-modulation. An additional cause which could aid in exciting them is required. If this cause is absent, the laser will operate in a stationary regime without spiking.

In a multimode laser two regimes of relaxation oscillations and two characteristic relaxation frequencies are possible. In the first regime all the modes oscillate in phase as a whole at the relaxation frequency of a single-mode laser, ω_1 (a signal with amplitude modulation). In the second regime the partial intensities of the modes oscillate, roughly speaking, independently at frequency $\omega_2 \sim \omega_1/\sqrt{N}$, where N is the number of modes generated (a signal with phase modulation). The relaxation frequency ω_2 for $\gamma_2 \gg \gamma_3$ appears in the analysis of the balance equations for a multimode laser [61]. Thus, as in the case of a single-mode laser, large-amplitude relaxation oscillations have the form of self-Q-switched giant pulses with width γ_3^{-1} and repetition rate γ_1.

Therefore, a qualitative analysis of Eq. (9) has yielded linear coupling (pulling) of the center of the frequency spectrum (10), the "diagonal" approximation and a region of anomalously close frequencies (11), coherent and noncoherent states, and has made it possible to introduce characteristic units for measuring the slow difference frequencies (Table 2).

We shall now develop this model in detail. Sections 2-4 are devoted to traveling-wave lasers (masers); $\Phi_\lambda = \dfrac{1}{\sqrt{L}} \exp\left(i\dfrac{\tilde{\omega}_\lambda}{c} r\right)$, where L is the cavity length.

TABLE 2

Difference frequencies	Anomalous effects	Relaxation effects		Nutation	Mode locking	Mode synchronization
		multimode laser	single-mode laser			
$\omega_m - \omega_n$	γ_1	ω_1, ω_2	ω_1	ω	0	$\tilde{\omega}_m - \tilde{\omega}_n$
$\tilde{\omega}_m - \tilde{\omega}_n$	—	—	—	—	γ_3	—

2. The Stationary States of Traveling-Wave Lasers

We now consider "single-mode" oscillations of the field. By "single-mode" oscillations we mean states such as (7) and (8) for which each frequency in the time expansion of the field corresponds to a mode in the space expansion of the field. The inverse is not generally true; i.e., $e_p = e_p^p \Phi_p$. From a study of Eq. (9) in a linear approximation in the field it follows that such states exist if the condition $\Omega_{th}^2 = \Omega^2/\gamma_2\gamma_3 > 1$ is satisfied, where Ω_{th}^2 is the ratio of the stationary inversion in the absence of a field to its threshold value for self-excitation of the laser. In the "diagonal" approximation the spectrum of the "single-mode" states is arranged either as single-frequency oscillations or two-frequency oscillations of the field. In the latter case the frequencies are placed symmetrically with respect to the center of the line, ω_0. The total intensity of the oscillations as a function of the frequency difference falls off parabolically,

$$|e_n|^2 + |e_m|^2 = \frac{\gamma_1}{4\gamma_3}\left\{1 - \Omega_{th}^{-2}\left[1 + \frac{(\omega_0 - \omega_n)^2}{\gamma_2^2}\right]\right\}, \tag{23}$$

with a simple dispersion law $\omega_k = \alpha\widetilde{\omega}_k + \beta\omega_0$, $k = n, m$ (linear pulling for each frequency and not only on the average). We shall examine the effect of oscillations in the inverted population using the example of two-frequency and three-frequency oscillations in the field.

In accordance with the conclusions of Section 1, Eq. (9) for the two-frequency field oscillations is independent of the phase and there are four equations for finding the frequencies ω_n, ω_m and the partial intensities $|e_n|^2$, $|e_m|^2$. It is convenient to introduce the new unknowns $\overline{\omega}$, $\omega_n - \omega_m$, $|e_n|^2 + |e_m|^2$, and $\Pi = (|e_n|^2 - |e_m|^2)(|e_n|^2 + |e_m|^2)^{-1}$. The real parts of Eqs. (9) yield a dispersion relation for the center of the frequency spectrum $\overline{\omega}$ (10) and the dependence of $\omega_n - \omega_m$ on $\widetilde{\omega}_n - \widetilde{\omega}_m$:

$$\alpha\left(\widetilde{\omega}_n - \widetilde{\omega}_m\right) = (\omega_n - \omega_m)\frac{(\omega_n - \omega_m)^2 + \gamma_1^2 - \frac{1}{2}\omega_1^2\left(1 - \frac{\gamma_1}{2\gamma_3}\right)}{(\omega_n - \omega_m)^2 + \gamma_1^2\left(1 + \frac{\omega^2}{2\gamma_1\gamma_2}\right)}. \tag{24}$$

This formula characterizes the beat frequency at the output of a two-mode laser. It is clear from this that if $\gamma_1 < 2\gamma_3$ then for pumping such that

$$\omega_1^2\left(1 - \frac{\gamma_1}{2\gamma_3}\right) > 2\gamma_1^2, \tag{25}$$

the dispersion curve (24) is nonmonotonic and is made up of three branches (Fig. 2). The branch 1-3, 6-8 describes the nonlinear decoupling between the field frequencies first noted in [62]; the branch 3-4, 5-6 shows the anomalous variation of the dispersion curve in the close frequency region (11).

The total intensity $|e_n|^2 + |e_m|^2$ generally depends in a fairly complicated fashion on $\omega_0 - \frac{1}{2}(\widetilde{\omega}_n + \widetilde{\omega}_m)$ and $\omega_n - \omega_m$. An idea of the dependence on $\omega_n - \omega_m$ is given by the case $\widetilde{\omega}_n + \widetilde{\omega}_m = 2\omega_0$, where for ω^2 we obtain the equation

$$3\omega^4 + 2\left[\gamma_1\gamma_2(4 - \Omega_{th}^2) + (\omega_n - \omega_m)^2\left(2\frac{\gamma_2}{\gamma_1} - 1\right)\right]\omega^2 +$$
$$+ [(\omega_n - \omega_m)^2 + \gamma_1^2][(\omega_n - \omega_m)^2 - 4\gamma_2^2(\Omega_{th}^2 - 1)] = 0. \tag{26}$$

This equation has a single positive root in the region $(\omega_n - \omega_m)^2 \leqslant 4\gamma_2^2(\Omega_{th}^2 - 1)$. Neglecting oscillations in the inverted population, the total intensity varies parabolically (23) as $|\omega_0 - \omega_n|$ increases. Including the oscillations in the inversion causes a dip in this curve in the region

Fig. 3. A stability diagram for the single-frequency lasing regime on the frequency difference plane (x_1, y_1) for $\gamma_2 \gg \gamma_3 \gg \gamma_1$ and $\Omega^2_{th} \gg 1$: I) the contour $\Pi = -1$; II) $\Pi = 0$; III) $1 < \Pi < 0$. The $\Pi > 0$ contours are obtained from the $\Pi < 0$ contours by the transformation $x'_1 = x_1 - y_1$, $y'_1 = -y_1$. In the region denoted by the index $D(n, m)$ the eigenvalue equation for the stability of the single-frequency regime has n roots in the left and m roots in the right complex half-planes.

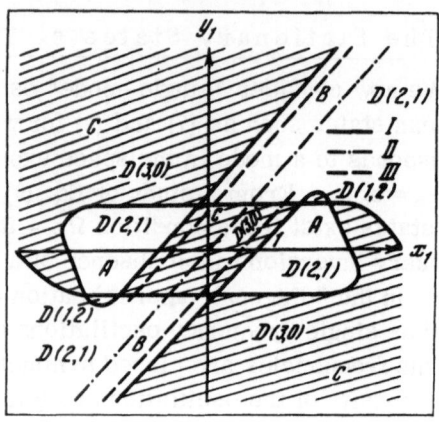

of anomalously close frequencies (11) with pumping such that

$$\Omega^2_{th} - 1 \gtrsim \gamma_1^2 \gamma_2^{-2}. \tag{27}$$

The width of the dip increases with pumping, and its depth saturates. If inequality (27) is satisfied with room to spare,[†] then

$$(|e_n|^2 + |e_m|^2)_{min} (|e_n|^2 + |e_m|^2)_{max}^{-1} = {}^2/_3. \tag{28}$$

The anomalous variation in the dispersion curve and the dip in the amplitude characteristic in the region of close frequencies (11) are analogous to the anomalies in the phase- and amplitude-frequency characteristics of a quantum amplifier [41]. In a plasma, anomalous dispersion leads to the formation of negative energy waves and the explosive instability [63]. One of L. I. Mandelshtam's last papers is devoted to a search for real media in which waves with a negative group velocity might propagate [64].

In order to fully characterize the two-frequency oscillations of the field we must find the distribution of partial intensities, which is described by the quantity Π. It is convenient to construct contours $\Pi = $ const on the frequency difference plane $(x_1 = \omega_0 - \tilde{\omega}_n, y_1 = \tilde{\omega}_m - \tilde{\omega}_n)$. Various relationships between the relaxation constants γ_1, γ_2, and γ_3 and different pumping levels have been examined. These lines have been plotted in Fig. 3 for a laser at high pump levels, i.e.,

$$\gamma_2 \gg \gamma_3 \gg \gamma_1, \quad \Omega^2_{th} \gg 1. \tag{29}$$

Two directions for the contour groups can be seen in the figures. The broad groups along the line $y_1 = 2x_1$ is described by the formula

$$\Pi = (2x_1 - y_1)(\omega_m - \omega_n) \, \omega^{-2}. \tag{30}$$

An extremely narrow group is directed along the lines

$$y_1 = \pm \, (\gamma_3 \gamma_2^{-1} - 1) \, \omega/\sqrt{2}, \tag{31}$$

which in all cases, that is, for arbitrary ratios of γ_3 to γ_2, corresponds to

$$\omega_m - \omega_n = \mp \omega/\sqrt{2}. \tag{32}$$

---

† This is typically the case for most masers and lasers (see Table 1).

Equations (30)-(32) show that nutation in the inverted population at frequency ω has an effect on the redistribution of the field energy among the separate components. This effect is seen especially clearly in the case of three-frequency field oscillations.

Because the equations are so complicated, we shall limit ourselves to three-frequency oscillations of the field with components e_n, e_m, and e_l under the condition

$$\omega_n = \frac{1}{2}(\omega_m + \omega_l) = \frac{1}{2}(\widetilde{\omega}_m + \widetilde{\omega}_l) = \widetilde{\omega}_n = \omega_0. \tag{33}$$

The system of equations (9) yields three complex equations. Since Eqs. (33) include the synchronization condition, these equations will depend on the phase of the components. The phases φ_k, $k = l$, m, n, enter into the equations in the combination $2\varphi_n - \varphi_m - \varphi_l$. The symmetry of the frequency configuration (33) implies that

$$2\varphi_n - \varphi_m - \varphi_l = 0. \tag{34}$$

Eliminating $|e_m| = |e_l|$ and $|e_n|$ from the remaining equations, we obtain a cubic equation for $(\widetilde{\omega}_n - \widetilde{\omega}_m) - (\omega_n - \omega_m)$ with coefficients which are polynomials in $\omega_l - \omega_m$. From the condition that the partial intensities $|e_m|$ and $|e_n|$ are positive, we find a physically meaningful solution. Two branches of the three-frequency field oscillations are thus isolated. For a laser with parameters γ_2, $\gamma_3 \gg \gamma_1$ at high pump levels $\Omega_{th}^2 \gg 1$ branches 1 and 2 lie in regions where the frequency differences $\omega_n - \omega_m$ satisfy the inequalities

$$(\omega_n - \omega_m)^2 < \omega^2, \tag{35}$$

$$\frac{1}{4}\omega^2 < (\omega_n - \omega_m)^2 < 2\omega^2, \tag{36}$$

respectively. Graphs of $|e_m|$, $|e_n|$, and $|v_{\omega_l - \omega_n} v_{\omega_l - \omega_m}^{-1}|$ as functions of $\omega_n - \omega_m$ for branches 1 and 2 are shown in Figs. 4 and 5, respectively. As follows from these graphs, at frequency differ-

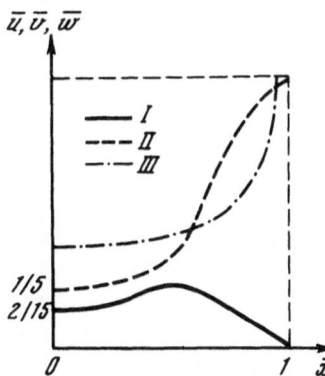

Fig. 4. The partial intensities $|e_m|^2$, $|e_n|^2$, and $v_{\omega_m - \omega_n}(v_{\omega_m - \omega_l})^{-1}$ for the second branch of the three-frequency field oscillations. I) $\bar{u} = (4\gamma_3/\gamma_1)|e_m|^2$; II) $\bar{v} = (4\gamma_3/\gamma_1)|e_n|^2$; III) $\bar{w} = |v_{\omega_m - \omega_n} \times (v_{\omega_m - \omega_n})^{-1}|$; $\bar{x} = (\omega_m - \omega_n)^2/\omega^2$.

Fig. 5. The partial intensities $|e_m|^2$, $|e_n|^2$, and $v_{\omega_m - \omega_n}(v_{\omega_m - \omega_l})^{-1}$ for the second branch of the three-frequency field oscillations. Same notation as in Fig. 4.

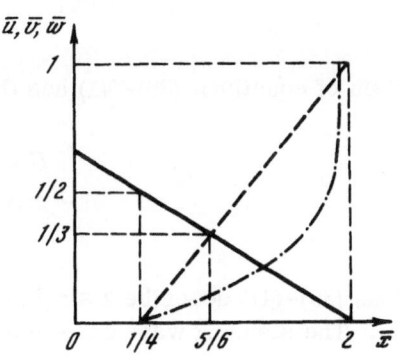

ences of the order of the nutation frequency ω, oscillations in the inverted population at frequency $\omega_m - \omega_n$ are amplified (branch 1) or excited (branch 2) by oscillations ν at the doubled frequency $\omega_l - \omega_m$. Such effects are typical of a parametric (here a self-parametric) resonance [65].

3. Single-Mode Laser

The single-mode oscillator is the simplest model for a laser. In this case it is possible to make a complete analysis of the phase space of the system. This analysis has been done in [66] for high pumping levels when the nutation frequency is much greater than all the relaxation parameters. The calculations done there showed that the laser is a complicated system in which, even when the stability condition for a spikeless regime is satisfied, the so-called non-linear or practical (operating) instability may occur; that is, a spiking regime develops strongly when the amplitude of the fluctuations in the dynamic parameters exceeds some critical value. Some calculations for customary pumping powers in a quasilinear approximation [66] justify saying that even at customary pumping levels complicated effects of this type may be seen in lasers. In this paragraph we shall study fully the phase space of a single-mode laser with the parameters

$$\gamma_3 \gg \gamma_2 \gg \gamma_1, \quad \mu \ll 1, \tag{37}$$

where $\mu = \mathring{\omega} \left[(\gamma_2 + \gamma_3) \gamma_3 \right]^{-\frac{1}{2}}$. A single-mode laser with arbitrary relationships among the relaxation constants at typical pumping powers,

$$\gamma_2, \ \gamma_3 \gg \gamma_1, \quad \mu \ll 1, \tag{38}$$

is also examined near the stability boundary for single-frequency operation in the relaxation approximation.

We shall write the equations for a single-mode laser tuned to the center of the luminescence line in the form of an abridged system of equations for the field U, the polarization V, and the population inversion W in the material [67][†] :

$$\mu \, \frac{dU}{d\tau'} = -U + V, \tag{39}$$

$$\beta\mu \, \frac{dV}{d\tau'} = \alpha(-V + UW), \tag{40}$$

$$\alpha \, \frac{dW}{d\tau'} = \mu \, [(1 - \Omega_{th}^{-2})^{-1} - (\Omega_{th}^2 - 1)^{-1} W - UV], \tag{41}$$

where

$$U = \left[\frac{4\gamma_3}{\gamma_1 (1 - \Omega_{th}^{-2}) \, 2\pi n_0 \hbar \omega_0} \right]^{\frac{1}{2}} E_0, \quad E = E_0 \exp(-i\omega_0 t) + \text{c.c.}, \quad \left| \frac{dE_0}{dt} \right| \ll \omega_0 |E_0|,$$

$$\tau' = \mu\gamma_3 t.$$

The system of equations (39)–(41) has three singular points:

$$U = V = 0, \quad W = \Omega_{th}^2, \tag{42}$$

$$\pm U = \pm V = W = 1. \tag{43}$$

[†] Equations (39)–(41) describe a single-mode traveling wave laser or a laser with a point active medium. The standing wave case is examined in Section 5, Paragraph 5.2.

Point (42) is a saddle. Points (43), which describe single-frequency lasing, are unstable if the inequality [67-69]

$$\dot{\omega}^2 (\gamma_3 - \gamma_2 - \gamma_1) > \gamma_1 (\gamma_2 + \gamma_3)(\gamma_1 + \gamma_2 + \gamma_3) \tag{44}$$

is valid. It is more difficult to study the critical case

$$\dot{\omega}^2 (\gamma_3 - \gamma_2 - \gamma_1) = \gamma_1 (\gamma_2 + \gamma_3)(\gamma_1 + \gamma_2 + \gamma_3). \tag{45}$$

A calculation shows that at the boundary (45) the points (43) are complicated 1-fold foci with a positive Lyapunov magnitude [70]. This means, first, that the single-frequency regime is practically (nonlinearly) unstable [50, p. 181] in the neighborhood of the boundary (45). Second, in the region of stability of the single-frequency regime there must be small unstable limiting cycles near the boundary (45). These cycles were calculated in [66]. The physical significance of the instability (linear in the Lyapunov sense or nonlinear in the practical sense) is as follows. Field fluctuations in the neighborhood of the single-frequency operating regime (43) produce relaxation oscillations in the inverted population at frequency ω_1. When the frequency of the relaxation oscillations approaches the optical nutation frequency ω, that is, $\beta \sim 1$ (see Fig. 1), relaxation oscillations may be excited and the single-frequency regime disrupted. The threshold conditions for this effect may be fulfilled either with infinitely small fluctuations [in region (44)] or with finite-amplitude fluctuations (in the practical instability).

Let us now examine the integral behavior of the trajectories of the system (37), (39)-(40). Every large-amplitude vibrational solution is characterized by an emission stage over times of order μ and a pumping stage with a time scale of order μ^{-1}. Harmonic linearization and averaging methods are of little use in analyzing such solutions. The method of point images [71, p. 258] is natural for problems of this type.

In the emission stage it is possible to neglect the effect of pumping and relaxation of the inversion on the behavior of the system; specifically, introducing the new variables

$$u = \mu U, \quad v = \mu V, \quad w = \alpha W, \quad \tau_i = \mu^{-1}\tau', \quad \dot{} = d/d\tau_i \tag{46}$$

and setting $\alpha = \mu = 0$ in view of Eq. (37), we find

$$\dot{u} = -u + v, \tag{47}$$

$$\dot{v} = uw, \tag{48}$$

$$\dot{w} = -uv. \tag{49}$$

The phase space of the system (47)-(49) is made up of cylindrical layers

$$v^2 + w^2 = C^2. \tag{50}$$

Motions on cylinder (50) are easily classified using the qualitative theory of differential equations [70].

The cylinder contains two stationary points: a saddle

$$u = v = 0, \quad w = C, \tag{51}$$

and a stable node (for $4C < 1$) or focus (for $4C > 1$)

$$u = v = 0, \quad w = -C. \tag{52}$$

There are no cycles, as Eqs. (47)–(49) yield the equation

$$d/d\tau_{i}\ (u^2 + 2w) = -2u^2. \tag{53}$$

Thus, choosing $u^2 + 2w$ as a Lyapunov function, we find from Eq. (53) that all the trajectories on the cylinder except the two separatrices in the saddle (51) tend toward a stable equilibrium position (52). It will be apparent from the following that the behavior of those trajectories which begin in the neighborhood of the saddle (51) on cylinders with radius C less than $^1/_4$ is important in constructing a point image. The integral (50) shows that these trajectories are described by a π-pulse.

In the pump stage it is possible to neglect the effect of the field on the population inversion in the material, that is, to eliminate the term UV in Eq. (41). Thus, we transform to the following variables and parameters:

$$|\,U\,| = \exp\,[-\mu^{-2}\bar{u}],\quad V = \bar{v}U,\quad \tau_{\text{H}} = \mu\tau',\quad a_1 = [\alpha\,(\Omega_{\text{th}}^2 - 1)]^{-1}, \tag{54}$$
$$' = d/d\tau_{\text{H}}$$

and in the limit $\alpha = \mu = 0$ we obtain

$$\mu^2\bar{u}'' = \bar{u}'^2 - \bar{u}' - w, \tag{55}$$

$$w' = (1 - \Omega_{\text{th}}^{-2})^{-1} - a_1w. \tag{56}$$

The phase plane $(\bar{u}', w)$ of Eqs. (55) and (56) is shown in Fig. 6. The complicated equilibrium state (42) of the initial system (39)–(41) has been broken down into some simple states as a result of transformation (54); These states are the saddle (3) and node (6) in Fig. 6 lying on the parabola of the slow motions [see Eq. (55)],

$$\bar{u}'^2 - \bar{u}' - w = 0. \tag{57}$$

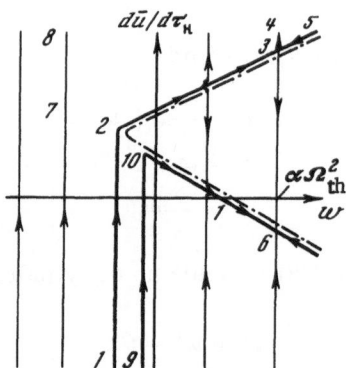

Fig. 6. The phase plane of the system of equations (55) and (56) which approximates the initial system (39)–(41) for a single-mode laser in the pump stage. The dot–dashed curve is the parabola of the slow motions (57); 1-2-3, 3-4, 3-5, and 3-6 are separatrices of the saddle 3; 6 is a stable node; 7-8-9-10 is the trajectory of a jump which is part of the cycle describing discontinuous oscillations in a single-mode laser.

The upper branch of the parabola [given by the larger root of Eq. (57)] is a discontinuity curve (in the terminology of the theory of discontinuous instabilities [48, p. 65; 72]) and the lower branch is the (stable) curve of sliding motions, often appearing in the theory of relay systems [71, p. 71]. The large cycle given by Eqs. (47)-(49), (55), and (56) may contain, in general, unstable sliding motion along the discontinuity curve followed by a jump and stable sliding motion along the sliding motion curve. It is clear that such a discontinuous oscillation with sliding is unstable with a characteristic time $\tau' \sim 1$. Thus, we shall study the case in which the initial data for Eqs. (55) and (56) are given on the sliding motion curve. Equations (56) and (57) are integrated. We then find the duration of this stage from Eq. (57) by equating the values of $\bar{u}$ at the beginning and end of the pump stage and find the inversion w at the end of the pump stage from Eq. (56). This program may be fulfilled if the initial value of w satisifes $-1/4 < w < 0$ [see Eq. (57)]; this justifies the previous assumption about the value of C in the emission stage.

We have thus constructed the point image

$$(1 + 4a\Omega_{th}^2)^{-\frac{1}{2}} = \frac{z_{2k} - z_{2k-1} - \text{th } z_{2k} + \text{th } z_{2k-1}}{\ln\left(\frac{\text{ch } z_{2k}}{\text{ch } z_{2k-1}}\right)} , \tag{58}†$$

$$(1 + 4a\Omega_{th}^2)^{-\frac{1}{2}} = \frac{1}{\sqrt{2}}[\text{th}^2 z_{2k} + \text{th}^2 z_{2k+1}]^{\frac{1}{2}}, \tag{59}$$

where $\text{ch}^2 z_{2k\mp1} = \frac{1 + 4a\Omega_{th}^2}{4(a\Omega_{th}^2 - w_{2k\mp1})}$, $\tau_{H, 2k} - \tau_{H, 2k-1} = a(\Omega_{th}^2 - 1)\ln\frac{\text{ch}^2 z_{2k}}{\text{ch}^2 z_{2k-1}}$; w_{2k-1} and w_{2k+1} are the initial values of the inversion in the k and $(k + 1)$-st pumping stages; and $\tau_{H, 2k} - \tau_{H, 2k-1}$ of the k-th pumping stage ($k = 1, 2, \ldots$).

The point image (58), (59) has a unique fixed point ($z_{2k+1} = z_{2k-1}$) if the pump power satisfies the conditions

$$^3/_4 \sim \varkappa_c < a\Omega_{th}^2 < 1. \tag{60}$$

This fixed point is unstable. According to [71, p. 189] the system of equations (39)-(41) has two unstable limiting cycles [for positive and negative U; cf. Eq. (54)]. At the upper boundary of the existence region $a\Omega_{th}^2 = 1$ the unstable cycles merge with the stationary points (43) for the single-frequency lasing regime, which is unstable at large pump powers $a\Omega_{th}^2 > 1$ [see Eqs. (37) and (44)]. At the lower boundary of the region (60) $a\Omega_{th}^2 = \varkappa_c$ the unstable cycles come close to the separatrix 2-3-6 of Fig. 6 and, ultimately, "get stuck" in it.‡

From Fig. 6 it is obvious that in the pumping region

$$\varkappa_c < a\Omega_{th}^2 \tag{61}$$

the laser must produce stable discontinuous oscillations. Actually, in the pump stage the image point of the system slides along the lower branch of the parabola (57) of Fig. 6, and in the emission stage a π-pulse is produced which transforms the inversion. The combined action of these

---

† Here and elsewhere "th" (except when used as a subscript), "cth," and "ch" refer to hyperbolic tangent (tanh), hyperbolic cotangent (coth), and hyperbolic cosine (cosh), respectively.
‡ From this comment it is clear that the boundaries of region (60) follow immediately from Eqs. (58) and (59) for the point image if we examine them in the neighborhood of the single-frequency regime $z_{2k\mp1} = z_{2k}$ and the separatrix $z_{2k-1} = 0$.

transformations (as shown above) causes a growth in the amplitude of the oscillations at an arbitrary [in region (44)] or sufficiently large [in region (60)] initial inversion. The amplitude of the oscillations grows until the initial value of the inversion at the k-th pumping stage reaches $w = -\frac{1}{4}$ which corresponds to the abscissa of the vertex of the parabola of the slow motions (57). After this the point jumps from the vertex of the parabola along the trajectory 7-8-9-10 of Fig. 6 with a change in sign of U. [We note again that according to Eq. (54) $\bar{u}$ gives the change in the absolute value of U.] The jump is completed by sliding along the sliding motion curve, emission of a π-pulse for negative values of U, and a jump with a change in the sign of the field strength U. After this the process is repeated. Thus, we obtain a stable discontinuous limiting cycle. In the pump region (44) the discontinuous oscillations are excited weakly and for pump intensities (60), strongly. The repetition period of the giant pulses is determined by the time required to slide along the lower branch of the parabola (57) and, in view of the preceding remarks, is given by Eq. (58) with $z_{2k-1} = 0$.†

The analysis of the general case (38) in the pumping stage differs in no way from the above analysis for a laser with parameters (37). Equation (58) remains in force with the change

$$\alpha\Omega_{th}^2 \to \alpha\beta\,(\Omega_{th}^2 - 1). \tag{62}$$

In the emission stage the system (47)-(49) must be replaced by the system

$$\dot{u} = -u + v, \tag{63}$$

$$\beta\dot{v} = \alpha\,(-v + uW), \tag{64}$$

$$\alpha\dot{W} = -uv. \tag{65}$$

These equations contain nonlinear terms that are quadratic in the dynamic variables. From them it is possible to obtain the equations of motion for quadratic quantities. They will contain higher-order terms in the field, polarization, and population inversion for which it is also possible to find equations of motion. In this way we obtain a chain of kinetic equations with ever higher powers of the dynamic variables. Near the stability boundary of the single-frequency regime (45) u, v, $|W - 1| \ll 1$ it is possible to close this chain by eliminating in the n-th step terms of order (n + 1) in the field, polarization, and deviation of the inversion from its stationary value. It is possible to integrate this system in this way. In particular, in the second step we have the integral relation

$$\alpha\,(2W - W^2)\Big|_{\tau_i=0}^{\tau_i=\infty} - \int_0^\infty u^4\,d\tau_i = 0, \tag{66}$$

if we consider that the polarization and field equal zero before and after the pulse is emitted. Leaving out the integral in this equation, we obtain a crude representation of the laser's operation in the emission stage. This approximation also follows directly from Eqs. (63)-(65) if we eliminate the polarization and inversion from them and use the condition $d/d\tau_i$, $u \ll 1$. Then we have

$$u = C/[\mathrm{ch}\,(C\tau_i - \psi_0)], \quad \alpha W = \alpha + (\dot{u}/u), \tag{67}$$

which yield Eq. (66) without the integral term for the inverted population before and after the

---

† This expression gives the repetition period of the giant pulses on the intensity curve $|U|^2$; if the field intensity is measured, then the pulse structure has a period of 2T.

output pulse. In order to improve the approximation we substitute Eq. (67) for the field in the integral. After an elementary integration and substitution of variables we find the following relation between the inversions at the beginning and end of the emission stage:

$$1 - \frac{1}{2}[1 + 4\alpha\beta(\Omega_{th}^2 - 1)](\text{th}^2 z_{2k} + \text{th}^2 z_{2k+1}) + \frac{\alpha}{48\beta}[1 + 4\alpha\beta(\Omega_{th}^2 - 1)]^2(\text{th}^2 z_{2k} - \text{th}^2 z_{2k+1})^2 = 0. \qquad (68)$$

Equations (58), (62), and (68) specify the point image. An investigation shows that it has a single fixed point if the laser parameters lie in the stability region of a single-frequency lasing regime. This point is unstable; thus, all the conclusions about the existence and weak and strong excitation of discontinuity oscillations drawn above for the case of a laser with conditions (37) transfer to the general case (38) near the stability boundary of a single-frequency lasing regime. The repetition period of the giant pulses is determined by Eq. (58) with the substitution (62) for $z_{2k-1} = 0$. This equation was derived in [73] assuming that discontinuity oscillations exist. This assumption is justified and considered in detail for a laser with parameters (37) and (38) in this article. Computer calculations were done for a laser with the parameters[†]

$$\gamma_1 = 10^2 \text{ sec}^{-1}, \quad \gamma_2 = 10^8 \text{ sec}^{-1}, \quad \gamma_3 = 4 \cdot 10^8 \text{ sec}^{-1}, \quad \Omega_{th}^2 = 10, \qquad (69)$$

and the following results were obtained: $U_{max} = 499$, the pulse width at a level of U = 1 is $\tau = \frac{0.033}{\sqrt{2\beta} \, \mathring{\omega}}$, and the repetition period of the pulses is $v^{-1} = \frac{1081}{\sqrt{2\beta} \, \mathring{\omega}}$ in accordance with Eq. (58) with the substitution (62).

Therefore, it has been shown that a single-mode laser may generate discontinuity oscillations with a pulse width $\sim\gamma_3^{-1}$, a repetition rate $\sim\gamma_1$, a field strength and polarization with a maximum of $U \sim V \sim \mu^{-1}$ and a minimum of $U \sim V \sim \exp[-\mu^{-2}]$, and oscillations in the inversion at an amplitude of $W \sim \alpha^{-1}$ [see Eqs. (46) and (54)]. The region for pulling of the giant limiting cycle and its bifurcation [Eqs. (58)-(60)] has been found. The discontinuity oscillations are a complicated, relatively little studied phenomenon [48, p. 65; 73-76]. Discontinuity oscillations have been studied in more detail on the phase plane of the Van der Pol equation [77]. A single-mode laser is an example of a physically significant system with a three-dimensional phase space which produces discontinuity oscillations that may be strongly or weakly excited. It may be of interest to calculate later approximations in μ for the pulse repetition period since these terms usually [75-77] have a complicated asymptotic character ($\sim\mu^{2/3}$, $\mu \ln\mu$) and slowly approach zero as $\mu \to 0$. Including noise in the pumping stage, as in [78, 79], does not lead to qualitatively new conclusions since the field strength at the spike maximum depends logarithmically on the field strength at the minimum $U_{min} \sim \exp[-U_{max}^2]$.

The effects described above may be observed in a maser, molecular generator, or semiconductor laser. In a ruby maser [13] spiking has been experimentally obtained with a spike repetition rate of about 100 Hz, in agreement with our calculations. It would be interesting to make a more detailed experimental verification of the basic conclusions of the theory. In doing this it would not be necessary to strive for fulfillment of the inequalities (37) or to work near the stability boundary for spikeless operation (45). Based on the results of [66], we might expect that a strong transition by the laser to the discontinuity oscillation regime is possible with much weaker restrictions on the parameters. It is shown rigorously in [66] for high pumping levels ($\mu \gg 1$) that a strong spike excitation regime is realized in lasers with relaxation

---

[†] These calculations were done by D. F. Davidenko.

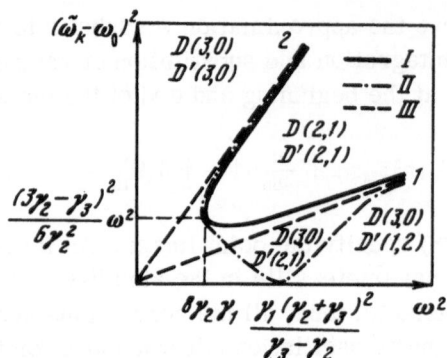

Fig. 7. The stability diagram for single-frequency operation at the frequency of the center of the luminescence line at all pumping levels above threshold for different values of the ratio γ_3/γ_2. I) $D(n,\ m)$, $\gamma_2 > \gamma_3$; II) $D'(n,\ m)$, $\gamma_2(1 + \sqrt{2}) > \gamma_3 > \gamma_2$; III): 1) $(\widetilde{\omega}_k - \omega_0)^2 = (\gamma_3/\gamma_2 - 1)^2\omega^2$; 2) $(\widetilde{\omega}_k - \omega_0)^2 = 2\omega^2$; the region denoted by $D(n,\ m)$ or $D'(n,\ m)$, $m \neq 0$, is the instability region for the single-frequency regime.

constants satisfying†

$$3\gamma_3 - 2\gamma_1 > \gamma_2 > \gamma_3 - \gamma_1. \qquad (70)$$

Such experiments would not only clarify some aspects of the physics of lasers and masers, but could aid in making a generator of discontinuity oscillations and a pulse discriminator in the optical range. At present giant pulses are usually obtained with the aid of passive or active modulation (Q-switching) of the cavity losses. This theory examines the possibility of producing such pulses by means of an autonomous system without a filter. In the strong pulse regime such a generator might find application in neuristor grids [81].

4. The Stability of the Stationary States of a Multimode Traveling-Wave Laser with Respect to Coherent Perturbations

4.1. Multiple Synchronization

Let one of the eigenfrequencies of a multimode cavity coincide with the center of the transition line ω_0 and $\widetilde{\omega}_{\bar{k}} + \widetilde{\omega}_k = 2\omega_0$. We shall examine the stability of the single-frequency regime $e = e_0 \exp\left(-i\omega_0 t + i\frac{\omega_0}{c}r\right)$ + c.c. with respect to coherent perturbations (14) in the $\widetilde{\omega}_{\bar{k}}$, $\widetilde{\omega}_k$ modes. This problem was solved for different relationships between the linewidth and the spectral width of the modes for $\gamma_2, \gamma_3 \gg \gamma_1$. Some typical stability diagrams are shown in Fig. 7. If $\gamma_3 > \gamma_2$ then the single-frequency operating regime may be unstable in a single-mode

---

† At high pumping levels the field oscillations are not oscillations of a relaxation discontinuity type, and the spikes repeat at the optical nutation frequency. This limiting case ($\mu \gg 1$) may be studied experimentally in molecular masers with superconducting cavities. Superconducting cavities made of niobium have a Q of 10^9-10^{11} in the frequency range 10^9-10^{10} Hz [80], which is five or six orders of magnitude greater than the Q of ordinary cavities.

laser. This question was analyzed in detail in Section 3. When $\gamma_2 > \gamma_3$ an instability in the single-mode regime develops in a multimode cavity at frequency differences $\tilde{\omega}_k - \omega_0$ of the order of the optical nutation frequency ω in the following way. Field fluctuations in the modes $\tilde{\omega}_k^-$, $\tilde{\omega}_k$ interfere with the strong field at frequency ω_0 and cause oscillations in the inverted population at frequency $\tilde{\omega}_k - \omega_0 = \omega_0 - \tilde{\omega}_k^-$. If this frequency is close to the optical nutation frequency, then oscillations may be driven and single-frequency operation may be cut off. A strong interaction between the branch describing output intensity oscillations at a period occurs when these branches intersect. For this it is necessary (see Fig. 1) that the frequency difference $\tilde{\omega}_k - \omega_0$ be greater than $(\gamma_1 \gamma_2)^{1/2}$ since otherwise, with a frequency difference of the order of the nutation frequency, the threshold is attained at pump levels such that ω^2 is greater than $\gamma_1 \gamma_2$. An exact calculation for the instability threshold (see Fig. 7) yields

$$\omega^2 > 8\gamma_1\gamma_2, \quad \text{or} \quad \Omega_{th}^2 > 9. \tag{71}$$

At high pumping levels $\Omega_{th}^2 \gg 1$ the mechanisms responsible for a growth in fluctuations may be classified in more detail. The dispersion curve for the three-frequency field oscillations, which relates $\omega_k - \omega_0$ to $\tilde{\omega}_k - \omega_0$, is similar to the dispersion curve for the two-frequency oscillations (See Fig. 2). The intersection of the dispersion curve with the line $\omega_k - \omega_0 = \tilde{\omega}_k - \omega_0$ has the same equation as the asymptote 2 in Fig. 7: $(\tilde{\omega}_k - \omega_0) = 2\omega^2$. Thus, for small frequency differences the phase velocity of the envelope, $c(\omega_k - \omega_0)/(\tilde{\omega}_k - \omega_0)$, is greater than the velocity of light in the material and an instability develops due to the Cerenkov effect [82]. The instability boundary coincides with the upper boundary for existence of the second branch of the three-frequency field oscillations; thus, it is a safe boundary and branch 2 of the three-frequency field oscillations corresponds to stable field oscillations [83], at least near the boundary. The instability in the single-frequency regime develops with a growth rate

$$\text{Im}\,(\omega_k - \omega_0) = -\gamma_3 \frac{2\omega^2}{\omega^2 + 2\,(\gamma_2 + 2\gamma_3)^2} \left[\frac{[\tilde{\omega}_k - \omega_0]^2}{2\omega^2} - 1 \right]. \tag{72}$$

According to the dispersion curve, asymptote 1 corresponds to field oscillations for which $(\omega_k - \omega_0)^2 = \omega^2$ with an arbitrary relationship between the constants γ_2 and γ_3. At these frequencies field oscillations with a period equal to the photon transit time across the cavity are in resonance with nutation oscillations in the inversion. As was shown in Section 2, an autoparametric resonance occurs: Oscillations in the population inversion at frequency ω are amplified due to oscillations in the inversion at twice this frequency $\omega_k - \omega_k^- = 2\omega$. A parametric resonance in a damped oscillator has a threshold [65]; thus, the instability is nonlinear and asymptote 1 is a hazardous boundary. Since asymptote 1 outlines the upper boundary for existence of the first branch of the three-frequency field oscillations (see Section 2), branch 1 describes unstable oscillations, at least near the boundary [83]. The autoparametric resonance gives an instability in the single-frequency regime if $^1/_4\omega^2 < (\omega_k - \omega_0)^2 < \omega^2$, where $(\omega_k - \omega_0)^2 = {}^1/_4\omega^2$ is the lower boundary for existence of the second (stable) branch of the three-frequency field oscillations.

Two-frequency field oscillations at the side modes (with frequencies $\tilde{\omega}_k$ and $\tilde{\omega}_k^-$) of a three-mode resonator are unstable with respect to amplitude perturbations in the central mode (with frequency $\tilde{\omega}_0 = \omega_0$ and phase such that $\varphi_k^- + \varphi_k - 2\varphi_0 = 0$) when

$$^1/_4\,\omega^2 < (\omega_k - \omega_0)^2. \tag{73}$$

At the instability boundary (73) the stable cycle which describes branch 2 of the three-frequency field oscillations merges with the unstable two-frequency cycle, and at smaller frequency differences the two-frequency field oscillations are stable with respect to amplitude perturbations in the central mode. This bifurcation in the stability at the boundary of region (73) has a physical basis: For smaller frequency differences the phase velocity of a wave with two components e_k and $e_{\bar{k}}$ is greater than the propagation velocity of light in the medium. Thus this modulation in the output intensity is stable because of the Cerenkov effect. However, for arbitrary frequency differences the two-frequency field oscillations at the side modes are unstable with respect to phase perturbations in the central mode $\varphi_{\bar{k}} + \varphi_k - 2\varphi_0 = \pi$.

Therefore, on the whole, lasing at the central mode is stable at large frequency differences[†] $(\tilde{\omega}_k - \omega_0)^2 > 2\omega^2$. For differences $\omega^2 < (\tilde{\omega}_k - \omega_0)^2 < 2\omega^2$ and $^1/_4\omega^2 < (\tilde{\omega}_k - \omega_0)^2 < \omega^2$ the laser goes weakly and strongly, respectively, into a spiking regime with a spike repetition rate of ω. For small differences $(\tilde{\omega}_k - \omega_0)^2 < ^1/_4\omega^2$ single-frequency operation at the central mode is stable on the whole but a strong cutoff of the spikeless regime is possible with entry into the unstable two-frequency cycle and later restoration to a single-frequency regime.

Similar effects were obtained in [84] for quantum amplifiers with $\gamma_2 \gg \gamma_3$. This problem for lasers was studied in another way in [85] but the authors of this article devoted most attention to analyzing the operation of a laser with a single relaxation time ($\gamma_1 \sim \gamma_2$). In recent years these results have been generalized to a laser with an inhomogeneous luminescence line [86], with lumped losses [87], and the cavity space factor [88] and the effect of differences between the frequency of the single-frequency regime and the center of the luminescence line [89] have been taken into account.

4.2. Mode Capture

Mode capture develops on coherent perturbations (13). Trapped waves with different eigenfrequencies have different phase velocities. The phases of the waves rapidly diverge and, since the traveling waves remove the inversion uniformly over the volume, the packet of traveling waves is unstable. Thus, traveling waves are stable with respect to formation of a packet of modes oscillating at a single frequency and the "single-mode" state apparently constitutes the entire class of stationary field states.

5. The Effect of Inhomogeneities in the Laser Parameters on Its Operation

In the preceding paragraphs we have examined traveling wave lasers whose operation is characterized by the following parameters: γ_1, γ_2, γ_3, Ω, Δ. A reduction in the number of parameters leads to a laser without relaxation constants ($\gamma_1 = \gamma_2 = \gamma_3 = 0$). In this model all effects associated with relaxation oscillations in the inversion are lost. The inversion may oscillate only at the optical nutation frequency. Thus spikes with a repetition rate of ω were first observed in a laser without relaxation parameters [18, 19]. This approximation gives a first representation of laser operation at high pumping levels when the optical nutation frequency is greater than any of the relaxation constants [66]. In Section 5 we shall expand this minimum of parameters, examine a point model of a multimode laser, and study the effect of spatial field inhomogeneities in the cavity, Q dispersion, saturable filters, and inhomogeneity in the luminescence line on the operating regime of a laser.

5.1. Multimode Laser with a Point Active Medium

Let the dimensions of the active medium, l, be much less than the wavelength λ. Then it is possible to neglect the sample dimensions and consider a uniform (for notational simplicity) cavity of length L such that $0 \le r \le L$, in which the material occupies a point at position a. Then the l in the curly brackets of Eq. (5) has to be replaced by the δ-function $\delta[(r - a)/l]$.

[†] For simplicity, all the following inequalities are written for the case $\gamma_2 \gg \gamma_3$.

We shall analyze the conditions for self-excitation of a multimode point oscillator. To do this we eliminate the nonlinear terms in Eq. (9) (with complex frequencies):

$$\frac{A_m B_m}{4\omega_0^2} e_m = - \Omega^2 \delta \left(\frac{r-a}{l} \right) e_m.$$

(74)

Multiplying Eq. (74) by Φ_λ^* and integrating over space, we find an infinite system of equations for e_m^λ and from the consistency conditions for this system we obtain the following characteristic equation:

$$1 = \frac{\Omega_{\text{th}}^2 \, l \gamma_2 \gamma_3}{\gamma + \gamma_2} \, G_a,$$

(75)

where $G_a = \sum_\lambda | \Phi_\lambda (a) |^2 (\gamma + \gamma_3 + i\Delta_\lambda \gamma_3)^{-1}$, $\gamma = -i (\omega_m - \omega_0)$, $\Delta_\lambda = \dfrac{\tilde{\omega}_\lambda - \omega_0}{\gamma_3}$. Without limiting the generality, we may apply periodic boundary conditions, that is, $\Phi_\lambda = \dfrac{1}{\sqrt{L}} \exp \left(i \dfrac{\tilde{\omega}_\lambda}{c} r \right)$, with the frequency of one of the modes $\tilde{\omega}_0$ equal to that of the center of the luminescence line ω_0. Then the series for G_a is summed as a Fourier series at the point $r = 0$ for the function $f = 2c^{-1} \left\{ \exp \left[-(\gamma + \gamma_3) \dfrac{L}{c} \right] - 1 \right\}^{-1} \exp \left[-(\gamma + \gamma_3) \dfrac{r}{c} \right]$, over the interval $[0, L]$. At the point $r = 0$ this series converges to $^1/_2 [f(0) + f(L)]$, that is,

$$G_a (\gamma) = c^{-1} \operatorname{cth} (\gamma + \gamma_3) \frac{L}{2c} \, .$$

(76)

Thus, the characteristic equation for self-excitation (75) [like the secular equation (83) for the stability of mode capture] is transcendental. A delta-function perturbation introduced by the active medium at point a involves the entire infinite number of degrees of freedom of the field to some extent. Nevertheless, the characteristic equation (75) for a laser with a finite linewidth γ_2 has a finite number of roots in the right-hand complex half-plane which approach ∞ as $\gamma_2 \to \infty$. We shall further limit ourselves to the next case; that is, we shall consider a laser with a very broad luminescence line. Then Eq. (75) will have one real root and a countable number of complex roots in the right-hand complex half plane if the pump power satisfies the inequalities (Fig. 8)

$$2c (\gamma_3 L)^{-1} \operatorname{th} \left(\frac{\gamma_3 L}{2c} \right) < \Omega_a^2,$$

(77)

$$2c (\gamma_3 L)^{-1} \operatorname{th} \left(\frac{\gamma_3 L}{2c} \right) < \Omega_a^2 < 2c (\gamma_3 L)^{-1} \operatorname{cth} \left(\frac{\gamma_3 L}{2c} \right),$$

(78)

respectively, where $\Omega_a^2 = \Omega_{\text{th}}^2 (2l/L)$ is the lasing parameter for a point active medium. (The coefficient 2 in the expression for Ω_a^2 accounts for traveling waves in two directions.) The real root for pumping levels (77) describes excitation of a single-frequency regime, and the complex roots for pumping levels (78) bring the oscillator into a spiking regime with a spike repetition rate of Δ near the threshold for self-excitation and $\Delta/2$ in the neighborhood of the upper boundary of region (78).

We shall study the stability of the single-frequency regime at the line frequency ($\omega_n = \omega_0$) with respect to coherent perturbations (14). In the linear approximation in the perturbation we find these equations for e_k, $k = n, m, l$:

$$\frac{A_n B_n}{4\omega_0^2} e_n = - f_1 e_n,$$

(79)

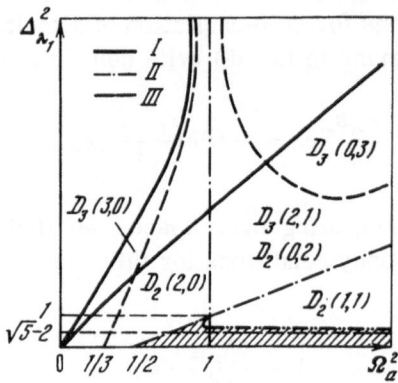

Fig. 8. A self-excitation and stability diagram
for the single-frequency regime of a laser with
a point-active medium. Shown are the self-
excitation characterizing: I) a single-mode oscillator; II) a $D_2(n, m)$ two-mode oscillator;
III) a $D_3(n, m)$ three-mode oscillator. The
index $D_k(n, m)$, $k = 2, 3$, means that in the corresponding region the characteristic equation
(75) has n roots in the left and m roots in the
right-hand complex half-planes. The equations
of curves II are $\Omega_a^2 = 1$. $\Delta_{\lambda}^2 = 2\Omega_a^2 - 1$, and of curves
III are $\Delta_{\lambda_1}^2 = (3\Omega_a^2 - 1)/(1 - \Omega_a^2)$ for $\Omega_a^2 < 1$, and $\Delta_{\lambda_1}^2 =$
$(3\Omega_a^2 - 2)^2/(\Omega_a^2 - 1)$ for $\Omega_a^2 > 1$. The shaded region
represents the region of stable mode locking in
a two-mode oscillator.

$$\frac{A_m B_m}{4\omega_0^2} e_m = -f_1[e_m - f_2(e_m + e_{-l})], \tag{80}$$

$$\frac{A_{-l} B_{-l}}{4\omega_0^2} e_{-l} = -f_1[e_{-l} - f_2(e_m + e_{-l})], \tag{81}$$

where $f_1 = \dfrac{\Omega^2\delta\left(\dfrac{r-a}{l}\right)}{1 + \dfrac{4\gamma_3}{\gamma_1}\Omega_{\text{th}}^2|e_n|^2}$ and $f_2 = \dfrac{\omega^2(\gamma + 2\gamma_2)}{2\gamma_2[(\gamma + \gamma_2)(\gamma + \gamma_1) + \omega^2]}$. Multiplying Eqs. (79)–(81) by Φ_λ^* and

integrating over space, we find an infinite system of equations for e_k^λ, $k \neq n, m, l$. These
equations are easily solved and, in particular, we find the optical nutation frequency $\omega^2 =$
$\gamma_1\gamma_2 [\Omega_{\text{th}}^2 l\gamma_3 G_a(0) - 1]$, where $G_a(0) = G_a$ for $\gamma = 0$. From the consistency conditions for the
equations for e_m^λ and e_{-l}^λ we derive the characteristic equation

$$1 = f_2\left\{\left[1 - \frac{G_a(0)(\gamma + \gamma_2)}{G_a(\gamma)\gamma_2}\right]^{-1} + \left[1 - \frac{G_a(0)(\gamma + \gamma_2)}{\bar{G}_a(\gamma)\gamma_2}\right]^{-1}\right\}, \tag{82}$$

where $\bar{G}_a(\gamma) = \sum_\lambda |\Phi_{\bar{\lambda}}(a)|^2(\gamma + \gamma_3 + i\Delta_\lambda\gamma_3)^{-1}$ and $\tilde{\omega}_{\bar{\lambda}} + \tilde{\omega}_\lambda = 2\omega_0$. For a multimode laser with period-
ic boundary conditions and a broad luminescence line Eq. (82) has the form

$$2\omega^2 \operatorname{sh}\frac{\gamma_3 L}{2c}\left(\operatorname{ch}\frac{\gamma_3 L}{2c}\operatorname{cth}\frac{\gamma L}{2c} + \operatorname{sh}\frac{\gamma_3 L}{2c}\right) = -(\gamma_2\gamma + \omega^2 + \gamma_1\gamma_2). \tag{83}$$

From the real part of this equation (83) it is clear that it has no solutions γ in the right-hand complex half plane since for Re $\gamma > 0$ the left side of the equation is positive while the right side is negative.

For two modes ($\Delta_{\lambda_1} = -\Delta_{\lambda_2}$, $\gamma_3 = \infty$, $\lambda \neq \lambda_1, \lambda_2$) Eqs. (75) and (82) yield the results of [20] and, in particular, an instability in the single-frequency regime under certain conditions for the frequency difference Δ and pump power Ω_a^2 (see Fig. 8). In a multimode cavity, as was shown above, the single-frequency regime is stable in a small region. It is easy to prove that the single-frequency regime is also stable in a three-mode laser ($\Delta_{\lambda_1} = -\Delta_{\lambda_2}$, $\Delta_{\lambda_3} = 0$) with a wide line.

We shall examine the operation of a two-mode laser ($\Delta_{\lambda_1} = -\Delta_{\lambda_2}$, $\gamma_2 \gg \gamma_3 \gg \gamma_1$) near the instability boundary of the single-frequency regime in the quasi-linear approximation. From the initial laser equations (1)-(3) we can obtain the following equations for a laser with a point active medium by averaging over the major frequency ω_0 and expanding the field in modes under conditions such that the polarization follows the field:

$$z_{\lambda}' = -l_{\lambda} z_{\lambda} + \int z y \Phi_{\lambda} \delta \left(\frac{r}{l}\right) dr, \tag{84}$$

$$y' = \varepsilon_1 (\Omega_{\text{th}}^2 - y - y|z|^2), \tag{85}$$

where z_{λ} is the field strength in mode Φ_{λ} and $z = \sum_{\lambda} z_{\lambda} \Phi_{\lambda}$;

$$e = z \left(\frac{\gamma_1}{4\gamma_3 \Omega_{\text{th}}^2}\right)^{\frac{1}{2}} \exp(-i\omega_0 t) + \text{c.c.}, \quad \left|\frac{dz}{dt}\right| \ll \omega_0 |z|;$$

$l_{\lambda} = 1 + i\Delta_{\lambda}$; $y = \Omega_{\text{th}}^2 \nu$ is the inverted population in the medium; $\varepsilon_1 = \gamma_1 \gamma_3^{-1}$; and differentiation is with respect to the reduced time $\tau = \gamma_3 t$.

Equating the derivative to zero, we find the singular points of the system (84), (85):

$$z = 0, \quad y = \Omega_{\text{th}}^2, \tag{86}$$

$$z_{\lambda} = z(0)\Phi_{\lambda}(0) y l l_{\lambda}^{-1}, \quad y = \Omega_{\text{th}}^2 [1 + |z(0)|^2]^{-1}, \quad |z(0)|^2 = \frac{2\Omega_a^2}{|l_{\lambda}|^2} - 1, \tag{87}$$

where $z(0)$, $\Phi_{\lambda}(0)$, and similar quantities denote the value of the functions at the point 0 where the active medium is located. The equilibrium state (86) is unstable when the condition for self-excitation of the laser is satisfied, i.e.,

$$\Omega_a^2 > \min\left\{1, \frac{1 + \Delta_{\lambda}^2}{2}\right\}. \tag{88}$$

For $\Delta_{\lambda}^2 > 1$ excitation is oscillatory and the laser output is modulated at the beat frequency between the cavity modes, and for $\Delta_{\lambda}^2 < 1$ the excitation is aperiodic and brings the laser into the single-frequency lasing regime (87).

The single-frequency regime (87) exists for $\Omega_a^2 > \frac{1}{2}(1 + \Delta_{\lambda}^2)$. It is unstable with respect to phase perturbations if $\Delta_{\lambda}^2 > 1$. The growth of such perturbations gives a phase-modulated signal and there is no interest in examining them since with such frequency differences the laser goes into a beating regime during the self-excitation process. In studying the amplitude perturbations we shall limit ourselves to the practically important case of $\varepsilon_1 \ll 1$. The amplitude perturbations in such a laser grow in the region which is limited below by an arc of the curve (see Fig. 8) [20]

$$\Omega_a^2 = \frac{\Delta_\lambda^2 (\Delta_\lambda^2 + 1)^2}{\Delta_\lambda^4 + 4\Delta_\lambda^2 - 1} . \tag{89}$$

To calculate the cycle in the neighborhood of the instability boundary (89) we decompose the amplitude and period of the cycle in a series in the square root of the parameter which specifies the deviation in $\varkappa$ and Δ_λ from their values on the boundary. This asymptotic behavior follows from the assumption (confirmed by the calculation) that the point (87) is a composite focus of unit multiplicity on the boundary (89). Calculations using the Poincaré small parameter method [90] yield

$$|z_{1;1}(0)|^2 = \frac{6\varepsilon_1 (1 - \Delta_\lambda^4)^2}{\gamma^4 F_1} \left[\frac{(\Omega_{th}^2)'(1 - \Delta_\lambda^2)}{y} - \frac{(\Delta_\lambda^2)' 2 |z(0)|^2}{1 - \Delta_\lambda^2} \frac{\Delta_\lambda^6 + \Delta_\lambda^4 7 - \Delta_\lambda^2 3 - 1}{\Delta_\lambda^4 + 4\Delta_\lambda^2 - 1} \right] \tag{90}$$

for the cycle amplitude $z_{1;1}$, where $F_1 = 8\Delta_\lambda^{10} - 109\Delta_\lambda^8 - 25\Delta_\lambda^6 + \Delta_\lambda^4 + \Delta_\lambda^2 - 32$; the primed quantities $(\Omega_{th}^2)'$ and $(\Delta_\lambda^2)'$ equal the deviations of the corresponding parameters from their values on the instability boundary and the unprimed quantities $(y, z(0), \Delta_\lambda)$ are fixed on the boundary; and $\gamma = \left[\frac{2 |l_\lambda|^2 \varepsilon_1 |z(0)|^2}{2 - |l_\lambda|^2} \right]^{\frac{1}{2}}$ is the angular frequency of the cycle.

From Eqs. (89) and (90) we find that the "small" cycle lies in the stability region of the single-frequency regime – the shaded region of Fig. 8; hence, it is unstable [83] and the stability boundary (89) is hazardous. In the stability region of the single-frequency regime strong excitation of a spiking regime is possible. The region in which pulling of the stable single-frequency regime occurs grows rapidly as ε_1^{-1} on moving away from the boundary. The development of an instability, linear or nonlinear, brings the laser into a self-Q-switch spiking regime with width $\sim \gamma_3^{-1}$ and repetition rate $\sim \gamma_1$. This result follows from the form of Eqs. (84) and (85). The presence of the small parameter ε_1 in Eq. (85) makes it possible (as in the theory of a single-mode laser in Section 3) to break the process of spike formation into a pumping state of duration γ_1^{-1} and an emission stage of duration γ_3^{-1}. The instability in the "small" cycle (85) ensures undamped giant pulses (also in complete analogy with the single-mode laser; see Section 3).

We now calculate the giant cycle for a two-mode laser near the instability boundary (89) in the relaxation approximation $(\varepsilon_1 \ll 1, \Delta_{\lambda_1} = -\Delta_{\lambda_2} = \Delta < 1)$. In the pumping stage the laser equations have the form

$$\dot{\zeta} = l_\lambda^\bullet + 2i\Delta\xi - 2y_a, \tag{91}$$

$$\varepsilon_1 \ddot{\zeta} = (\dot{\zeta} - 1)^2 + (\dot{\zeta} - 1) 2y_a + \Delta^2, \tag{92}$$

$$\dot{y}_a = \Omega_a^2 - y_a, \tag{93}$$

where the variables ξ, ζ, y_a are given by the equalities $|\Phi_{\lambda_1}(0)| z_{\lambda_1} = \xi \exp\left(-\frac{\zeta}{\varepsilon_1} \right)$, $|\Phi_{\lambda_2}(0)| z_{\lambda_2} = (1 - \xi) \exp\left(-\frac{\zeta}{\varepsilon_1} \right)$, and $y_a = yl |\Phi_{\lambda_1}(0)^2|$, and for simplicity we write $|\Phi_{\lambda_1}(0)| = |\Phi_{\lambda_2}(0)|$ and $\tau_H = \varepsilon_1 \gamma_3 t$. The dot over the symbols denotes differentiation with respect to τ. The phase plane of Eqs. (92) and (93) is analogous to the phase space of a single-mode laser in the pumping stage (Fig. 6). Using the same arguments as in the theory of a single-mode laser (Section 3), we find the relation between the inversion at the beginning and end of the pumping stage:

$$(\text{ch}\,\theta_1 - \text{ch}\,\theta_2) \ln \frac{\text{ch}\,\theta_1 - x_{2k}}{\text{ch}\,\theta_1 - x_{2k-1}} + f_{2k} - f_{2k-1} + \text{ch}\,\theta_1 \ln \frac{f_{2k}}{f_{2k-1}} - (\theta_{2k} - \theta_{2k-1}) \text{sh}\,\theta_1 = 0, \tag{94}$$

where the unknowns and parameters are defined as follows: $x = \frac{1}{2}(f + f^{-1}) = y_a \Delta^{-1}$, $\mathrm{th}\,\theta = \frac{\sqrt{x^2 - 1}\,\mathrm{sh}\,\theta_1}{x\,\mathrm{ch}\,\theta_1 - 1}$, $0 < \theta < \infty$, $\mathrm{ch}\,\theta_1 = \Omega_a^2 \Delta^{-1}$, $\mathrm{ch}\,\theta_2 = \Delta^{-1}$. The duration of the pumping state is given by

$$\tau_{\mathrm{H},\,2k} - \tau_{\mathrm{H},\,2k-1} = \ln \frac{\mathrm{ch}\,\theta_1 - x_{2k-1}}{\mathrm{ch}\,\theta_1 - x_{2k}} \,. \tag{95}$$

In the emission stage the laser obeys the equations

$$\hat{z}_\lambda' = -l_\lambda \tilde{z}_\lambda + \tilde{z} y_a, \tag{96}$$

$$y_a' = -y_a |\hat{z}|^2, \tag{97}$$

where $\tilde{z}_\lambda = \sqrt{\varepsilon_1}|\Phi_\lambda(0)|z_\lambda$, $\tilde{z} = \tilde{z}_{\lambda_1} + \tilde{z}_{\lambda_2}$. These equations can be integrated near the instability boundary (89) following the scheme given in detail for a single-mode laser with the parameters (38). In fact, from Eqs. (96) and (97) together with the boundary conditions (the field in modes $\tilde{z}_\lambda$ is zero before and after the output pulse) we find the integral equation

$$(-\ln y_a^2 + 4 y_a)\big|_{\tau=0}^{\tau=\infty} = -\Delta^2 \int_0^\infty \int_0^\infty \hat{z}(t)\,\hat{z}^*(s)\exp\left[-|t-s|\right]\,dt\,ds. \tag{98}$$

The left side of this equation corresponds to the conservative approximation which follows from Eqs. (96) and (97) with $d/d\tau$, $\tilde{z}_\lambda \ll 1$ and gives the field in the form

$$\hat{z}(\tau) = \frac{C}{\mathrm{ch}\left[C\sqrt{\dfrac{1+\Delta^2}{1-\Delta^2}}\,\tau - \psi_0\right]}\,, \qquad -4\sqrt{\frac{1-\Delta^2}{1+\Delta^2}}\,C = \ln y_a^2\big|_{\tau=0}^{\tau=\infty}. \tag{99}$$

Substituting this in the right-hand side of Eq. (98), integrating, and substituting the variables, we obtain the following relation between the inversion before and after the emission stage:

$$4(x_{2k+1} - x_{2k}) - \mathrm{ch}\,\theta_2 \ln(x_{2k+1}^2 x_{2k}^{-2}) = -2\,\mathrm{ch}^{-1}\theta_2\,\frac{\mathrm{ch}\,\theta_2 - \mathrm{ch}^{-1}\theta_2}{\mathrm{ch}\,\theta_2 + \mathrm{ch}^{-1}\theta_2}\,\zeta\left\{2,\ \frac{1}{2}\left[4\,\frac{\mathrm{ch}\,\theta_2 - \mathrm{ch}^{-1}\theta_2}{(\mathrm{ch}\,\theta_2 + \mathrm{ch}^{-1}\theta_2)\ln(x_{2k}^2 x_{2k+1}^{-2})} + 1\right]\right\}, \tag{100}$$

where $\zeta(s, q)$ is the generalized Riemann ζ-function.

Equations (94) and (100) define the point image. An investigation shows that it has a unique fixed point if the parameters of the laser lie in the instability region of the single-frequency regime. This point is stable. Thus, a two-mode laser in the instability region (see Fig. 8) goes into a stable giant limiting cycle and emits pulses of width γ_3^{-1} with a repetition rate γ_1 in accordance with the assumption used as a basis for the quasilinear analysis. This cycle collapses to a point describing a spikeless regime on the stability boundary (89). This does not contradict the quasilinear approximation since the relaxation approximation fixes the boundary for existence of a cycle to within some small quantities which approach zero as $\varepsilon_1 \to 0$. A comparison of analyses of laser operation in the relaxation and quasilinear approximations allows us to state that a strong transition to a giant pulse regime can occur only in a narrow band (of width $\sim \varepsilon_1$) lying in the stability region for spikeless lasing near the boundary (89) (cf. Section 3). The large cycle, which according to its parameters belongs in an intermediate region between the relaxation and quasilinear approximation regions, has been found in [20] with some additional assumptions.

Therefore, a two-mode laser with a point active medium has a stable mode capture region. In three- and multimode lasers mode capture is stable for arbitrary frequency differ-

ences between the modes but strong excitation of self-Q-switched spikes with mode synchronization (locking) is possible. In fact, in the $D_3(0,3)$ region for a three-mode laser and in region
(78) for a multimode laser oscillatory self-excitation with frequencies $\sim \gamma_3 \Delta_\lambda$ can occur; thus,
in these regions it is possible to excite mode locking. The envelope of the ultrashort pulses
will be a giant pulse, as our considerations and calculations for a two-mode laser imply.

5.2. Spatial Inhomogeneity in the Field and Dispersion in Q

Let the cavity modes Φ_λ have a spatial structure (for example, standing waves in a one-
dimensional resonator) and cavity losses which depend on the frequency and wave number of
the wave; that is, γ_3 is generally an operator $\hat{\gamma}_3$ such that $\hat{\gamma}_3 e_p^\lambda = \gamma_3(\widetilde{\omega}_\lambda, \omega_p) e_p^\lambda$. The stationary
"single-mode" oscillations of the field in the "diagonal" approximation are described by a
system of integral equations (101) which may be derived from Eqs. (9):

$$f_p^{-1} = \int \frac{\Phi_p^2 dr}{1 + \sum\limits_q 4\gamma_{3,q} \gamma_1^{-1} \hat{f}_q |e_q|^2} , \tag{101}$$

where $\hat{f}_p = \Omega^2 \gamma_2^{-1} \gamma_{3,p}^{-1} [1 + (\omega_0 - \omega_p)^2 \gamma_2^{-2}]^{-1}$ and $\gamma_{3,p} = \gamma_3(\widetilde{\omega}_p, \omega_q = \alpha\widetilde{\omega}_p + \beta\omega_0)$. Equations (101)
for $\gamma_2 \gg \gamma_3$ follow from the kinetic equations [91] and have been studied in detail in [92-94]. To
analyze the reasons for the appearance of spiking we have to go beyond the confines of the
"diagonal" approximation. Proceeding as in our study of the traveling wave laser we can show
that a spatial inhomogeneity in the field and a dispersion γ_3 have no significant effect on the dispersion equation (24) and the amplitude characteristic (26). However, an inhomogeneity in the
parameters may strongly change the partial intensities of the field components and, therefore,
the regions of stable operation in the single-frequency regime. From Eqs. (101) it follows that
in the framework of a study of the stability of the single-frequency regime, the effect of a spatial inhomogeneity in the field and dispersion in the cavity Q on the spectrum of the "single-
mode" field oscillations is described by a single parameter, the growth rate (measured in units
of the mode width) $\varkappa' = \dfrac{\gamma_{3,n} - \gamma_{3,m}}{\gamma_{3,n}} + \dfrac{4\gamma_{3,n}}{\gamma_1} \hat{f}_n |e_n^n|^2 (I - I')$, where

$$I = \int \hat{f}_n \Phi_n^4 \left[1 + \frac{4\gamma_{3,n}}{\gamma_1} \hat{f}_n |e_n^n|^2 \Phi_n^2\right]^{-1} dr,$$

$$I' = \int \hat{f}_n \Phi_n^2 \Phi_m^2 \left[1 + \frac{4\gamma_{3,n}}{\gamma_1} \hat{f}_n |e_n^n|^2 \Phi_n^2\right]^{-1} dr.$$

A spatial inhomogeneity in the field inside the cavity broadens the spectrum of the modes in the
output since $I > I'$. The dispersion γ_3 may either broaden the mode spectrum of the output (if
the maximum of the cavity Q is for modes which are far from the line center, i.e., if $\gamma_{3,n} > \gamma_{3,m}$)
or narrow it (if $\gamma_{3,n} < \gamma_{3,m}$). For $\varkappa' = 0$ the spectrum broadening due to spatial inhomogeneity
in the field is compensated by its narrowing due to Q dispersion. Thus, spatial effects usually
have a very strong effect on the stability. The growth rates of the instabilities which develop
in a traveling wave laser are of order γ_1 and the spatial inhomogeneity of the field in the cavity
and the dispersion γ_3 give instabilities with a growth rate $\sim \gamma_3 \varkappa'$. Thus, in the transition to a
traveling wave regime it is necessary to have the ratio of the amplitudes of the waves travel-
ing in opposite directions and the loss dispersion less than the ratio $\gamma_1 \gamma_3^{-1}$. Only in that case
is it possible to quantitatively verify a stability diagram of the type of Fig. 7. This fact has
been noted in theoretical [95, 96] and experimental [97] papers. Nevertheless, as will be shown
here, the developed instability regimes in systems with a spatial wave structure and in traveling wave lasers have much in common. In the following we shall assume that $\varkappa' \neq 0$, but we
shall not limit ourselves to "single-mode" oscillations.

The following discussion applies to a one-dimensional cavity filled completely with the active medium and having the eigenfunctions

$$\Phi_\lambda = \left(\frac{2}{L}\right)^{\frac{1}{2}} \sin\frac{\widetilde{\omega}_\lambda}{c} r. \tag{102}$$

As in Paragraphs 4.1 and 5.1, $\widetilde{\omega}_0 = \omega_0$ and $\widetilde{\omega}_{\overline{\lambda}} + \widetilde{\omega}_\lambda = 2\omega_0$. We also assume that

$$\gamma_2, \gamma_3 \gg \gamma_1, \qquad \gamma_3\Delta_\lambda \gg \gamma_1. \tag{103}$$

We now examine the stability of a single-frequency laser regime with respect to the coherent perturbations (13) and (14). In a single-mode oscillator at the mode Φ_0 the perturbations are damped as

$$-i(\omega_m - \omega_n) = -\frac{3}{8}\gamma_1\frac{(z_1+1)(z_1+2)^2}{z_1+3}\left[1 - \frac{2\beta(3-2\beta)z_1(z_1+3)^2}{3(z_1+1)(z_1+2)^2}\right] \pm i\left[\frac{\gamma_1\gamma_2\gamma_3 z_1(z_1+3)}{\gamma_2+\gamma_3}\right]^{\frac{1}{2}}, \tag{104}$$

where $z_1 = \sqrt{1+\lambda} - 1$, $\lambda = \frac{1}{2}[4\Omega_{th}^2 - 1 - (8\Omega_{th}^2 + 1)^{\frac{1}{2}}]$, and $(\Omega_{th}^2 - 1) \gg \max\left\{\frac{\gamma_1}{\gamma_3}, \frac{\gamma_1}{\gamma_2}\right\}$. Thus, the spatial structure of the field in the cavity removes the instability in the single-frequency regime (44) which is characteristic of a traveling wave laser or a laser with a point active medium. When $\gamma_2 \gg \gamma_3$, as is assumed later in this paragraph, Eq. (104) gives an increased damping constant without changing the relaxation oscillation frequencies compared to a traveling wave laser for which [see Eq. (17)]

$$-i(\omega_m - \omega_n) = -\tfrac{1}{2}\gamma_1\Omega_{th}^2 \pm i[2\gamma_1\gamma_3(\Omega_{th}^2 - 1)]^{\frac{1}{2}}. \tag{105}$$

This effect was noted in [98] where the damping constant of the relaxation oscillations for a mode with transverse structure (TEM_{00q}) was evaluated on a computer. The physical reason for the increase in the damping constant is multiphoton processes due to which the inversion is burnt out near the nodes of the standing wave at high intensities. This gives increased damping compared with a traveling wave laser in which the inversion is removed uniformly over the volume.

A study of perturbations (13) in a multimode laser without Q dispersion gives the following formula for the mode capture region,

$$(\widetilde{\omega}_\lambda - \omega_0)^2 = \gamma_3^2 z_1^2 (2z_1 + 3)(z_1 + 1)^{-2}, \tag{106}$$

which has been found by another method in [99]. In [100] it is shown that mode capture in a multimode laser is unstable. This proof was done for a one-dimensional dispersion-free cavity with ideally reflecting walls for which the frequency of one mode lay at the center of the luminescence line. This result, however, is easily generalized to a cavity with Q dispersion, an arbitrary wall reflection coefficient R, and detuning relative to the line center. The nature of the instability is basically that mode capture in a packet is energetically unfavorable and fails when

$$\frac{\Omega_{th}^2}{N}\sum_\lambda\int\frac{\Phi_\lambda^2\,dr}{1+|z|^2} > 1, \tag{107}$$

where N is the number of modes in the cavity. When $R \neq 1$ the modes Φ_λ include both sines

(102) and cosines. This condition resembles the condition for self-excitation of a laser: If we neglect saturation then the factor $(1 + |z|^2) \simeq 1$, and the inequality (107) becomes a condition for lasing. Some calculations as in [100] show that the inequality (107) is satisfied for an arbitrary packet and that the smaller the wall reflection coefficient R the better it is satisfied. This is understandable because mode capture in a traveling wave laser (R → 0) is generally impossible (Section 4, Paragraph 4.2). For a laser with ideally reflecting walls (R ≃ 1) and a smaller number of modes a region of frequency differences between modes may be found in which the difference between the left and right sides of Eq. (90) is of the order of the inequality $\gamma_1 \gamma_3^{-1}$. In such a region it is necessary to take into account the oscillations in the population inversion and associated subtle effects.

A calculation shows that this situation does not occur for a three-mode laser. Figure 9 shows the characteristics of single-frequency two- and three-mode field states in a three-mode resonator. From the figure it is clear how quickly the phase space of the laser becomes more complicated as the number of modes is increased. In a two-mode oscillator only two stationary states for mode locking are possible (Curve V of Fig. 9). In a three-mode cavity the number of stationary states corresponding to different variants of mode capture is much greater and they (as well as the two-mode cases) are all unstable with large damping constants $\sim \gamma_3(\Omega_{th}^2 - 1)$ for arbitrary frequency differences between the modes. In a two-mode cavity mode capture is

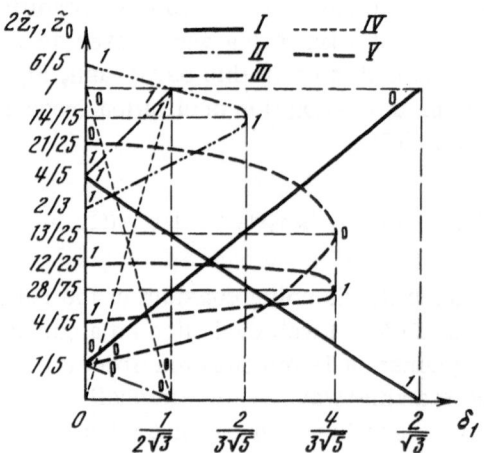

Fig. 9. The dependence of the partial intensities of single-frequency (non-single-mode) field oscillations on the frequency difference $\delta_1 = \tilde{\omega}_1 - \omega_0 / [\gamma_3 (\Omega_{th}^2 - 1)]$ near the lasing threshold $(\Omega_{th}^2 - 1 \ll 1)$ in a three-mode cavity: $\tilde{\omega}_0 = \omega_0$, $\tilde{\omega}_{\bar{1}} + \tilde{\omega}_1 = 2\omega_0$, $\Phi_\lambda = (2/L)^{\frac{1}{2}} \sin \tilde{\omega}_\lambda r/c$, $\lambda = \bar{1}, 0, 1$. [The behavior of $\tilde{z}_\lambda = 6\gamma_3 |e_n^\lambda|^2 / [\gamma_1 L (1 - \Omega_{th}^2)]$, $\lambda = \bar{1}, 0, 1, \tilde{z}_1 = \tilde{z}_{\bar{1}}$, is determined by the curve with the index $(0, 1)$.] $\varphi_1 + \varphi_{\bar{1}} - 2\varphi_0 = 2l\pi$: I) $\sin 2(\varphi_1 - \varphi_0) > 0$, $\cos 2(\varphi_1 - \varphi_0) = -\frac{1}{2}$; II) $\sin 2(\varphi_1 - \varphi_0) < 0$, $\cos 2(\varphi_1 - \varphi_0) = -\frac{1}{2}$; III) $\sin 2(\varphi_1 - \varphi_0) > 0$, $\cos 2(\varphi_1 - \varphi_0) \to \mp 1$ for $\delta_1 \to 0$ for the upper (lower) branch; $\varphi_1 + \varphi_{\bar{1}} - 2\varphi_0 = (2l + 1)\pi$; IV) $\cos 2(\varphi_1 - \varphi_0) = \pm\frac{1}{2}$; $x_0 = 0$; V) $\cos 2(\varphi_1 - \varphi_{\bar{1}}) \gtrless 0$ for the upper (lower) branch.

unstable for frequency differences such that [3]

$$\Delta^2 > \frac{8}{25} \gamma_1 \gamma_3^{-1} (\Omega_{\text{th}}^2 - 1). \tag{108}$$

Inequality (107) is proven in [96] in the framework of the kinetic equations for a two-mode cavity, and mode capture is found to be unstable for arbitrary frequency differences. This conclusion is unjustified for frequency differences in the neighborhood of the relaxation frequency (108) when effects not related to the energy balance and not described by the kinetic equations must be taken into account. Thus, the emission intensity of a non-single-mode laser must be amplitude modulated as a rule. The type of modulation may be determined by an analysis of perturbations (14). Such perturbations grow in a $(2q + 1)$ mode cavity at the relaxation frequency ω_1 if [100]

$$3[4q^2 (z_1 + 3)]^{-1} \gamma_1 \gamma_3^{-1} z_1 (z_1 + 1) (z_1 + 2)^2 \leqslant \Delta^2 \leqslant 2 [\gamma_1 \gamma_3^{-1} z_1^3 (z_1 + 3)]^{\frac{1}{2}}. \tag{109}$$

We now consider the operation of a two-mode laser $(\gamma_2 \gg \gamma_3 \gg \gamma_1,\ \Delta_{\lambda_1} = -\Delta_{\lambda_2})$ near the stability boundary (108) in a quasilinear approximation. The truncated equations for standing mode lasers have the form

$$z_\lambda' = - l_\lambda z_\lambda + \int zy\Phi_\lambda \, dr, \tag{110}$$

$$y' = \varepsilon_1 (\Omega_{\text{th}}^2 - y - y\,|\,z\,|^2), \tag{111}$$

and are derived in the same way as the analogous equations (84) and (85) for a laser with a point active medium.

The equilibrium points for the system (110) and (111) are given by

$$z = 0, \qquad y = \Omega_{\text{th}}^2, \tag{112}$$

$$\Omega_{\text{th}}^2 = [\eta_{12} \cos (\varphi_{\lambda_1} - \varphi_{\lambda_2}) + \eta_{11}]^{-1}, \qquad \Delta_{\lambda_1} = - \Omega_{\text{th}}^2 \eta_{12} \sin (\varphi_{\lambda_1} - \varphi_{\lambda_2}), \tag{113}$$

where $\eta_{ij} = \int \Phi_i \Phi_j (1 + |z|^2)^{-1}\, dr$, and φ_λ is the phase of the field component z_λ. The state (112) is unstable when the self-excitation condition $\Omega_{\text{th}}^2 > 1$ is satisfied. The instability is oscillatory with frequency Δ_λ and a damping constant $\sim(\Omega_{\text{th}}^2 - 1)$ and causes the laser output intensity to be modulated at a frequency equal to the frequency difference between the modes Δ.

An investigation of the single-frequency laser regime yields more information. The single-frequency regime is implicitly described by Eqs. (113). In the following we shall limit ourselves to near-threshold pumping, such that $\Omega_{\text{th}}^2 - 1 \ll 1$. In this case, Eqs. (113) may be inverted to give curve V of Fig. 9 for mode capture in a two-mode laser. The stability diagram for the single-frequency regime is shown in Fig. 10 in the coordinates $\mathcal{I}'^2 = \frac{4}{L}\,|\,z_\lambda\,|^2 \sin^2 (\varphi_{\lambda_1} - \varphi_{\lambda_2})$ and $\mathcal{I}''^2 = \frac{4}{L}\,|\,z_\lambda\,|^2 \cos^2 (\varphi_{\lambda_1} - \varphi_{\lambda_2})$. An aperiodic instability in the mode capture develops on the radial rays (Fig. 10). For phase perturbations it is energy based, i.e., the total intensity of the field of the modes for mode capture parameters in the sector below the dot-dashed curve is less than the field intensity of a single-mode standing wave laser. This means that for mode capture parameters lying on the dot-dashed ray, the mode capture point moves in the phase space of a two-mode laser through a separatrix surface of the same type as the separatrix in the phase space of a system with two potential wells. In the latter case the separatrix separates trajectories enveloping both wells from trajectories localized to one well. Two equilibrium states describing mode capture are merged on the smooth radial ray of Fig. 10; that is,

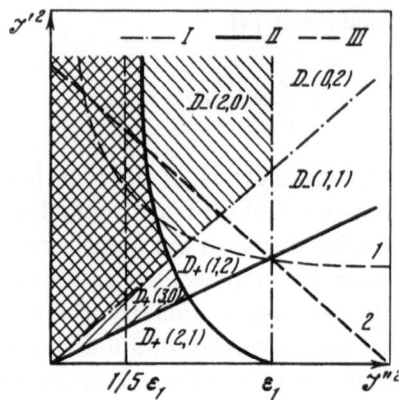

Fig. 10. The mode-capture-stability diagram for a standing-wave two-mode laser. (I) $D_-(n, m)$ phase perturbations; (II) $D_+(n, m)$ amplitude perturbations; the left (right) shading covers the region for mode capture that is stable relative to phase (amplitude) perturbations; in the doubly shaded region mode capture is stable. The equation of the radius of curve I is $\tan^2(\varphi_{\lambda_1} - \varphi_{\lambda_2}) = 3$, and of curve II, $\tan^2(\varphi_{\lambda_1} - \varphi_{\lambda_2}) = 9/5$; curve III is the contour for mode capture, that is, a graph of (113) with the left sides of the equations held constant: 1) $\Delta_\lambda^2 = \text{const}$; 2) $\Omega_{\text{th}}^2 = \text{const}$.

the points on curve V of Fig. 9 with a vertical tangent lie on this ray and the points on the upper (lower) branch of this curve are stable (unstable) with respect to aperiodic amplitude perturbations. For typical pump powers such that $\varepsilon_1 \ll (\Omega_{\text{th}}^2 - 1)$ an oscillatory instability with respect to amplitude perturbations develops when

$$y''^2 > \tfrac{1}{5}\,\varepsilon_1, \tag{114}$$

which is equivalent to the inequality (108), and is the only type of instability hazardous for capture.

The amplitude and frequency of the limit cycle which splits off from point (113) in the neighborhood of the instability boundary (114) decompose into a series in the square root of a parameter which characterizes the deviation in the laser parameters from their values on the boundary.[†] Calculations yield the following expressions for the partial intensities of the cycle:

$$|z_{1\lambda_j; 1}|, \quad j = 1.2:$$

$$|z_{1\lambda_1; 1}|^2\,|z_{\lambda_1}|^{-2} \simeq -\,0.211\,\frac{|\Delta_\lambda|'}{\gamma} + 0.0378\,\frac{\varepsilon_1(\Omega_{\text{th}}^2)'}{\gamma^2}, \tag{115}$$

$$z_{1\lambda_2; 1}\,z_{1\lambda_1; 1}^{-1} \simeq -i\,0.283\,\sin(\varphi_{\lambda_1} - \varphi_{\lambda_2}),$$

where $z_1 - z_0 = z_{1;1}\exp(i\gamma\gamma_3 t) + \text{c.c.}$, $z_{1;1} = z_{1\lambda_1; 1}\Phi_{\lambda_1} + z_{1\lambda_2; 1}\Phi_{\lambda_2}$, $\gamma\gamma_3 = \left[\tfrac{8}{5}\,\gamma_1\gamma_3(\Omega_{\text{th}}^2 - 1)\right]^{\frac{1}{2}}$, and the frequency of the cycle z_0 is determined by Eq. (113) at the boundary of Eq. (114).

[†] This is proven as in the case of a laser with a point active medium (Section 5, Paragraph 5.1).

From Eq. (115) it is clear that a small limiting cycle exists in the stable region for single-frequency operation; thus, the instability boundary (114) or (108) is hazardous and the width of the pulling region for the stable single-frequency regime grows slowly with distance from the boundary. From the nature of Eqs. (110) and (111) it also follows (as in Paragraph 5.1) that the development of a linear or nonlinear instability brings the laser into a self-Q-switched regime.

We now find the giant cycle for a two-mode laser near the threshold

$$\Omega^2_{\text{th}} - 1 \ll 1 \tag{116}$$

in the relaxation approximation $(\varepsilon_1 \ll 1, \Delta_{\lambda_1} = - \Delta_{\lambda_2} = \Delta)$. The inverted population enters Eq. (110) through $y_{\lambda_1\lambda_2} = \int y\Phi_{\lambda_1}\Phi_{\lambda_2}dr$, $\lambda_1, \lambda_2 = 1, 2$. Multiplying Eq. (111) by $\Phi_{\lambda_1}\Phi_{\lambda_2}$ and integrating over space we can obtain the equations of motion for $y_{\lambda_1\lambda_2}$ which will contain the "quadruple" term $y_{\lambda_1\lambda_2\lambda_3\lambda_4} = \int y\Phi_{\lambda_1}\Phi_{\lambda_2}\Phi_{\lambda_3}\Phi_{\lambda_4}dr$. Continuing this process we obtain a chain of kinetic equations for the "multiples" $y_{\lambda_1\lambda_2\ldots\lambda_n}$. This chain may be closed near the lasing threshold (116) since the $(n + 1)$-tuple term $y_{\lambda_1\lambda_2\ldots\lambda_{n+1}}$ enters in the equation for the n-th approximation with a small term of order $(\Omega^2_{\text{th}} - 1)$. In Eqs. (111) the nonlinear term has the form $y|z|^2$ and $|z|^2 \sim (\Omega^2_{\text{th}} - 1)$. If we neglect the "sextuples" and assume, in view of the symmetry of the frequency configuration $(\Delta_{\lambda_1} = -\Delta_{\lambda_2} = \Delta)$, that stable lasing is possible only for $|z_{\lambda_1}| = |z_{\lambda_2}|$ and $y_{\lambda_1\lambda_1} = y_{\lambda_2\lambda_2}$, then some simple calculations yield the following equations for the field and the population inversion:

$$y'_{\lambda_1\lambda_1} = \varepsilon_1 [\Omega^2_{\text{th}} - y_{\lambda_1\lambda_1} - |\bar{z}_1|^2 - |\bar{z}_2|^2], \tag{117}$$

$$y'_{\lambda_1\lambda_2} = \varepsilon_1 [- y_{\lambda_1\lambda_2} - 2I_2(I_1 + I_2)^{-1}(|\bar{z}_1|^2 - |\bar{z}_2|^2)], \tag{118}$$

$$\bar{z}'_1 = (-1 + y_{\lambda_1\lambda_1} + y_{\lambda_1\lambda_2})\bar{z}_1 + \Delta\bar{z}_2, \tag{119}$$

$$\bar{z}'_2 = (-1 + y_{\lambda_1\lambda_1} - y_{\lambda_1\lambda_2})\bar{z}_2 - \Delta\bar{z}_1, \tag{120}$$

where

$$\bar{z}_1 = \frac{1}{2} \sqrt{\Omega^2_{\text{th}}(I_1 + I_2)}(z_{\lambda_1} + z_{\lambda_2}),$$

$$\bar{z}_2 = \frac{1}{2i} \sqrt{\Omega^2_{\text{th}}(I_1 + I_2)}(z_{\lambda_1} - z_{\lambda_2}),$$

$$I_1 = \int \Phi^4_{\lambda_1}dr = \int \Phi^4_{\lambda_2}dr, \quad I_2 = \int \Phi^2_{\lambda_1}\Phi^2_{\lambda_2}dr.$$

In the pumping stage the variables $\tau_{\text{H}} = \varepsilon_1\gamma_3 t \left(\cdot = \frac{d}{d\tau_{\text{H}}} \right)$, $\bar{z}_1 = \exp\left(-\frac{\xi}{\varepsilon_1}\right)$, $\bar{z}_2 = \xi\Delta^{-1}\bar{z}_1$ may be used to reduce the basic system of equations (117)–(120) to

$$\dot{\bar{\zeta}} = -\xi + 1 - y_{\lambda_1\lambda_1} - y_{\lambda_1\lambda_2}, \tag{121}$$

$$\varepsilon_1\ddot{\bar{\zeta}} = (\dot{\bar{\zeta}} - 1 + y_{\lambda_1\lambda_1} - y_{\lambda_1\lambda_2})(\dot{\bar{\zeta}} - 1 + y_{\lambda_1\lambda_1} + y_{\lambda_1\lambda_2}), \tag{122}$$

$$\dot{y}_{\lambda_1\lambda_1} = \Omega^2_{\text{th}} - y_{\lambda_1\lambda_1}, \tag{123}$$

$$\dot{y}_{\lambda_1\lambda_2} = - y_{\lambda_1\lambda_2} \tag{124}$$

to within exponentially small terms. The phase space of the system (121)–(124) contains two planes of sliding motions:

$$\dot{\bar{\zeta}} = 1 - y_{\lambda_1\lambda_1} + y_{\lambda_1\lambda_2}, \tag{125}$$

$$\dot{\bar{\zeta}} = 1 - y_{\lambda_1\lambda_1} - y_{\lambda_1\lambda_2}. \tag{126}$$

If $y_{\lambda_1\lambda_2} > 0$ $(y_{\lambda_1\lambda_2} < 0)$, then the plane (125) [plane (126)] is unstable for sliding motions, or a discontinuity plane and plane (126) [plane (125)] is stable for sliding motions. Rapid motions bring the system into one of (from one of) these planes. Integrating Eq. (125) or (126) as in Section 3, we obtain the following relationship between the inversions before and after the pumping stage:

$$\widetilde{w}_{2k} = \widetilde{w}_{2k-1}(\overline{w}_{2k} - 1)(\overline{w}_{2k-1} - 1)^{-1}, \tag{127}$$

$$\mathrm{ch}^{-2}\theta_3 \ln\left(\frac{\overline{w}_{2k} - 1}{\overline{w}_{2k-1} - 1}\right) - (1 - \overline{w}_{2k-1} \mp \widetilde{w}_{2k-1})\frac{\overline{w}_{2k} - \overline{w}_{2k-1}}{\overline{w}_{2k-1} - 1} = 0, \tag{128}$$

where $\widetilde{w} = y_{\lambda_1\lambda_1}\Omega_{\mathrm{th}}^{-2}$, $\overline{w} = y_{\lambda_1\lambda_1}\Omega_{\mathrm{th}}^{-2}$, $\mathrm{th}^2\theta_3 = \Omega_{\mathrm{th}}^{-2}$, and the upper (lower) sign in Eq. (128) refers to sliding along plane (125) [along plane (126)]. The duration of the pumping stage is given by

$$\tau_{\mathrm{H}, 2k} - \tau_{\mathrm{H}, 2k-1} = \ln\left(\frac{1 - \overline{w}_{2k-1}}{1 - \overline{w}_{2k}}\right). \tag{129}$$

In the emission stage the laser obeys the equations

$$y'_{\lambda_1\lambda_1} = -\frac{1}{\widetilde{\delta}^2}|\hat{z}_2|^2, \tag{130}$$

$$y'_{\lambda_1\lambda_2} = 2I_2(I_1 + I_2)^{-1}\widetilde{\delta}^{-2}|\hat{z}_2|^2, \tag{131}$$

$$\hat{z}'_2 = (-1 + y_{\lambda_1\lambda_1} - y_{\lambda_1\lambda_2})\widetilde{z}_2, \tag{132}$$

$$\widetilde{z}'_1 = (-1 + y_{\lambda_1\lambda_1} + y_{\lambda_1\lambda_2})\widetilde{z}_1 + \hat{z}_2, \tag{133}$$

where $\widetilde{\delta}^2 = \Delta^2\varepsilon_1^{-1}$, $\hat{z}_2 = \sqrt{\varepsilon_1}\widetilde{z}_2$. Equations (130)-(132) have two first integrals from which we obtain the following relationship between the inversions before and after the emission stage:

$$\widetilde{w}_{2k+1} - \widetilde{w}_{2k} = 2I_2(I_1 + I_2)^{-1}(-\overline{w}_{2k+1} + \overline{w}_{2k}), \tag{134}$$

$$\overline{w}_{2k+1} + \overline{w}_{2k} - 2\,\mathrm{th}^2\theta_3 - \widetilde{w}_{2k+1} - \widetilde{w}_{2k} = 0. \tag{135}$$

Equations (127), (128), (134), and (135) determine the point image. This image has a single fixed point for the upper and lower signs in Eq. (128). The fixed point of the image, including sliding in the plane (125) in the pump stage, is unstable with a large growth rate $\sim\gamma_3$, since $\widetilde{w}_{2k-1} = \widetilde{w}_{2k+1} > 0$ for a fixed point and the plane (125) is a discontinuity plane. The fixed point of the image with the lower sign in Eq. (128) is stable, as can be shown by linearizing the equations near the fixed point and examining the resulting finite difference equation for stability in the usual manner.

Therefore, the equations for a two-mode laser have a stable giant limiting cycle which describes spiking with spikes of width γ_3^{-1} and a repetition rate $\sim\gamma_1$. The cycle exists for all frequency differences $\Delta \sim \sqrt{\varepsilon_1}$ and does not shrink to the single-frequency regime at the instability boundary (114). This distinguishes the two-mode standing wave laser from the single-mode (Section 3) and two-mode (Section 5, Paragraph 5.1) lasers with point active media. Formally this is a consequence of the slow growth of the unstable cycle found in the stable spikeless regime in the quasilinear approximation. Physically, the giant pulses are due to inhomogeneous removal of the population inversion where the inhomogeneity is described by the cross term

$y_{\lambda_1 \lambda_2}$. In the pumping stage the action of the pump radiation equalizes the spatial distribution of the population inversion and $y_{\lambda_1 \lambda_2}$ damps out according to Eq. (124). In the emission stage the standing wave field nonuniformly burns out the inversion and $y_{\lambda_1 \lambda_2}$ grows according to Eq. (131). In the stability (instability) region for a single-frequency regime the laser goes strongly (weakly) into a self-Q-switched giant pulse emission regime.

In three- and multimode lasers mode capture is, as shown above, unstable with a high growth rate $\sim \gamma_3 (\Omega_{th}^2 - 1)$. As this instability develops it forms the mode spectrum of the laser output. In addition, at the boundary of the oscillatory instability a "small" limiting cycle splits off from the single-frequency regime with the relaxation frequency [see Eq. (109)]

$$\Delta^2 = \frac{3\gamma_1 \gamma_3^{-1}}{4q^2(z_1 + 3)} z_1 (z_1 + 1)(z_1 + 2)^2.$$

(136)

The calculations done for a two-mode laser give reason to suppose that this cycle is unstable. This instability forms a giant pulse weakly in the case of frequency differences greater than the boundary value (136), and strongly otherwise.

The region of growing relaxation spikes (109) is considerably changed if we take Q-dispersion into account. This may be proved for the example of a three-mode laser for which $\gamma_{3,\lambda} = \gamma_{3,\bar{\lambda}} > \gamma_{3,0}$. We shall study the instability of single-frequency operation at the highest-Q mode Φ_0. We seek complex roots of the characteristic equation whose imaginary parts are of the order of the relaxation frequency ω_1. The characteristic equation has two complex roots,

$$- i(\omega_m - \omega_n) = -\frac{1}{2}\left[\gamma_{3,\lambda} - \gamma_{3,0} - \frac{\Delta_\lambda^2 \gamma_{3,0}}{2z_1}\right] \pm i \left\{\gamma_{3,0}\gamma_1 z_1 (z_1 + 3) - \frac{1}{4}\left[\gamma_{3,0} - \gamma_{3,\lambda} - \frac{\Delta_\lambda^2 \gamma_{3,0}}{z_1}\right]^2\right\}^{\frac{1}{2}}.$$

(137)

From the conditions $\mathrm{Re}\,(\omega_m - \omega_n) \neq 0$ and $\mathrm{Im}\,(\omega_m - \omega_n) > 0$ we obtain an estimate of the range of frequency differences where relaxation peaks exist with a positive growth rate

$$0 < \gamma_{3,0}^2 \Delta_\lambda^2 - \gamma_{3,0}(\gamma_{3,\lambda} - \gamma_{3,0}) z_1 < 2\,[\gamma_{3,0}^3 \gamma_1 z_1^3 (z_1 + 3)]^{\frac{1}{2}}.$$

(138)

In deriving Eqs. (137) and (138) it is assumed that the inequalities $\gamma_2 \gg \gamma_{3,\lambda} \gg \gamma_{3,\lambda} - \gamma_{3,0} \gg \gamma_1$, $\gamma_2 \gg \Delta_\lambda \gamma_{3,0}$, and $\Omega_{th}^2 - 1 \gg (\gamma_{3,\lambda} - \gamma_{3,0})\gamma_{3,0}^{-1}$ are satisfied, as is usually the case for solid-state lasers. We note that undamped spikes at the relaxation frequency ω_1 appear in modes whose frequencies, in accordance with condition (138), are separated from one another by a distance $[\gamma_{3,0}(\gamma_{3,\lambda} - \gamma_{3,0})]^{\frac{1}{2}}$ much greater than the repetition frequency of the spikes. A cavity with Q-dispersion differs in this respect from a cavity with equal Q for all modes. The reason is that damping of the spikes in a dispersion cavity is determined by a term in an exponent containing $\gamma_{3,\lambda} - \gamma_{3,0}$ while in a dispersionless cavity it is determined by a term containing γ_1. The growth of the spikes, however, is characterized by terms of a single order $\Delta_\lambda^2 \gamma_{3,0}$. Accordingly, the growth rate in a dispersive cavity appears at frequency differences $\Delta_\lambda^2 \sim \gamma_{3,0}^{-1} \times (\gamma_{3,\lambda} - \gamma_{3,0})$, and in a cavity without dispersion, at differences $\Delta_\lambda^2 \sim \gamma_1 \gamma_{3,0}^{-1}$. Thus, even a slight difference in the Q of the modes, due, for example, to different diffraction losses of modes with different tranverse indices, strongly displaces the region of growing relaxation spikes.

The output intensity of a laser, as noted several times previously, is modulated at the frequency of the intermode beating. This kind of modulation is distinct if the modes are resolved (that is, if the distance between the mode frequencies is greater than the spectral width of a mode). In this case the solutions for the field are "single-mode." Single-mode perturbations of the form (14) are characterized by the phase relationships

$$\varphi_{\bar{\lambda}} + \varphi_\lambda - 2\varphi_0 = n\pi$$

(139)

and have a growth rate $\sim \gamma_3 (\Omega_{th}^2 - 1)$ if

$$\Delta_\lambda^2 \leqslant \frac{1}{2} \gamma_2^2 \gamma_3^{-2} z_1. \tag{140}$$

As the perturbations (14), (139) with even (odd) n develop, they amplitude (phase) modulate the signal. In a three-mode cavity near the lasing threshold the stability of the modulation is determined by

$$\frac{d}{dt} (\varphi_1 + \varphi_{\bar{1}} - 2\varphi_0) = b \sin (\varphi_1 + \varphi_{\bar{1}} - 2\varphi_0), \tag{141}$$

where

$$b = \frac{1}{7} \gamma_1 \gamma_3 \gamma_2^{-1} \bar{\delta}^{-2} (\Omega_{th}^2 - 1) [2\gamma_1 \gamma_2^{-1} (\Omega_{th}^2 - 1)^{-1} (3 - 2\bar{\delta}^2) - \bar{\delta}^2 (5 + 6\bar{\delta}^2)],$$
$$\bar{\delta}^2 = \gamma_3^2 \gamma_2^{-2} (\Omega_{th}^2 - 1)^{-1} \Delta_1^2. \tag{142}$$

If the intermode frequency differences satisfy

$$\Delta_1^2 > \frac{6}{5} \gamma_1 \gamma_2 \gamma_3^{-2}, \tag{143}$$

which ensures efficient interaction of the output intensity oscillations having a period equal to the transit time of a photon through the cavity with the nutation oscillations in the inversion (see Fig. 1), then the amplitude modulation is stable and, in the opposite case, the phase modulation is stable.[†]   The damping decrement b which characterizes the phase stability is of the same order of magnitude as the growth rate (72) for establishing (multiple) mode synchronization in a traveling wave laser. For stable mode locking in a free-running laser it is necessary that the damping decrement (142) be greater than the phase diffusion coefficient D_θ associated with various noise sources which destroy the stationary phase relations between the modes[‡]; that is,

$$b \gtrsim D_\theta. \tag{144}$$

Since the decrement b is usually small, mode locking is not observed in a free lasing regime [105, 106] with rare exceptions [102-104]. It follows from Eqs. (142)-(144) that for mode locking it is preferable to have lasers with active media having small upper state lifetimes γ_1^{-1} (dyes, semiconductors, gases) and working at high pump powers [103] with a high density of modes in the cavity [102] and a low noise level. In general the signal will have amplitude and phase modulation with the width of the spectrum (140). The limiting value of the frequency difference (143) which divides the regions of amplitude and phase modulation is determined by the internal properties of the active medium and is independent of the cavity Q. The analogy between Eqs. (142) and (72) indicates that the nature of mode locking in traveling and standing wave lasers is the same. The difference is that in a standing wave laser the slow stage of synchronization of the mode phases precedes the rapid stage in which multiple-frequency lasing appears. The nonlinear stage in which ultrashort pulses develop, when the width is much less

---

[†] Equations (141)-(143) differ somewhat from those in [102] because in this paper the case of a very long cavity is being discussed (L ≫ l). An equation of the type (141) for a gas laser has been studied in [27].

[‡] These estimates follow from similar calculations for relaxation spikes. See Paragraph 7.1.

than the characteristic dimensions of the active medium, must occur in the same way in travel-ing and standing wave lasers (and in lasers with homogeneous and inhomogeneous luminescence lines). Ultrashort pulses develop due to Cerenkov radiation when the optical nutation frequency DE/ℏ (where E is the field strength in the pulse) is of the order of the spike repetition rate. Then it is possible to choose the laser parameters and pump power so that nonlinear growth is possible only for sufficiently powerful spikes (see Sections 4 and Fig. 7) and the active medi-um acts as a pulse discriminator.

These conclusions may be compared with the data of [107] which deal with the conditions for self-mode locking in a gas laser. Uchida and Ueki [107] plotted the region for stable mode synchronization in coordinates of the energy of the laser pulses versus the distance between the cavity mirrors. If we replot this diagram in coordinates of the energy of the oscillations E^2 versus the square of the spike repetition rate, then the upper boundaries of the synchroniza-tion regions for spikes with repetition rates $\Delta\gamma_3$ and $2\Delta\gamma_3$ at high oscillation energies fit a straight line well, in accordance with the theory (Fig. 11). The slope of the curve gives the correct order of magnitude of the dipole moment of the working transition in a He−Ne laser (about 1 Debye unit). Thus, the main conclusion of the theory, that the upper boundary of the synchronization region must go asymptotically at large pulse energies into a straight line of the type

$$L^2 E^2 m^{-2} = K, \tag{145}$$

where K is a constant dependent on the fundamental constants and dipole moment of the working transition and $m\Delta\gamma_3$ is the spike repetition rate, is apparently confirmed experimentally.

The most widespread method of mode locking is passive Q-switching [108]. In this re-gime the laser emits a train of ultrashort pulses of duration $\sim\gamma_2^{-1}$ whose envelope is a giant pulse of duration γ_3^{-1}. The technique for analyzing passive mode locking is a semistochastic approach developed in [109, 110]. In this approach the initial fluctuating output, whose probabil-ity is calculated subject to certain assumptions about the statistics of the laser output, evolves further according to the dynamic equations. First the filter saturates and ultrashort pulses are formed. The subsequent saturation of the active medium determines the pulse envelope.

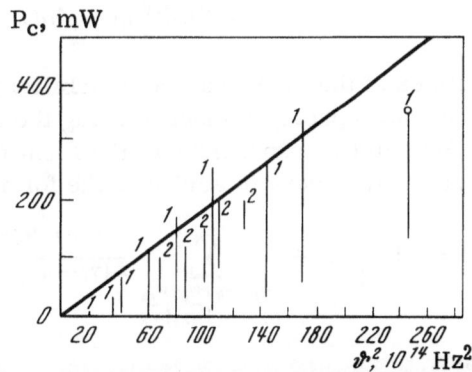

Fig. 11. The regions for stable synchronization of modes with equidistant frequencies in coor-dinates of wave power traveling in one direc-tion, P_i, versus the square of the repetition rate of the spikes, ν^2 (data from [107]). 1 (2) corre-sponds to spikes with rate Δ (2Δ). The circle denotes the maximum wave power obtained in the experiment.

The mechanisms which ensure stability of mode locking do not work in these times and the trains of ultrashort pulses are not always reproduced from shot to shot. Another experimental setup, however, is possible, in which the field configuration in the cavity is chosen so that the active medium saturates first and the laser goes into a single-mode free lasing regime for a time $\sim \gamma_3^{-1}$. Then an ultrashort pulse of duration $\sim \gamma_2^{-1}$ develops over a time $\sim b^{-1}$. In this stage the effect of the passive filter is to increase[†] $\gamma_1 = \tau_1^{-1} + P$ and ω so that higher pump powers P are to saturate the working transition. Thus, the time to form the ultrashort pulses is reduced [see Eq. (142)], conditions (144) and (145) may be satisfied, and stable mode locking and rapid nonlinear growth of the ultrashort pulses are possible. Bleaching of the filter determines the final form of the pulse, whose duration must be less than in the nonstationary case (with the usual method). The nonstationarity of the process reduces the pulse spectrum by about a factor of 10 [109, 110]. These considerations may explain the results of [111] in which this method yielded ultrashort pulses in a YAG : Nd^{3+} laser with a contrast of better than 10^3 for pump intensities up to 3% above the threshold level. With the usual method the contrast was less than 10^2 with an accuracy of maintaining near-threshold pumping of 1%. However, in this case damping of the less intense pulses in the nonlinear stage of ultrashort pulse development was noticed [112]. This fact, not included in previous theoretical representations [109, 110], is in agreement with the above theory. The role of dynamic effects in mode locking in media with short upper working level lifetimes has been noted by various authors [85-89], [113], but this type of Cerenkov instability behavior, general for all types of lasers and leading to the formation of ultrashort pulses, has not been reported in the literature as far as we know.

5.3. Inhomogeneous Luminescence Line

We shall not analyze the operation of a solid-state laser with an inhomogeneous luminescence line. This sort of laser obeys the following system of equations which is a generalization of Eqs. (1)-(3) to an ensemble of molecules with a distribution $\nu_0(v)$ of transition frequencies $\omega_v = \omega_0(1 + v/c)$:

$$\dot{\nu} = \gamma_1 (\nu_0 - \nu) + 2i\Omega e (\rho^* - \rho), \tag{146}$$

$$\dot{\rho} = - (i\omega_v + \gamma_2)\rho - i\Omega ev, \tag{147}$$

$$\ddot{e} + 2\gamma_3 \dot{e} - c^2 \nabla^2 e = - 2\Omega \omega_0^{-1} \int (\ddot{\rho} + \rho^*)\, dv. \tag{148}$$

For studies of the stationary states of the laser and their stability the field is decomposed into Fourier series (7) and (8) in time and space. Without limiting the generally, we may assume that Φ_λ are traveling waves. Substituting expansions of the form (7) and (8) in Eqs. (146)-(148) and eliminating the required terms, it is easy to arrive at the following system of equations:

$$\frac{A_{\lambda p} B_p}{4\omega_p^2} e_{pv}^\lambda = - \Omega^2 \left\{ \nu_0 e_p^\lambda + \sum_{\substack{\omega_p = \omega_k + \omega_l + \omega_m \\ \tilde{\omega}_\lambda = \tilde{\omega}_{\lambda_1} + \tilde{\omega}_{\lambda_2} + \tilde{\omega}_{\lambda_3}}} \frac{e_k^{\lambda_1} e_l^{\lambda_2} \check{e}_{mv}^{\lambda_3}}{L\left[\gamma_1 - i(\omega_l + \omega_m)\right]} \right\}, \tag{149}$$

$$e_p^\lambda = \int e_{pv}^\lambda\, dv, \tag{150}$$

where $e_{pv}^\lambda = \frac{4\Omega\omega_p^2}{A_{\lambda p}\omega_0} (\operatorname{Re}\rho)_p^\lambda$, $\check{e}_{pv}^\lambda = \frac{(\gamma_2 - i\omega_p) A_{\lambda p}}{\omega_p^2} e_{pv}^\lambda$, $A_{\lambda p}$, and $A_{\lambda p}$ is the operator A_p with $c^2\nabla^2$ replaced by $\tilde{\omega}_\lambda^2$. We now write Eq. (149) in matrix form. To do this we introduce the infinite-dimensional columns (e) and e_v with components e_p^λ and e_{pv}^λ, respectively, and the infinite dimensional

---

[†] τ_1 is the lifetime of the upper working level with respect to spontaneous emission.

matrices Φ_v and F with elements

$$\Phi_{v\lambda_1 p_1,\ \lambda_2 p_2} = -\delta_{\lambda_1\lambda_2}\delta_{p_1 p_2}\Omega^2 4\omega_{p_1}^2 v_0\,(A_{\lambda_1 p_1}B_{p_1})^{-1}, \tag{151}$$

$$F_{\lambda_1 p_1,\ \lambda_2 p_2} = \Omega^2 \sum_{\lambda_3,\ p_3} 4\omega_{p_3}^2(\gamma_2 - i\omega_{p_3})\,A_{\lambda_2 p_2}e_{p_3}^{\lambda_3}e_{p_4}^{\lambda_4}\{A_{\lambda_1 p_1}B_1[\gamma_1 - i(\omega_{p_1}+\omega_{p_3})]\omega_{p_3}^2 L\}^{-1}. \tag{152}$$

$$\omega_{p_4} = \omega_{p_1} + \omega_{p_3} - \omega_{p_2}, \qquad \widetilde{\omega}_{\lambda_4} = \widetilde{\omega}_{\lambda_1} + \widetilde{\omega}_{\lambda_3} - \widetilde{\omega}_{\lambda_2}.$$

Then the system of Eqs. (149) goes into the matrix equation

$$e_v = (1 + F)^{-1}\,\Phi_v\,(e). \tag{153}$$

Combining Eqs. (150) and (153), we arrive at the closed equation for the field components:

$$(e) = \int (1 + F)^{-1}\,\Phi_v\,(e)\,dv. \tag{154}$$

In the "diagonal" approximation $F_{\lambda_1 p_1,\ \lambda_2 p_2}$ breaks down into the product of two cofactors,

$$F_{\lambda_1 p_1,\ \lambda_2 p_2} = F_{\lambda_1 p_1}F'_{\lambda_2 p_2}, \qquad F_{\lambda_1 p_1} = \gamma_1^{-1}v_0^{-1}e_{p_1}^{\lambda_1}\Phi_{v\lambda_1 p_1,\ \lambda_1 p_1}. \tag{155}$$

Thus we shall alter the derivation of Eq. (154) somewhat. Rewriting Eq. (153) in the form

$$e_{pv}^{\lambda} = \gamma_1 v_0 F_{\lambda p} - F_{\lambda p}\sum_{\lambda_2 p_2}F'_{\lambda_2 p_2}e_{p_2 v}^{\lambda_2}, \tag{156}$$

we multiply it by $F'_{\lambda p}$ and sum over λ and p,

$$\sum_{\lambda_2 p_2}F'_{\lambda_2 p_2}e_{p_2 v}^{\lambda_2} = \gamma_1 v_0 \sum_{\lambda p}F_{\lambda p,\ \lambda p}\Big(1 + \sum_{\lambda p}F_{\lambda p,\ \lambda p}\Big)^{-1}. \tag{157}$$

From Eqs. (156) and (157) we find an explicit form for the inverse operator:

$$(1 + F)_{\lambda_1 p_1,\ \lambda_2 p_2}^{-1} = \delta_{\lambda_1\lambda_2}\delta_{p_1 p_2}\Big(1 + \sum_{\lambda p}F_{\lambda p,\ \lambda p}\Big)^{-1}. \tag{158}$$

In particular, if the transition frequencies have a Gaussian distribution $v_0(v) = \dfrac{1}{\sqrt{2\pi}u}\exp\left(-\dfrac{v^2}{2u^2}\right)$, then the stationary regime of a solid-state laser with two traveling waves whose frequencies are placed symmetrically with respect to the line center ($\widetilde{\omega}_{\bar{k}} + \widetilde{\omega}_k = 2\omega_0$) is described by the equation

$$\frac{\widetilde{\omega}_k - \omega_k}{\gamma_3} - i = \frac{\overline{\Omega}^2}{\pi}\int_{-\infty}^{+\infty}\frac{\exp(-\zeta^2)\,d\zeta}{(\xi_n + \zeta + i\eta)\left\{1 + \dfrac{4\eta\gamma_3\overline{\Omega}^2|e_k^k|^2}{\gamma_1\sqrt{\pi}L}\left[\dfrac{1}{(\xi_n+\zeta)^2+\eta^2} + \dfrac{1}{(\xi_n-\zeta)^2+\eta^2}\right]\right\}}, \tag{159}$$

where $\xi_k = (\omega_k - \omega_0)\delta^{-1}$, $\eta = \gamma_2\delta^{-1}$, $\zeta = \dfrac{1}{\sqrt{2}}vu^{-1}$, $\delta = \omega_0 uc^{-1}\sqrt{2}$ is the inhomogeneous broadening of the line, and $\overline{\Omega}^2 = \sqrt{\pi}\Omega^2(\gamma_3\delta)^{-1}$ is the laser parameter for an inhomogeneous luminescence line with $\omega_{\bar{k}} + \omega_k = 2\omega_0$ and $|e_{\bar{k}}^k| = |e_k^{\bar{k}}|$. Equations analogous to (159) for a gas laser have been derived in [114, 115]. In the region $\xi_n \sim \eta$, Eq. (159) gives the Lamb dip. Near the self-excitation threshold of the laser we obtain Lamb's formula for the dip [27]:

$$|e_k^k|^2 = \frac{\sqrt{\pi}\gamma_1 \eta L}{2\gamma_3 \overline{\Omega}^2}(\overline{\Omega}^2 - 1)\frac{\xi_n^2 + \eta^2}{\xi_n^2 + 2\eta^2},$$

(160)

and in the limit of large pumping $\overline{\Omega}^4 \gg 1$, the following expression:

$$\frac{\overline{\Omega}^2 \gamma_1 \sqrt{\pi}\eta L}{4\gamma_3 |e_k^k|^2} - 2 = \Xi\{-(3 + 6\Xi + 2\Xi^2) + 2[\Xi(\Xi + 2)^3]^{\frac{1}{2}}\},$$

(161)

where $\Xi = \xi_n^2 \eta^{-2}\left[1 + \frac{4\gamma_3\overline{\Omega}^2 |e_k^k|^2}{\sqrt{\pi}\gamma_1\eta L}\right]^{-1}$. An equation of the type (161) for a gas laser is found in [116] where a bibliography is given on this question.

A study of the condition for spiking in solid-state lasers with inhomogeneously lumines-cent lines requires going beyond the "diagonal" approximation. This has been done for a single-mode laser tuned to the line center with a Gaussian form-function $\nu_0(\nu)$. For amplitude perturbations (14) $(2\varphi_n - \varphi_m - \varphi_l = 0)$ the matrix $(1 + F)^{-1}$ in Eq. (154) has been calculated assuming $\eta\overline{\Omega}^2 \ll 1$, $\gamma_1\overline{\Omega}^2 \ll \gamma_2$, and $\gamma_2 \gg |\omega_m - \omega_n| \gg \gamma_1\overline{\Omega}^4$. This yields the stability region in coordinates $\gamma_1\gamma_3\gamma_2^{-2}\eta^{-1}$, $\overline{\Omega}^2 - 1$:

$$\frac{\gamma_1\gamma_3\sqrt{\pi}}{\gamma_2^2\eta} > (\overline{\Omega}^2 - 1)^{-2}\left[\left(1 - \frac{\gamma_3}{\gamma_2}\right)(\overline{\Omega}^2 - 1)^3 + \left(5 - 3\frac{\gamma_3}{\gamma_2}\right)(\overline{\Omega}^2 - 1)^2 + 2\left(4 - \frac{\gamma_3}{\gamma_2}\right)(\overline{\Omega}^2 - 1) + 4\right].$$

(162)

If $\gamma_3 > \gamma_2$ then the right side of this inequality is a monotonically decreasing function of the pump power and spiking is possible for pumping above the boundary value. If $\gamma_3 < \gamma_2$ then the right side has a minimum as a function of the pumping level and spikes are generated in a limited pumping range. Instability arises in the single-frequency lasing regime due to the fact that when inequality (162) is satisfied the frequency of the relaxation oscillations in the inverted pop-ulation (of order[†] $[\gamma_1\gamma_3\eta^{-1}]^{1/2}$) is comparable to the width of the hole (of order γ_2) burned by the strong field in the inhomogeneous luminescence line of the material; thus, amplification at the combination frequencies exceeds the losses. For a ruby laser at low temperatures ($T^0 < 80°K$) the emission line is inhomogeneously broadened (see Table 1), and if $\gamma_3 < \gamma_2$ then Eq. (162) yields an instability in the single-frequency regime for $\gamma_2 < 10^7 \sec^{-1}$. A single-mode ruby laser is discussed in [117] in which the sample was cooled by liquid nitrogen or hydrogen and was pumped with monochromatic radia-tion from an argon laser. The ruby laser operated in a spikeless regime. This is in agreement with criterion (162) since according to the authors' calculations [117] the sample temperature always exceeded 25°K so the homogeneous line broadening was always greater than $10^7 \sec^{-1}$. Instability of the single-frequency regime of a single-mode laser with an inhomogeneous line was theoretically examined in [118-121] as well. There qualitatively similar results are ob-tained with simpler models. An inhomogeneous Lorentz line with $\gamma_1 = \gamma_2$ is discussed in [118]; a laser with two types of luminescence centers whose working transition frequencies differ by δ was analyzed in [119]; and a spin laser ($\gamma_3 \gg \delta$, $\gamma_1 = \gamma_2$) is studied in [120-121].

We now study the operation of a laser with an inhomogeneous luminescence line near the stability of the single-frequency regime in the quasilinear approximation. We shall examine

---

[†] The factor η^{-1} appears under the root because the excess pumping beyond the threshold for molecules which participate in lasing and whose fraction of the whole is η is in fact η^{-1} when a laser with an inhomogeneous line operates in the neighborhood of the threshold $\overline{\Omega}^2 \sim 1$.

a single-mode laser tuned to the center of the luminescence line, ω_0. We assume that the field in the cavity follows the polarization; that is, the laser parameters satisfy

$$\omega_0 \gg \gamma_3 \gg \delta, \gamma_2, \gamma_1. \tag{163}$$

Condition (163) may be realized experimentally in spin and molecular masers, lasers, and semiconductor lasers. With these limitations we obtain from Eqs. (146)–(148) the following equations for the dimensionless polarization Z_v and the inverted population y_λ of a sub-ensemble of molecules in the interval $(v, v + dv)$:

$$z_v' = -(1 + i\delta_v) z_v + y_v z, \tag{164}$$

$$y_v' = a_2 [\Omega_{th}^2 - y_v - 2 \operatorname{Re}(z \overset{*}{z_v})], \tag{165}$$

where $Z_v = \dfrac{1}{\sqrt{2}} \operatorname{Re}[z_v \exp(-i\omega_0 t)]$, $\delta_v = (\omega_v - \omega_0)\gamma_2^{-1}$, z is the polarization of the material averaged over the ensemble, and $a_2 = \gamma_1 \gamma_2^{-1}$. The unit chosen for measuring the time is the transverse relaxation time of the material γ_2^{-1}.

All the subsequent calculations may be done formally for an arbitrary form-function $v_0(v)$; however, discernible results can be obtained for the simplest cases. We shall consider a system with a discrete distribution function of the form

$$v_0(v) = \frac{1}{2} [\delta(v - v_1) + \delta(v + v_1)], \quad \delta_{v_1} = -\delta_{-v_1} = \delta. \tag{166}$$

The system of Eqs. (164) and (165) has the stationary state

$$z_v = z = 0, \quad y_v = \Omega_{th}^2, \tag{167}$$

$$z_v = y_v z_0 (1 + i\delta_v)^{-1}, \quad y_v = \frac{\Omega_{th}^2 (1 + \delta^2)}{1 + \delta^2 + 2z_0^2}, \tag{168}$$

$$2z_0^2 = \Omega_{th}^2 - (1 + \delta^2).$$

State (167) is unstable when one of the conditions for self-excitation of the laser,

$$\Omega_{th}^2 > 2, \tag{169}$$

$$\Omega_{th}^2 > 1 + \delta^2, \tag{170}$$

is satisfied. Equation (169) places weaker limitations on the pump power if $\delta > 1$. In this case the luminescence line of different centers do not intersect, the excitation of the laser is oscillatory (for the slow variables z_v, y_v), and the laser emits a band spectrum. This is the case of strong inhomogeneity of the line, and the output intensity is modulated at a frequency $\sim \delta$. The other case, $\delta < 1$, is also of interest. Then Eq. (170) is weaker and its overfulfillment brings the laser into the single-frequency regime (168).

An investigation of the stability of the single-frequency regime (168) yields a pair of characteristic equations for the amplitude

$$\gamma (\gamma + a_2)(\gamma + 1 - \delta^2) + 4a_2 z_0^2 (\gamma + 1) = 0 \tag{171}$$

and phase

$$\gamma [(\gamma + a_2)(\gamma + 1 - \delta^2) + 2a_2 z_0^2] = 0 \tag{172}$$

perturbations. An analysis of the distribution of the roots of Eq. (172) in the complex γ plane shows that $\delta > 1$ is required for the phase perturbations to grow (Fig. 12). Thus, in the following we shall dwell in detail on the behavior of the amplitude perturbations.

The Hurwitz−Routh criterion gives the following formula for the instability boundary of the single-frequency regime with respect to amplitude perturbations:

$$\Omega^2_{th} > 2 + \tfrac{1}{2} x (x + 3a_2 - 2)(1 + x)^{-1}, \tag{173}$$

where x is determined by $1 - \delta^2 = (1 - a_2)x \, (x + 1)^{-1}, \, 0 \leqslant x < \infty$. Amplitude perturbations near the instability boundary (173) grow with a rate γ' and a frequency γ'' given by

$$\gamma' = a_2(1 - a_2)(1 + x)(\Omega^2_{th})' \, [a_2 x (1 + x)^2 + (a_2 + x)^2]^{-1}, \tag{174}$$

$$\gamma''^2 = a_2 x. \tag{175}$$

According to these equations the growth rate decreases and the frequency increases with increasing x, and for $x \sim 1$, $\gamma' \sim \gamma''^2 \sim a_2$. Equations (171)-(175) have been derived in [119, 120] for $a_2 \simeq 1$ and $a_2 \ll 1$.

We now analyze the nature of the bifurcation in the stationary point (168) at the instability boundary (173). A calculation shows that the first Lyapunov quantity of point (168) at the instability boundary is nonzero everywhere except possibly at isolated points; thus, the Poincaré small parameter method [90] yields the following expression for the amplitude of the limiting cycle:

$$z - z_0 = z_1 \exp(\gamma'' \gamma_2 t) + \text{c.c.}:$$

$$|z_1|^2 = - \frac{[2 + (a_2 + 1) x] (a_2 + x) [4a_2 x (1 + x)^2 + (a_2 + x)^2] (1 - a_2) f_1}{2a_2 y_y (1 + x)^2 f}, \tag{176}$$

where $f_1 = - \dfrac{y_y |3(a_2 - 1) + (1 + x)^2| x'}{2 [2 + (a_2 + 1) x](1 + x)} + (\Omega^2_{th})'$, $f = 2a_2 x^4 - (6a_2^2 - 15a_2 + 2)x^3 + (14a_2^2 - 7a_2 + 2)x^2$

$- a_2(9a_2^2 - 38a_2 + 24)x + a_2^2 (3a_2 - 2)$, and the unprimed quantities are fixed at the instability boundary while the primed variables x' and $(\Omega^2_{th})'$ represent the deviations in x and Ω^2_{th} from their boundary values. The denominator of Eq. (176) includes the first Lyapunov quantity,

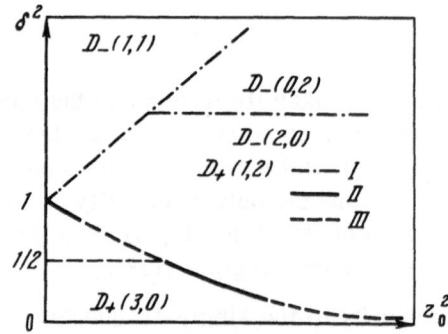

Fig. 12. A stability diagram of the single-frequency regime in a single-mode laser with two types of luminescence centers: I) $D_-(n, m)$ phase perturbations; II, III) $D_+(n, m)$ amplitude perturbations (II is a safe boundary and III, a hazardous boundary).

which is proportional to $+f$. Thus [83], if $f < 0$, then the single-frequency regime at the instability boundary goes weakly into a spiking regime and this part of the boundary is safe; $f > 0$ corresponds to a hazardous boundary; that is, there is a "small" unstable limiting cycle in the region $(\Omega_{th}^2)' < 0$ and hysteresis and strong excitations of spiking may occur.

In general the polynomial f has three positive roots x. If the transverse relaxation time of the material is less than the longitudinal (i.e., $a_2 \ll 1$), then

$$f < 0, \quad x < x_1, \qquad x_1 = a_2 (6 + \sqrt{37}), \tag{177}$$

$$f > 0, \quad x_1 < x < x_2, \ x_2 = 1, \tag{178}$$

$$f < 0, \quad x_2 < x < x_3, \ x_3 = a_2^{-1}, \tag{179}$$

$$f > 0, \quad x_3 < x. \tag{180}$$

In region (177) (see Fig. 12) the emission line is strongly inhomogeneous ($\delta \sim 1$), and a limiting cycle that appears weakly for $(\Omega_{th}^2)' > 0$ describes the band spectrum of the laser. An analogous situation occurs for $\delta > 1$ which was mentioned above. No luminescence lines of the individual centers are allowed in the neighborhood of a weakly inhomogeneous emission line (179) since $\delta^2 < 1/2$. Near the instability boundary the spikes repeat at a frequency of the order of the optical nutation frequency $\sim \sqrt{a_2}$. The amplitude of the spikes increases as $a_2^{-1}(\delta^2)'$ or $a_2^{-1}(\Omega_{th}^2)'$ with distance from the instability boundary. The physical picture of the breakup of the single-frequency regime is as follows. For small deviations from the stationary values (168) the polarization and inversion undergo relaxation oscillations at a frequency which for certain values of the laser parameters (163) is of the order of the optical nutation frequency. Thus, even a weak inhomogeneity of the line $\delta > \sqrt{2a_2}$ [see Eq. (179)] ensures sufficient conditions for growth of the perturbations. If we exclude pumping levels close to the instability boundary (to within a_2 of the boundary) then the final operating regime is determined by energy factors. The laser emits spikes with amplitude a_2^{-1}, unit width, and repetition rate a_2 [see Eqs. (164) and (165)]. For an inhomogeneous line with parameters (178) and (180) the instability boundary is hazardous. Region (178) is a buffer between the regions for a strongly (177) and a weakly (179) inhomogeneous line. In region (180) the behavior of lasers with inhomogeneous and homogeneous (see Section 3) luminescence lines is similar. As a_2 is increased with $0 < a_2 < 1$ the regions (177)–(180) are deformed in a rather complicated way.† In the limit $1 - a_2 \ll 1$ Eq. (176) yields the unstable cycle

$$| z_1 |^2 = -1/4 \, (4x + 1) \, (x + 1) \, (2x + 1)^{-1} \, (\delta^2 - 1). \tag{181}$$

We now study a single-mode laser with two types of luminescence centers near the instability boundary of the single-frequency regime (173) in the relaxation approximation ($a_2 \ll 1$, $\delta_{v_1} = -\delta_{v_2} = \delta < 1$). From the symmetry of the frequency configuration (166) it is clear that a lasing regime with $y_{v_1} = y_{v_2}$ and $z_{v_1} = z_{v_2}^*$ is established as a result of self-excitation. Thus, it is necessary to consider the system of three real equations

$$z' = -z + \delta z_- + yz,$$
$$z_-' = -z_- - \delta z,$$
$$y' = a_2 (\Omega_{th}^2 - y - 2 \, | z \, |^2),$$

---

† For $a_2 = a_* < a_{(1)}$ and $6a_{(1)}^2 - 15a_{(1)} + 2 = 0$, the roots x_1 and x_2 disappear; for $a_2 = 2/3$, the root $x = 0$ appears and increases with increasing a_2; and for $a_2 = a_{**} < a_{(2)}$ and $9a_{(2)}^2 - 38a_{(2)} + 24 = 0$ x_3 is "destroyed." If $a_2 > a_{**}$, then $f > 0$ and $x \geq 0$.

where $z_- = \frac{1}{2}i\,(z_{v_1} - z_{v_2})$ and $y = \frac{1}{2}(y_{v_1} + y_{v_2})$. The subsequent calculations are completely analogous to those done in Section 5, Paragraph 5.1 for a two-mode laser with a point active medium; hence, we present only the final results. The relationship between the inversions before and after the pump stage is given by Eq. (94) and, in the emission stage, by Eqs. (182),

$$x_{2k+1} + x_{2k} = \operatorname{ch}\theta_2 + \frac{1 - \operatorname{ch}^{-2}\theta_2}{2\,(x_{2k} - x_{2k+1})}\,\zeta\left\{2, \frac{1}{2}\left[1 + \frac{\operatorname{ch}\theta_2 - \operatorname{ch}^{-1}\theta_2}{x_{2k} - x_{2k+1}}\right]\right\}, \tag{182}$$

in which the parameters $\operatorname{ch}\theta_1$ and $\operatorname{ch}\theta_2$ must be redefined using the substitution

$$2\Omega_a^2 \to \Omega_{\mathrm{th}}^2, \quad 2y_a \to y, \quad \Delta^2 \to \delta^2. \tag{183}$$

The point image (94), (182), and (183) in the neighborhood of the instability boundary (173) has a unique fixed point which is stable (unstable) for

$$\operatorname{sh}^2\theta_2 \gtrless \frac{1}{\sqrt{10}}\left(\operatorname{sh}^2\theta_2 \lessgtr \frac{1}{\sqrt{10}}\right) \tag{184}$$

and lies in the instability (stability) region for single-frequency lasing. Thus, in this plane the degree of inhomogeneity of the line δ^2 (the pump intensity Ω_{th}^2) may be divided into four regions with different structures for the phase space of the laser. For a weakly inhomogeneous line in the instability region of the spikeless regime [see Fig. 12 and inequality (184)] the laser goes weakly into the giant pulse regime with pulses of width $\sim \gamma_2^{-1}$ and repetition rate γ_1, and in the stability region of the spikeless regime lasing is stable on the whole. For a strongly inhomogeneous luminescence line in the spikeless stability (instability) region the laser goes strongly (weakly) into discontinuity oscillations with a pulse repetition rate given by Eq. (94) with the substitution (183) and $x_{2k-1} = 1$. These conclusions agree with the quasilinear analysis if we note that in the relaxation approximation the instability boundary for the spikeless regime is fixed to within small quantities which approach zero as $a_2 \to 0$, so the case of differences between the luminescence center frequencies (177) and (180) is not described by these formulas.

Therefore, it has been shown with a simple model that even a weak inhomogeneity in the luminescence line can cause the laser to emit large pulses with a frequency of the order of the lifetime of the upper working level γ_1^{-1} and a width inversely proportional to the homogeneous linewidth γ_2. When the line has a large inhomogeneity each pulse will be modulated at a frequency $\sim \gamma_2\delta$. Both weak and strong transitions to spiking and various hysteresis effects are possible. The limiting cases $a \ll 1$ [Eqs. (176)-(180)] and $a \simeq 1$ [Eq. (181)] illustrate the general dependences (171)-(176) and are experimentally realized in lasers and masers and in spin and molecular oscillators, respectively. The theory can be verified quantitatively by artificially creating and controlling inhomogeneous line broadening by the Zeeman effect, for example. An experimental realization of this model would be a device which generates powerful pulses whose parameters are determined by the properties of the active medium, independently of the cavity characteristics; that is, the pulses would have the repetition rate of self-Q-switched spikes and the width of ultrashort pulses in a laser with a homogeneous line. In the strong spike excitation regime we may obtain powerful trains of pulses with the high contrast required for thermonuclear applications. Such a device might also be useful in neuristor networks. Studying the pulse parameters gives information on the relaxation constants of the material and is a technique for high-resolution spectroscopy.

The dispersion relation (24) for the difference frequencies in a laser with an inhomogeneously broadened line has the form

$$\widetilde{\omega}_m - \widetilde{\omega}_n = (\omega_m - \omega_n)\left\{1 - \omega_g^2\,\frac{(\omega_m - \omega_n)^2 + 2\gamma_1(\gamma_1 + \gamma_2)}{[(\omega_m - \omega_n)^2 + 4\gamma_2^2]\,[(\omega_m - \omega_n)^2 + \gamma_1^2]}\right\}, \tag{185}$$

where $\omega_g^2 = 4\gamma_2\gamma_3 \dfrac{\overline{\Omega}^2\gamma_3}{\sqrt{\pi}\eta L\gamma_1}[|e_n^n|^2 + |e_m^m|^2]$, if we limit ourselves to the Doppler limit $(\delta \gg \gamma_2, \gamma_3)$ and pumping near threshold. The second term in the curly brackets describes frequency pushing. For difference frequencies $|\omega_m - \omega_n| \gg \gamma_1\gamma_2$ the frequency pushing is due to the inhomogeneity of the line, is characterized by the frequency ω_g, and yields a branch with anomalous dispersion for $\omega_g^2 > 4\gamma_2^2$:

$$(\omega_m - \omega_n)^2 = \omega_g^2 - 4\gamma_2^2, \qquad \widetilde{\omega}_m = \widetilde{\omega}_n. \tag{186}$$

The relaxation frequency ω_g is unrelated to the oscillations in the inverted population, is of order $(\gamma_2\gamma_3)^{1/2}$, and appears when the spectral width of the cavity mode γ_3 is greater than the homogeneous line broadening γ_2. In this case, when the field oscillates with a large amplitude and the lasing process consists of a pump stage of duration γ_1^{-1} and an emission stage of duration γ_3^{-1}, the giant pulse decays into smaller pulses whose number is determined by the number of unphased radiators and is proportional to the ratio of the mode width to the homogeneous line broadening. For small amplitude field oscillations this subdivision of the pulse yields a fast relaxation frequency ω_g (cf. Section 1). These effects may be observed in a laser with a low-Q cavity and small homogeneous line broadening. For typical solid-state lasers the inequality $\omega_g \ll \gamma_2$ is usually satisfied and the dispersion relation (185) has the form (24) with the obvious redefinition of the relaxation frequency ω_1 typical for a laser with a homogeneously broadened line.

The dependence of the total intensity of the two-frequency regime on the difference frequency is of interest. If the frequencies ω_m and ω_n are located symmetrically relative to the line center ω_0, i.e., $|e_n^n| = |e_n^m|$, then near the lasing threshold $(\overline{\Omega}^2 - 1 \ll 1)$ we have

$$\overline{\Omega}^2 - (1 + \xi_n^2) = |e_n^n|^2 \frac{2\gamma_3\overline{\Omega}^2}{\sqrt{\pi}\gamma_1\eta L} \frac{(\omega_m - \omega_n)^4 + (\omega_m - \omega_n)^2(8\gamma_2^2 + \gamma_1^2 - 2\gamma_1\gamma_2) + 12\gamma_1^2\gamma_2^2}{[(\omega_m - \omega_n)^2 + 4\gamma_2^2][(\omega_m - \omega_n)^2 + \gamma_1^2]}. \tag{187}$$

From this equation it is clear that at small difference frequencies the monotonicity of the function $|e_n^n|^2$ of $\omega_m - \omega_n$ is disrupted and Eq. (187) describes the superposition of two dips — the Lamb dip, of width γ_2 [see Eq. (160)], and a narrow dip, of width γ_1 which is characteristic of a homogeneous line [see Eq. (26)]. These dips are distinctly separated for $\gamma_1 \ll \gamma_2 \ll \delta(\overline{\Omega}^2 - 1)$ into

$$|e_n^n|^2 = \frac{\sqrt{\pi}\gamma_1\eta L(1 - \overline{\Omega}^{-2})}{2\gamma_3} \frac{(\omega_m - \omega_n)^2 + 4\gamma_2^2}{(\omega_m - \omega_n)^2 + 8\gamma_2^2}, \qquad |\omega_m - \omega_n| \sim \gamma_2, \tag{188}$$

$$|e_n^n|^2 = \frac{\sqrt{\pi}\gamma_1\eta L(1 - \overline{\Omega}^{-2})}{2\gamma_3} \frac{(\omega_m - \omega_n)^2 + \gamma_1^2}{2(\omega_m - \omega_n)^2 + 3\gamma_1^2}, \qquad |\omega_m - \omega_n| \sim \gamma_1. \tag{189}$$

The nature of these dips is different. The Lamb dip originates in the overlapping of the holes burnt by different modes in an inhomogeneous luminescence line if the difference between the frequencies of the modes is less than the homogeneous line width. The narrow dip (189) is due to relaxation oscillations in the inversion which follows the field in this frequency range. As a result, part of the field energy is scattered in other low-Q modes because of the generation of combination frequencies.

We have just considered a solid-state laser with an inhomogeneous luminescence line. Many of these results can be carried over to a gas laser almost without changes. In a gas laser the inhomogeneity of the line is due to Doppler broadening of the emission line of moving active molecules. In Eqs. (146)-(148) this is taken into account by replacing the derivatives $\dot{\nu}$ and $\dot{\rho}$ by the substantial derivatives $\dfrac{\partial\nu}{\partial t} + v\dfrac{\partial\nu}{\partial r}$ and $\dfrac{\partial\rho}{\partial t} + v\dfrac{\partial\rho}{\partial r}$. It is usually assumed that there are no other sources of inhomogeneous broadening and that $\omega_v = \omega_0$.[†] The subsequent calculations for a gas laser do not differ from the analogous ones for a solid-state laser with an inhomogeneous line. One feature of the Doppler effect — the dependence of the shift in the frequency of the working transition of the active molecule on the angle between the propagation direction of the electromagnetic wave and the direction of motion of the molecule — leads to differences in the behavior of gas and solid-state lasers with the same relaxation parameters. For example, the Lamb dip can be observed in a single-mode standing wave gas laser. In a solid laser with an inhomogeneous line this effect is possible only in a two-mode cavity with different eigenfrequencies. Keeping this obvious observation in mind, we can interpret these equations for a gaseous active medium. In particular, Eqs. (187)-(189) are best verified in a He—Ne laser. By changing the pressure of the gaseous mixture in such a laser it is possible to change the relation between the constants γ_1 and γ_2 and, therefore, to separate the Lamb dip in the amplitude characteristic of the laser from the dip characteristic of a homogeneous line. A gas laser also has the correct dispersion since $\gamma_1 \gg \gamma_3$ [see Table 1, Eq. (185), and Section 2]; thus, it is possible to extract experimentally the entire dip of width $\sim \gamma_1$.

6. The Output Regimes of Lasers with

Unstable Parameters

In recent years it has been demonstrated experimentally that in a number of cases an instability in the parameters of a laser is a cause of spiking. In this section we examine the effect of instabilities in the cavity (6.1, 6.2) and in the pump (6.3) on the operation of a laser.

6.1. The Output Regimes of a Single-Mode Laser with an

Unstable Cavity

It has been found [17, 21, 122] that spiking may occur in the output of solid-state lasers because of an instability in the resonator. The mechanism for parametric instability in the spikeless regime is as follows. An instability in the cavity causes modulation in the losses with frequency ω_3 and depth $\varkappa$. For certain ω_3 and $\varkappa$ a nonlinear resonance sets in which leads to driving of relaxation oscillations in the intensity and to disruption of the spikeless regime. The operating regimes of a single-mode laser described by the kinetic equations with sinusoidal modulation of the losses have been studied with a computer in [21, 123, 124]. The resonance characteristics of such lasers have been studied analytically in [3, 125, 126]. Our problem is to analyze a single-mode laser with an unstable resonator in more detail.

We write the initial equations of the laser in the following way:

$$\dot{U}^2 = 2\gamma_3 U^2 \left[-(1 + \varkappa \sin \omega_3 t) + W \right], \tag{190}$$

$$\dot{W} = \gamma_1 \left[\Omega_{th}^2 - W - (\Omega_{th}^2 - 1) U^2 W \right]. \tag{191}$$

If $\varkappa = 0$ then Eqs. (190) and (191) are the ordinary kinetic equations for a laser, which follow, for example, from the system (39)-(41) when $\gamma_2 \gg \gamma_3$.

[†] In Eqs. (146)-(148) it is possible to take the difference in the lifetimes of the upper (γ_a) and lower (γ_b) working levels (typical in gas lasers) into account. This does not lead to serious changes in the computational scheme and the final results. In Eqs. (159)-(161) and (185)-(189) it is necessary to replace $|e_k^k|^2$ by $|e_k^k|^2 4\gamma_1^2 [4\gamma_1^2 - (\gamma_a - \gamma_b)^2]^{-1}$ and to set $\gamma_1 = \frac{1}{2}(\gamma_a + \gamma_b)$.

From Eqs. (190) and (191) it is easy to obtain a closed equation for the field $U^2 = \exp X$:

$$X'' + 2\mu_1^2 (\exp X - 1) = -\mu_1 \sqrt{\varepsilon_2}\{(1 + \varepsilon_1\varepsilon_2^{-1} \exp X) X' + 2\varkappa_1 [\cos\tau + \mu_1 \sqrt{\varepsilon_2}(1 + \varepsilon_1\varepsilon_2^{-1} \exp X) \sin\tau]\},$$

(192)

where $\mu_1 = \sqrt{\beta}\, \overset{\circ}{\omega}\omega_3^{-1}$, $\varepsilon_2 = \varepsilon_1(\Omega_{th}^2 - 1)^{-1}$, $\varkappa_1 = \varkappa\varepsilon_1^{-1}$, $' = d/d\tau$, and $\tau = \omega_3 t$. For a solid-state laser inequalities

$$\varepsilon_1 \ll \sqrt{\varepsilon_2} \ll 1$$

(193)

usually hold. We shall also assume that the amplitude of the modulations in the cavity Q is not very large, i.e.,

$$\varkappa_1\sqrt{\varepsilon_2} \ll 1.$$

(194)

When conditions (193) and (194) are satisfied the term with $\sin\tau$ may be omitted in Eq. (192). This sort of equation is known [127] to have stable periodic solutions with either the period of an external interaction (harmonic oscillations) or a period $2\pi n$ (subharmonic oscillations of order $1/n$).

We shall study the conditions for the existence of harmonic oscillations. The second-order equation (192) is close to a system of Lyapunov equations. The generating solution of this system for harmonic oscillations may be either the equilibrium position $X = 0$ or a closed trajectory $X = \xi^{(m)}$ along which the inversion period is 2π. Using the method given in [54, p. 456] we find in the first case that

$$X^{(0)} = -\frac{2\mu_1\varkappa_1\sqrt{\varepsilon_2}}{2\mu_1^2 - 1}\cos\tau + O(\varepsilon_2), \quad 2\mu_1^2 - 1 \gg \sqrt{\varepsilon_2},$$

(195)

$$X^{(p)} = (12\sqrt{2}\varkappa_1\sqrt{\varepsilon_2})^{\frac{1}{3}}\cos\tau + O\left(\varepsilon_2^{\frac{1}{3}}\right), \quad |2\mu_1^2 - 1| \sim \sqrt{\varepsilon_2}.$$

(196)

In the second case

$$X^{(m)} = \xi^{(m)}(\tau + h) + O(\sqrt{\varepsilon_2}),$$

(197)

where the phase h is found from the defining equation

$$\sin h = \Omega_{th}^2 \sum_{n=0}^{\infty} n^2\xi_n^{(m)2}(2\varkappa_1\xi_1^{(m)})^{-1}, \quad \xi^{(m)} = \sum_{n=0}^{\infty} \xi_n^{(m)}\cos n(\tau + h),$$

(198)

which has a solution only for $\varkappa_1 > \varkappa_{1cr}$ [3, 126]. We note that the threshold amplitude of the modulation in the losses $\varkappa_{1cr}\gamma_3$ is of the same order of magnitude ($\sim \gamma_1\Omega_{th}^2$) as the dispersion $\gamma_{3n} - \gamma_{3m}$ that leads to a substantial change in the region of growing relaxation oscillations in the laser (see Section 5, Paragraph 5.2). Of the two solutions (197) and (198) one (X_+^m) is stable and the other (X_-^m) is unstable. The resonance solution $X^{(p)}$ is stable. The solution $X^{(0)}$ is stable everywhere if condition (194) is satisfied except in the narrow region $\left|2\mu_1^2 - \frac{1}{4}\right| \sim \sqrt{\varepsilon_2}$, where the parametric resonance [65]

$$\left(2\mu_1^2 - \frac{1}{4}\right)^2 \leqslant \frac{\varepsilon_2}{72}\left(\varkappa_1^2 - \frac{9}{4}\Omega_{th}^4\right)$$

(199)

occurs.

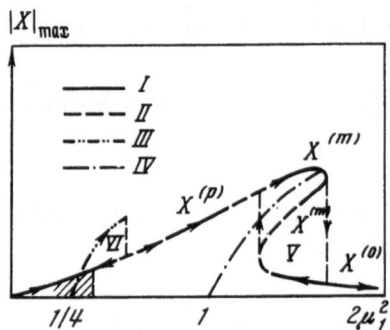

Fig. 13. The amplitude−frequency characteristic for harmonic (I, II) and subharmonic of order 1/2 (III) oscillations. I) Lines plotted from Eqs. (195)−(197); II) the assumed resonance curve; IV) a skeleton trace of the harmonic oscillations; the shading covers the region of instability in the harmonic oscillations; V, VI) hysteresis loops produced when the frequency of the cavity loss modulations is changed.

If we do some calculations similar to (197) and (198) it is possible to show [126] that subharmonic oscillations of order $^1/_2$ exist in the frequency region†

$$0 < \sqrt{2}\mu_1 - \frac{1}{2} < \frac{1}{3}\left(\frac{2}{3}\frac{\varkappa_1}{\Omega^2_{\text{th}}} - 1\right). \qquad (200)$$

Therefore the amplitude−frequency characteristics of the harmonic and half subharmonic oscillations have the form shown in Fig. 13. The appearance of a nonlinear resonance leads, first, to a displacement of the resonance along the curve toward lower frequencies $2\mu^2_1 > 1$. This occurs because the characteristic of the restoring force $f(X) = \exp X - 1$ is nonsymmetric. Secondly, hysteresis loops V and VI appear which can be observed when the loss modulation frequency ω_3 or the pump power Ω^2_{th} is changed. All these effects can occur for $\varkappa_1 \gtrless \gamma_1\gamma^{-1}_3\Omega^2_{\text{th}}$ and disappear when the amplitude $\varkappa$ of the modulations is reduced and the pump power Ω^2_{th} is increased. If $\varkappa_1 \ll \varepsilon^{1/4}_2$ then the resonance curve takes a form typical of linear systems [125].

6.2. Multimode Solid−State Laser with an Unstable Cavity

Theoretical studies of the effect of cavity instabilities on the operating regime of solid-state lasers are usually limited to a single-mode [3, 21, 125, 126] or two-mode [17, 123, 124] kinetic model. We shall consider effects due to modulation of the cavity losses in a solid-state laser with N ≫ 1 operating modes. The kinetic equations for a single-mode solid−state laser with a homogenous luminescence line and loss modulation may be written in the form

$$\dot{X}_n = 2\gamma_3 X_n\left[-(1 + \varkappa_n \bar{f}_n) + \chi_n \int v\Phi^2_n dr\right], \qquad (201)$$

$$\dot{v} = \gamma_1\left(1 - v - v\sum_n X_n \frac{\Phi^2_n}{\int \Phi^4_n dr}\right), \qquad (202)$$

---

† Equation (200) is true for $^2/_3\varkappa_1/\Omega^2_{\text{th}} - 1 \ll 1$.

where X_n are the partial field intensities normalized so that in a single-mode laser operating at a single frequency $\omega_0 \sum_n X_n = 1 - \Omega_{th}^{-2}$, $\chi_n = \Omega_{th}^2 \gamma_2^2 [(\omega_0 - \tilde{\omega}_n)^2 + \gamma_2^2]^{-1}$ takes the shape of the luminescence line into account, and the periodic time-dependent function $\varkappa_n \bar{f}_n$ gives the modulation of the losses from mode n at frequency ω_3 and depth $\varkappa_n$. If $\varkappa_n = 0$, then Eqs. (201) and (202) are equivalent to the kinetic equations in [91].

If $\varkappa = 0$, then Eqs. (201) and (202) have stationary states with $\sum_n X_n \neq 0$. The state corresponding to the maximum number of modes for a given pump level is asymptotically stable [128]. We shall choose it as a generating solution. According to the Poincaré assumption [54, p. 377] for sufficiently small $\varkappa_n \neq 0$ the system (201) and (202) allows one and only one periodic solution with the frequency for the external interaction ω_3 which becomes the generating solution for $\varkappa_n = 0$. It depends analytically on $\varkappa_n$ and is asymptotically stable. In the linear approximation in $\varkappa_n$ we may assume that $f_n = \cos(\omega_3 t + \varphi_n)$, where φ_n is the phase of the loss modulation in mode n. Then we find for the relative amplitudes of the variations in the total intensity $\sum_n x_n \left(\sum_n X_n\right)^{-1}$ and the partial intensities $x_n X^{-1}$ near the generating solution X_n close to the lasing threshold ($\Omega_{th}^2 - 1 \ll 1$) that

$$\frac{\sum_n x_n}{\sum_n X_n} = \left| \frac{(-i\omega_3 + \gamma_1) I_1 \sum_\kappa \tilde{\varkappa}_k f_k^{-1}}{\sum_n X_n \gamma_1 I_2 \left(1 + \sum_\kappa f_k^{-1}\right)} \right|, \tag{203}$$

and

$$\left| \frac{x_n}{X_n} \right| = \left| \frac{(-i\omega_3 + \gamma_1) I_1 \left[\sum_k f_k^{-1} (\tilde{\varkappa}_k - \tilde{\varkappa}_n) - \tilde{\varkappa}_n \right]}{X_n \gamma_1 I_2 f_n \left(1 + \sum_k f_k^{-1}\right)} \right|, \tag{204}$$

where $f_k = \dfrac{-i\omega_3(-i\omega_3 + \gamma_1) I_1}{2\gamma_1\gamma_3 \chi_k I_2 X_k} + \dfrac{I_1 - I_2}{I_2}$; it is assumed that the integrals I_1 and I_2 are independent of the mode indices n and k; and $\tilde{\varkappa}_k = \dfrac{\varkappa_k}{2\chi_k} \exp(-i\varphi_k)$.

The resonance curve for the total intensity (203) has a maximum at the modulation frequency

$$\omega_3^2 = -\gamma_1^2 + \left[4\gamma_1\gamma_3 \sum_n \chi_n X_n \frac{I_2}{I_1} \left(\gamma_1\gamma_3 \sum_k \chi_k X_k \frac{I_2}{I_1} + \gamma_1^2 \right) \right]^{\frac{1}{2}} \tag{205}$$

if the inequality

$$\frac{2 I_2 \gamma_3 \sum_k \chi_k X_k}{I_1 \gamma_1} > \sqrt{2} - 1 \tag{206}$$

is satisfied. When $\varkappa_k \sim \varkappa$ and the inequality (206) is satisfied with room to spare, we have

$$\left| \frac{\sum_n x_n}{\sum_n X_n} \right|_{max} \sim \gamma_3 \gamma_1^{-1} \varkappa. \tag{207}$$

It is natural to analyze the resonance curves for the partial intensities (204) in the case

$$\sum_k \varkappa_k f_k^{-1} = 0, \tag{208}$$

since a new effect, which disappears completely when the loss modulation is synchronous ($\widetilde{\varkappa}_k = \text{const}$), then appears clearly. When condition (208) is satisfied the behavior of the resonance curves for the partial intensities (204) is characterized by Eqs. (205)–(207) with the substitution

$$\frac{x_n}{X_n} \to \frac{\sum\limits_n x_n}{\sum\limits_n X_n}, \qquad \chi_n X_n \frac{I_1 - I_2}{I_1} \to \sum_n \chi_n X_n I_2 I_1^{-1}. \tag{209}$$

Thus, during loss modulation in a multimode laser it is possible to excite oscillations in the partial intensities at frequencies $\sim \omega_2$ [see Eqs. (205) and (209) and Section 1]. These frequencies are much less than the frequency ω_1 at which the total intensity oscillates. The condition for buildup of a low-frequency modulation in the modes closest to the luminescence line center in the case (208) follows from Eqs. (205)–(209):

$$\Delta > \frac{(-1 + \sqrt{2})^{\frac{3}{2}} 2^{\frac{3}{2}}}{3} \left(\frac{\gamma_1}{\gamma_3}\right)^{\frac{3}{2}} \frac{\gamma_2 \gamma_3^{-1}}{(\Omega_{\text{th}}^2 - 1)}, \tag{210}$$

$$\omega_3^2 \sim \frac{1}{2} \gamma_1 \gamma_3 \left[3 \Delta \gamma_3 \gamma_2^{-1} (\Omega_{\text{th}}^2 - 1)\right]^{\frac{2}{3}}, \tag{211}$$

$$\varkappa \gtrsim \gamma_1 \gamma_3^{-1}, \tag{212}$$

where it is assumed that all the modes are axial (102) with $N \gg 1$ and $\Omega_{\text{th}}^2 - 1 \ll 1$. If we go beyond the linear approximation in the amplitude of the loss modulation, then low-frequency oscillations in the partial intensities must be observed even during synchronous modulation of the losses and will lead to low-frequency oscillations in the total intensity. In addition, the low-frequency intensity oscillations will be accompanied by the usual relaxation oscillations with frequency ω_1. Low-frequency oscillations in the field intensity have been recorded experimentally [17]. In experimental studies of the low-frequency modulation in the output it is best to modulate the reactive component of the dielectric constant, that is, the refractive index $\varepsilon'(\omega_0)$, at frequency ω_2. Then, according to the Kramers–Kronig relations

$$\varepsilon''(\omega) = -\frac{1}{\pi} \oint \frac{\varepsilon'(x)\,dx}{x - \omega} \sim -\frac{1}{\pi} \frac{\varepsilon'(\omega_0)}{\omega_0 - \omega}$$

the losses from the various modes are modulated in opposite phase and Eq. (208) is satisfied. Relaxation oscillations in the intensity at frequency ω_1 are more easily obtained by modulating the real part of the dielectric constant, that is, the losses ε''. The refractive index and losses may be modulated with the aid of modulators employing the linear electro-optical effect [129, p. 82].

6.3. The Effect of Polarization on the Dynamics of a Single-Mode

Solid-State Laser with Modulated Pumping

When the polarization is included the small deviations in the field from the single-frequency regime oscillate at two characteristic frequencies [see Eqs. (17)–(20)]. At small pump levels, when the kinetic equations are valid, the field oscillates at the relaxation frequency ω_1. At sufficiently large pump levels the field oscillates at the optical nutation frequency. It is of interest to excite undamped oscillations in the field at this frequency by means of some external interaction. Modulating the cavity losses in the linear approximation in the amplitude $\varkappa \gamma_3$ does not give a resonance in the amplitude–frequency characteristic of the laser because at large pumping levels ω is the eigenfrequency of the inversion, the loss modulation acts on the field, and the inversion then oscillates with a large amplitude while the field gives almost no reaction to the external influence. In this way the oscillations of a damper on a ship dampen the tossing of the ship in the sea's swell.

A single-mode solid-state laser with a three-level scheme and modulated pumping obeys the following system of equations:

$$\dot{U} = \gamma_3 \left(- U + V \right), \tag{213}$$

$$\dot{V} = \gamma_2 \left(-V + UW \right), \tag{214}$$

$$\dot{W} = \gamma_1 \left[\Omega_{\text{th}}^2 - W - (\Omega_{\text{th}}^2 - 1) UV + P\gamma_1^{-1}\varkappa \bar{f} \left(\Omega_{\text{th}}^2 \frac{P\tau_1 + 1}{P\tau_1 - 1} - W \right) \right], \tag{215}$$

where the function $\varkappa \bar{f}$ gives the modulation in the pump power at frequency ω_3 and depth $\varkappa$. These equations are analogous to the equations used in [3] to study the effect of modulations in the number of active particles entering the cavity on the operating regimes of a molecular maser.

In solid-state lasers the relaxation constants usually satisfy the inequality

$$\gamma_2 \gg \gamma_3 \gg \gamma_1. \tag{216}$$

When $\varkappa = 0$ the single-frequency regime of a laser with parameters (216) is asymptotically stable. Thus [54, p. 377], for sufficiently small $\varkappa \neq 0$ Eqs. (213)-(215) have a unique periodic solution with frequency ω_3 that transforms to the generating solution $U = V = W = 1$ if $\varkappa \to 0$ and is asymptotically stable. In the linear approximation in $\varkappa$ we may assume that $f = \cos \omega_3 t$. Then for deviations in the field from the single-frequency regime $\delta U = u_1 \exp(-i\omega_3 t) + \text{c.c.}$ we obtain

$$u_1 = \frac{\gamma_1\gamma_2\gamma_3\bar{\varkappa}}{i\omega_3 \{\omega_3^2 - [\gamma_1(\gamma_2 + \gamma_3) + \overset{\circ}{\omega}{}^2]\} + [2\gamma_3\overset{\circ}{\omega}{}^2 - (\gamma_1 + \gamma_2 + \gamma_3)\omega_3^2]}, \tag{217}$$

where $\bar{\varkappa} = \frac{1}{2} P\gamma_1^{-1}\varkappa \left(\Omega_{\text{th}}^2 \frac{P\tau_1 + 1}{P\tau_1 - 1} - 1 \right)$.

A study of the amplitude−frequency characteristics given by the absolute value of Eq. (217) yields four pumping regions in the limit.

If

$$\overset{\circ}{\omega}{}^2 < \frac{1}{4} \gamma_1^2\gamma_2\gamma_3^{-1}, \tag{218}$$

then the resonance curve decreases monotonically with increasing ω_3.

In the pumping region

$$\frac{1}{4} \gamma_1^2\gamma_2\gamma_3^{-1} < \overset{\circ}{\omega}{}^2 < 4\gamma_2\gamma_3 \tag{219}$$

the amplitude−frequency characteristic has the resonance

$$\omega_{3\,\text{max}}^2 = 2\gamma_3\gamma_2^{-1}\overset{\circ}{\omega}{}^2 - \frac{1}{2} \gamma_1^2\Omega_{\text{th}}^4, \quad |u_1|_{\text{max}}^2 = \frac{\gamma_3^2\bar{\varkappa}^2}{\Omega_{\text{th}}^4 \left(2\gamma_3\gamma_2^{-1}\overset{\circ}{\omega}{}^2 - \frac{1}{4} \gamma_1^2\Omega_{\text{th}}^4 \right)}, \tag{220}$$

whose width at half maximum [if inequality (219) is satisfied with a margin] is

$$| \omega_3^2 - 2\gamma_3\gamma_2^{-1}\mathring{\omega}^2 | \leqslant (2\gamma_3\gamma_2^{-1})^{\frac{1}{2}} \mathring{\omega}\gamma_1\Omega_{th}^2. \qquad (221)$$

If

$$4\gamma_2\gamma_3 < \mathring{\omega}^2 < \gamma_2^2 (2 + \sqrt{3}), \qquad (222)$$

then the resonance curve again decreases monotonically as ω_3 increases.

Finally, at pump levels such that

$$\gamma_2^2 (2 + \sqrt{3}) < \mathring{\omega}^2, \qquad (223)$$

the amplitude–frequency characteristic has a local maximum. If inequality (223) is true with a margin then the resonance has coordinates

$$\omega_{3\,max}^2 = \mathring{\omega}^2, \qquad | u_1 |_{max}^2 = \frac{\gamma_3^2\bar{\varkappa}^2}{\gamma_2^2 (\Omega_{th}^2 - 1)^2} \qquad (224)$$

and a width at half maximum of

$$| \omega_3^2 - \mathring{\omega}^2 | \leqslant \gamma_2\mathring{\omega}. \qquad (225)$$

If the pump power in the free lasing regime satisfies conditions (218) or (222) then the deviations in the inverted population from the stationary state W = 1 are damped aperiodically. For pump powers lying within the limits of inequalities (219) or (223) the inverted population oscillates at the relaxation frequency ω_1 or at the optical nutation frequency $\mathring{\omega}$, respectively. This observation explains the form of the amplitude–frequency characteristic at the corresponding pump levels, in particular, the existence of a resonance at frequency ω_1 or ω in the pumping regions (219) or (223). The amplitude–frequency characteristic in the pumping region (219) has been studied in experiments with a Nd : YAG laser [130]. Active media with a narrow luminescence line and a small effective lifetime γ_1^{-1} of the upper level are to be preferred for experimental studies of the resonance at ω. From this standpoint $Dy^{2+} : CaF_2$ (see Table 1) is promising as with it inequality (223) may be satisfied in a sufficiently high-Q cavity. For modulation of the pumping it is best to use wide-band modulators with bandwidths satisfying condition (225). A mode-locked laser may be tried as a pump source. Such experiments have been done with a multimode laser: A dye laser was pumped with ultrashort pulses from a neodymium glass laser [131]. The proposed method of generating spikes has certain advantages over the multimode method since cavity lengths of the pump and subject laser do not have to have the same length [131] and it may be realized in small-sized apparatus.

7. Some Stochastic Problems in the Theory of Solid-State Lasers

If the parameters of a laser are modulated with large amplitude and a broad spectrum, then the resonances in the amplitude-freqency characteristics will be nonlinear, will broaden, and will multiply (Section 6). The emission takes on a random character. In the resonance, randomization takes place due to hysteresis phenomena associated with the nonlinear nature of the resonance. The intersection of resonances causes the system to go from one resonance to another and increases the randomness in the emission. The appearance of randomness in nonlinear dynamic systems when resonances intersect is treated theoretically in model conservative systems [132].

Randomization in lasers has been studied experimentally [133] and on computers [124, 134]. The basic conclusion of these articles is as follows: Irregular large amplitude spikes appear in the laser when the cavity Q is modulated at a large amplitude, with $\varkappa \sim 10^{-2}$ and $\omega_3 \sim \omega_1$ [133]. This irregularity appears distinctly if the phases of the modes are "disconnected." For example, one mode enters into lasing while the other leaves [124]. The theory of random processes is an adequate mathematical description of the random lasing regime. The appearance of randomness in the emission of a single-mode laser is examined analytically in Paragraph 7.1.

7.1. A Single-Mode Solid-State Laser with Noise Pumping

When ordinary nonmonochromatic sources are used the pump level may undergo significant fluctuations. To describe the operation of a solid-state laser with a three-level pump scheme we take the usual equations for a single-mode laser and add a noise term:

$$\frac{dU}{dt} = \gamma_3 (-U + V), \tag{226}$$

$$\frac{dV}{dt} = \gamma_2 (-V + UW), \tag{227}$$

$$\frac{dW}{dt} = \gamma_1 \left[\Omega_{\text{th}}^2 - W - (\Omega_{\text{th}}^2 - 1) UV + P\gamma_1^{-1} \varkappa \left(\Omega_{\text{th}}^2 \frac{P\tau_1 + 1}{P\tau_1 - 1} - W \right) \dot{\xi} \right], \tag{228}$$

where $\varkappa^2$ is the noise power, $\dot{\xi}$ is the white noise of unit intensity, and the stochastic differential equations (226)-(228) are understood in the Stratonovich sense.

We shall investigate the operation of a single-mode solid-state laser ($\gamma_2 \gg \gamma_3$) with noise pumping. Then the system of Eqs. (226)-(228) becomes a system of kinetic equations with a noise term:

$$U^{2\prime} = \varepsilon^{-1} [2 (\Omega_{\text{th}}^2 - 1)^{-1}]^{1/2} U^2 (-1 + W), \tag{229}$$

$$W' = \varepsilon [2^{-1}(\Omega_{\text{th}}^2 - 1)^{-1}]^{\frac{1}{2}} \left[\Omega_{\text{th}}^2 - W - (\Omega_{\text{th}}^2 - 1) U^2 W + \frac{P\varkappa}{\gamma_1 \sqrt{\mu\gamma_3}} \left(\Omega_{\text{th}}^2 \frac{P\tau_1 + 1}{P\tau_1 - 1} - W \right) \xi' \right], \tag{230}$$

where $\varepsilon = (\gamma_1 \gamma_3^{-1})^2$, $' = d/d\tau'$. Eliminating the population inversion from Eqs. (229)-(230), we find a closed equation for the field $U^2 = \exp \varepsilon X_1$:

$$X_1'' + X_1 = -\varepsilon^{-1} \sum_{n=2}^{\infty} {}' \frac{1}{n!} (\varepsilon X_1)^n - \varepsilon \left[(\Omega_{\text{th}}^2 - 1)^{-\frac{1}{2}} + (\Omega_{\text{th}}^2 - 1)^{\frac{1}{2}} \exp \varepsilon X_1 \right] X_1' +$$

$$+ \frac{P\varkappa}{\gamma_1 \sqrt{\mu\gamma_3}} \xi' \left[\frac{\Omega_{\text{th}}^2 (P\tau_1 + 1) (P\tau_1 - 1)^{-1} - 1}{(\Omega_{\text{th}}^2 - 1) \varepsilon} - \varepsilon (\Omega_{\text{th}}^2 - 1)^{-\frac{1}{2}} X_1' \right]. \tag{231}$$

Since the inequalities

$$\varepsilon \ll 1, \quad \varepsilon \ll (\Omega_{\text{th}}^2 - 1) \ll \varepsilon^{-1} \tag{232}$$

usually are satisfied in a solid-state laser, we shall make the standard substitution

$$X_1 = a \cos \psi, \quad X_1' = -a \sin \psi, \quad \psi = \tau' + \theta \tag{233}$$

in Eq. (231). The new variables, the amplitude a and phase θ, satisfy the following system of equations in which (since an averaging method will be used later on) terms of higher order in

ε than those shown here have been eliminated:

$$da = \varepsilon a \left[a \cos^2 \psi - \Omega_{th}^2 \left(\Omega_{th}^2 - 1 \right)^{-\frac{1}{2}} \sin \psi \right] \sin \psi \, d\tau' - \sqrt{\varepsilon} \, \widetilde{\varkappa} \sin \psi d^*\xi, \qquad (234)$$

$$d\theta = \varepsilon \left[a \cos^2 \psi - \Omega_{th}^2 \left(\Omega_{th}^2 - 1 \right)^{-\frac{1}{2}} \sin \psi \right] \cos \psi \, d\tau' - \sqrt{\varepsilon} \, a^{-1} \widetilde{\varkappa} \cos \psi d^*\xi, \qquad (235)$$

where $\widetilde{\varkappa} = \dfrac{P\varkappa \left[\Omega_{th}^2 \left(P\tau_1 + 1 \right) \left(P\tau_1 - 1 \right)^{-1} - 1 \right]}{\gamma_1 \left(\Omega_{th}^2 - 1 \right) \varepsilon^{3/2} \sqrt{\mu \gamma_3}}$.   From the Stratonovich stochastic differential equations (234) and (235) we transform to the Ito stochastic equations [135, p. 218]:

$$da = \varepsilon \left\{ a \left[a \cos^2 \psi - \Omega_{th}^2 \left(\Omega_{th}^2 - 1 \right)^{-\frac{1}{2}} \sin \psi \right] \sin \psi + \frac{1}{2} \widetilde{\varkappa}^2 a^{-1} \cos^2 \psi \right\} d\tau' - \sqrt{\varepsilon} \, \widetilde{\varkappa} \sin \psi \, d\xi, \qquad (236)$$

$$d\theta = \varepsilon \left\{ a \left[a \cos^2 \psi - \Omega_{th}^2 \left(\Omega_{th}^2 - 1 \right)^{-\frac{1}{2}} \sin \psi \right] \cos \psi - \frac{1}{2} \widetilde{\varkappa}^2 a^{-2} \sin 2\psi \right\} d\tau' - \sqrt{\varepsilon} \, \widetilde{\varkappa} a^{-1} \cos \psi \, d\xi. \qquad (237)$$

The solution of the stochastic differential equations (236) and (237) is a diffusive Markov process whose transient probabilities satisfy the Kolmogorov−Fokker−Planck (KFP) equation

$$\frac{\partial f}{\partial \tau'} + \frac{\partial (K_a f)}{\partial a} + \frac{\partial (K_\theta f)}{\partial \theta} = \frac{1}{2} \left[\frac{\partial^2 (D_a f)}{\partial a^2} + 2 \frac{\partial^2 (D_{a\theta} f)}{\partial a \, \partial \theta} + \frac{\partial^2 (D_\theta f)}{\partial \theta^2} \right], \qquad (238)$$

where the transport coefficients K_a and K_θ and diffusion coefficients D_a, $D_{a\theta}$, and D_θ are easily written in an explicit form. Averaging them over ψ [53], we obtain

$$K_a = -K_1 a + \frac{D_a}{2a}, \qquad K_1 = \frac{1}{2} \varepsilon \Omega_{th}^2 \left(\Omega_{th}^2 - 1 \right)^{-\frac{1}{2}}, \qquad K_\theta = 0,$$

$$D_a = \frac{1}{2} \varepsilon \widetilde{\varkappa}^2, \qquad D_{a\theta} = 0, \qquad D_\theta = D_a a^{-2}.$$

The Green's function of Eq. (238) is given by

$$G(a, \theta, a_0, \theta_0, \tau') = \frac{K_1 a \exp (2K_1 \tau')}{\pi D_a \left[\exp (2K_1 \tau') - 1 \right]} \times$$

$$\times \exp \left\{ - \frac{K_1 \left[a^2 \exp (2K_1 \tau') + a_0^2 - 2a_0 a \exp (K_1 \tau') \cos (\theta - \theta_0) \right]}{D_a \left[\exp (2K_1 \tau') - 1 \right]} \right\}, \qquad \tau' > 0; \qquad (239)$$

$$G = 0, \quad \tau' < 0.$$

Since the solution of Eq. (238) with an arbitrary initial distribution $f_0(a_0, \theta_0)$ has the form

$$f(a, \theta, \tau') = \int_0^{2\pi} \int_0^{\infty} G(a, \theta, a_0, \theta_0, \tau') f_0 (a_0, \theta_0) \, da_0 \, d\theta_0, \qquad (240)$$

it follows from the Green's function in Eq. (239) that any solution of Eq. (238) tends with the passage of time $(\tau' \to +\infty)$ to the distribution

$$f = \frac{K_1 a}{\pi D_a} \exp \left(- \frac{K_1}{D_a} a^2 \right). \qquad (241)$$

The stationary distribution (241) is independent of the phase and is a Rayleigh distribution for the amplitude a with an average value

$$\bar{a} = \left(\frac{D_a \pi}{4K_1} \right)^{\frac{1}{2}}$$

and dispersion

$$\overline{a^2} - \bar{a}^2 = D_a K_1^{-1}\left(1 - \frac{\pi}{4}\right).$$

The expressions for the average time for a transition from location a to $q \le a M_q(a)$ and from a to $p \ge a M_p(a)$ are of some interest. Proceeding as in [48, p. 142], we find

$$M_q = K_1^{-1}\ln\frac{a}{q},$$

$$M_p = \frac{1}{2D_a}\sum_{m=1}^{\infty}\frac{K_1^{m-1}}{D_a^{m-1}}\frac{p^{2m} - a^{2m}}{\lfloor m!\, m}. \tag{242}$$

Thus, the output of a solid-state laser with noise pumping at a sufficiently low noise amplitude will be narrow-band noise (233) with a carrier frequency of the order of the relaxation frequency ω_1, an envelope a that is distributed according to a Rayleigh law (241), and a uniform phase θ. The modulation in the intensity is greater the greater the diffusion coefficient D_a (i.e., the noise power) and the smaller the transport coefficient (i.e., the stability of the single-frequency regime). The inequality

$$D_a \gg K_1 \tag{243}$$

is the condition for a random lasing regime with a noticeable intensity in its fluctuating component. There are several experimental indications that pumping noise plays a controllable role in the appearance of spiking in solid-state lasers. A single-mode dysprosium laser was studied in [136, 137]. With nonmonochromatic pumping [136] the laser emitted random spikes while with monochromatic pumping by a ruby laser [137] the dysprosium laser operated without spiking.

The correlation function of the random process X_1 has been calculated in [138] in the linear approximation in the noise power $\varkappa^2$. The spectrum of the correlation function of the field in solid-state lasers has a narrow maximum at the relaxation oscillation frequency ω_1, as does the amplitude–frequency characteristic of a solid-state laser with sinusoidal modulation of its parameters (Section 6).

This randomness in the laser output is a consequence of the indeterminacy in the effect on the laser parameters. At large noise powers $\varkappa^2 \sim (\gamma_1\gamma_3^{-1})^2$ (see Section 6), the resonance becomes nonlinear and, since the noise power exceeds the stability of the resonant and non-resonant limit cycles [because condition (243) is satisfied], the output is further randomized due to the internal hysteresis properties of the system. If the noise power is increased further, then at a noise power level such that the width of the basically nonlinear resonance $\sim\gamma_3\varkappa$ equals the distance between the resonances $\sim\omega_1$, that is, such that

$$\varkappa \sim (\gamma_1\gamma_3^{-1})^{\frac{1}{2}}, \tag{244}$$

the resonances become collective and the randomization introduced into the signal by the system is so large that the lasing regime may become random even with a determinate external effect. Equation (244) yields $\varkappa \sim 10^{-2}$ for the boundary at which random spikes appear in typical solid-state lasers (see Table 1) in accordance with experiment [133]. A condition similar to Eq. (244) was obtained in [12, 124] from empirical considerations. According to the criteria given there the random regime develops in lasers for which the spatial dispersion of the losses $\Delta\gamma_3$ is greater than the frequency ω_1, that is, $(\Delta\gamma_3)\gamma_3^{-1} \gtrsim (\gamma_1\gamma_3^{-1})^{\frac{1}{2}}$.

7.2. The Spectral Analysis and Statistics of Burst in Random Spiking

To study the random emission in detail we need kinetic equations for the correlation function of the field or for the spectral energy density of the field. In deriving the kinetic equations the random phase approximation is always used in one form or another. The following hierarchy of times is introduced [132]:

$$t_{in} \ll t_{ph}, \quad t_a \ll t_t. \tag{245}$$

Here t_{in} is the time over which the laser parameters change significantly. The initial equations (1)-(3) are macroscopic and they include averaging over infinitesimally small volumes containing many molecules and over time intervals much greater than the characteristic intermolecular interaction time $t_m \sim l v^{-1} \sim 10^{-13}$ sec (where l is the distance between the molecules and v is the thermal velocity of the molecules). The laser parameters in the macroscopic equations may, however, change considerably due to slower, large-scale processes. The active medium of a laser has many degrees of freedom. At ordinary temperatures density oscillations (acoustic phonons) are strongly excited in it. The phonons scatter on one another and on lattice defects. These processes cause changes in the laser parameters over a time $\sim K t_m$, where K is the percent defect content in the active medium. In ruby this time is inversely proportional to the inhomogeneous line broadening, $\sim 10^{-10}$ sec (see Table 1). When optical pumping is done with gaseous discharge lamps and when an inversion is created by an electrical discharge or hydrodynamic methods over a wide frequency range as white noise, the pump circuit is noisy. The mirrors vibrate and the dimensions of the active medium change at low kilohertz frequencies ($\sim v L^{-1}$). For these reasons, and others which may be listed, the dynamic quantities describing the operation of the laser undergo strong fluctuations as in the case of the turbulent flow of a fluid. Thus, as in the theory of hydrodynamic and plasma turbulence [63, 139] a second stage of averaging† must be completed in order to obtain a kinetic equation for times of the order of the time to establish the turbulence spectrum, t_t; t_t is inversely proportional to the growth rate of the instability, and in a single-mode laser $t_t \sim \gamma_1^{-1}$. To average over the phases it is necessary that the time for the turbulence spectrum to develop, t_t, be much greater than the time for phase mixing, t_{ph}. In a single-mode laser with noise pumping (Paragraph 7.1) t_{ph} is inversely proportional to the diffusion coefficient for the phase, $t_a \sim D_a^{-1}$. Mixing of the amplitudes also occurs over a time D_θ in a layer whose thickness is directly proportional to the amplitude diffusion coefficient D_a and inversely proportional to the stability of the limiting cycle. These considerations allow us to obtain a kinetic equation for a laser in the random regime in the following way. We expand the field in a time-dependent Fourier series with slowly varying coefficients, and in the nonlinear terms of Eq. (9) we eliminate all terms which depend on the phase differences between the field components. Thus, all the methods developed in Section 1 for studying the stationary states of lasers and their stability and transition processes may be transferred to stochastic systems. All effects which do not depend on phase relationships are then preserved.

In certain cases it has been demonstrated experimentally that the output of a laser in the free lasing regime is similar (perhaps, identical) to amplified Gaussian noise. The spatial correlation functions of the field and intensity of a Nd^{3+}:YAG laser have been studied experimentally [142], and it was found that when a large number of transverse modes are generated ($N \sim 10^4$) the statistics of the output is approximately Gaussian. A study of self-mode-locking in a neodymium laser using two-photon fluorescence [106] showed that the picosecond structure

---

† In recent years articles [140, 141] have appeared in which the kinetic equations for turbulent motion are obtained from first principles, that is, averaging is done over the whole volume in the initial microscopic equations.

in the laser output is similar to Gaussian noise in many respects. It was shown in [143] that the emission of different microsecond spikes is incoherent. Thus, the problem of describing the field in terms of Gaussian stochastic processes arises. We shall expand some realization of a stochastic process in the Fourier integral

$$e(t) = \int_{-\infty}^{+\infty} e_\omega \exp(-i\omega t) \, d\omega. \tag{246}$$

It is known [144] that e_ω are complex Gaussian random quantities which are uncorrelated in a stationary process and, thus, independent. Hence, the transition to a statistical description in the dynamic equations for stationary states may be made rigorously as follows: Multiply Eq. (9) by e_{-p} and average the resulting equation over the ensemble. As a consequence of the averaging all phase-dependent terms go to zero since the phases are distributed independently and uniformly. We then come to consider the incoherent states of the field. This sort of transition has been discussed qualitatively above.

We shall solve this stochastic problem for a single-mode laser. Let the cavity be tuned to the line center and $\gamma_2, \gamma_3 \gg \gamma_1$. For an even number of components in the spectrum ($N_1 = 2m$) a solution is constructed such that $\omega_p + \omega_{\bar{p}} = 2\omega_0$, $p = 1, 2, ..., m$, and for an uneven number of components ($N_1 = 2m + 1$) a central component $\omega_n = \omega_0$ is added. We seek the state with the most components in its spectrum, $N_{1\max}$. The number of states can be found from the equation

$$2(\omega_m - \omega_n)^2 [1 - 2\beta(N_1 - 2)] = \omega_1^2 \frac{|e_m^m|^2}{\sum\limits_{p>0} |e_p^p|^2} \left[2\sum_{q=1}^{m-1} \frac{(\omega_m - \omega_n)^2}{(\omega_m - \omega_n)^2 - (\omega_q - \omega_n)^2} + \frac{1}{2} + \theta(|e_n^n|) \right], \tag{247}$$

where $\theta(|e_n^n|) = 1$, $|e_n^n| > 0$ and $\theta(|e_n^n|) = 0$, $|e_n^n| = 0$. From Eq. (247) it follows that if $2\beta = (N_2 - 2)^{-1}$, $N_2 = 3, 4, ...$, then the field oscillations with the largest number of components, $N_{1\max}$, are such that $|e_m^m| = 0$, $N_{1\max} = N_2$. If $\beta > [2(N_2 - 2)]^{-1}$, then $N < N_2$. In particular, when $\beta > 1/2$ stationary field oscillations do not occur in a single-mode cavity at more than two frequencies, and when $1/4 < \beta < 1/2$ they do not occur at more than three frequencies, and so on† (that is, as β is decreased, the greatest possible N_1 becomes larger). The width of the spectrum then decreases as

$$\frac{4\beta(\omega_m - \omega_n)^2}{\omega_1^2} + \frac{|e_m^m|^2}{\sum\limits_{p>0} |e_p^p|^2} = \frac{1+\beta}{N[1 - \beta(N-2)]}. \tag{248}$$

If $\beta \ll 1$, then the formulas

$$N_{1\max} \simeq \frac{1}{2\beta} \quad \text{and} \quad (\omega_m - \omega_n)^2 \simeq \omega_1^2 \tag{249}$$

are asymptotically valid. Equations (247)–(249) are obtained from Eq. (9) for sufficiently high pumping levels when

$$|\omega_p - \omega_n| \gg (\gamma_1 \gamma_3)^{\frac{1}{2}}, \; \gamma_1. \tag{250}$$

---

† It is shown in [66] that the dynamic model of a single-mode laser also has a unique limiting cycle for $\beta > 1/2$, while for $1/4 < \beta < 1/2$ it has two limiting cycles, one of which is unstable. True, these are no longer two- or three-frequency states — their spectra contain many components.

If this inequality holds in the opposite sense, then $m_{max} = 1$, that is, $N_1 = 2, 3$. These kinds of states were studied in Section 2, and it was established that they are possible if condition (25) is satisfied.

We have just made a correlation analysis of a random process which describes the field and have found its spectrum. More detailed characteristics of the process may be derived, for example, the average number of positive bursts per unit time, $\bar{N}$, the average duration of the bursts, $\bar{\tau}$, and the average interval $\bar{\theta}$, between their occurrence at some level C. In general in the optical range the laser field is the sum of a determined signal $s(t) = E_0(t) \cos[\omega_0 t + \varphi_0(t)]$ and a narrow-band (normal under our assumptions) noise $\xi_r(t) = E_r(t) \cos[\omega_0 t + \varphi(t)]$; that is,

$$\xi_n(t) = \xi_r + s = E(t) \cos[\omega_0 t + \psi(t)]. \tag{251}$$

Process (251) has a probability density

$$f(E) = \frac{E}{\sigma^2} \exp\left(-\frac{E^2 + E_0^2}{2\sigma^2}\right) I_0\left(\frac{EE_0}{\sigma^2}\right), \qquad E \geqslant 0, \tag{252}$$

where σ^2 is the dispersion of the noise ξ_r, and I_0 is the modified Bessel function of order zero. Limiting ourselves to the case in which the laser operates in a single-frequency regime in the absence of noise (i.e., E_0 and φ are time dependent), we have the following formulas for $\bar{N}$, $\bar{\tau}$, and $\bar{\theta}$ at a level C [145][†]:

$$\bar{N} = \frac{C}{\sigma}\sqrt{\frac{-\ddot{\rho}(0)}{2\pi}} \exp\left(-\frac{C^2 + E_0^2}{2\sigma^2}\right) I_0\left(\frac{CE_0}{\sigma^2}\right), \tag{253}$$

$$\bar{\tau} = \frac{\sigma}{C}\sqrt{\frac{2\pi}{-\ddot{\rho}(0)}} \exp\left(\frac{C^2 + E_0^2}{2\sigma^2}\right) \frac{1 - \mathcal{I}}{I_0\left(\frac{CE_0}{\sigma^2}\right)}, \tag{254}$$

$$\bar{\theta} = \frac{\sigma}{C}\sqrt{\frac{2\pi}{-\ddot{\rho}(0)}} \exp\left(\frac{C^2 + E_0^2}{2\sigma^2}\right) \frac{\mathcal{I}}{I_0\left(\frac{CE_0}{\sigma^2}\right)}, \tag{255}$$

where $\rho(0)$ is the second derivative of the correlation coefficient of the envelope $E_r(t)$ at zero and is related to the correlation function of the noise $\xi_r(t)$ by $2\langle \xi_r(t + \tau)$, where $\xi_r(t)\rangle = \sigma^2 \rho(\tau) \cos \omega_0 \tau$ and $\mathcal{I} = \sigma^{-2} \int_0^C z \exp\left(-\frac{z^2 + C^2}{2\sigma^4}\right) I_0\left(\frac{CE_0}{\sigma^4}\right) dz$. Therefore, one realization of the process is a sequence of bursts, both positive and negative, of average duration of the order of $[-\ddot{\rho}(0)]^{\frac{1}{2}} \sim \omega_1^{-1}$. The ratio $\bar{\tau}\bar{\theta}^{-1}$ is independent of the correlation function [see Eqs. (254) and (255)]; i.e.,

$$\bar{\tau}\bar{\theta}^{-1} = (1 - \mathcal{I})\mathcal{I}^{-1}, \tag{256}$$

and decreases as the level C is increased. The average characteristics of the process have a period with frequency $\sim \omega_1 N_1^{-1}$ [see Eq. (249)]. Hence the turbulence is quasiperiodic and for $\beta \ll 1$ a developed turbulence with a continuous spectrum is possible. The dependence of the average number of positive bursts $\bar{N}$ on the level C repeats the dependence of f on E to within a proportionality coefficient and has a maximum at $C \sim E_0$ of width $\sim \sigma$. Thus, by studying this dependence experimentally it is possible to determined the characteristics of the envelope of

---

† The effect of noise on the relaxation spikes when E_0 depends on time has been studied in Paragraph 7.1 and [79]. Besides assuming stationarity we have assumed that the process $\xi_r(t)$ is ergodic when we derived Eqs. (254) and (255).

the determined signal E_0 and the noise E_r. This theoretical conclusion that the turbulence in the output of a single-mode laser operating in a random regime is quasi-periodic has been verified experimentally in a single-mode dysprosium laser [136]. If a single mode is not isolated, then [146, 147] the output of a dysprosium laser does not have these features. Observation of this effect in other types of lasers is made difficult by the problem of selecting a single mode in the cavity.

The appearance of spiking may be explained physically in the following way: Let a single-mode laser operate in a single-frequency regime. The field fluctuations in this regime are damped out over a time of order γ_1^{-1} and oscillate at the relaxation frequency ω_1 if it is greater than γ_1 [see Eqs. (104) and (105)]. If a new fluctuation in the field appears over a time $\sim \gamma_1^{-1}$, then the laser output will consist of a sequence of undamped random spikes. Thus the condition for spiking in a single-mode laser with constantly acting perturbations is $\omega_1 \gtrsim \gamma_1$. This is a well-known effect. Thus, in the theory of stochastic differential equations, the equation $\ddot{x} + \dot{x}(R + \sigma_1 \xi) + \omega_1^2 x = 0$, where ξ is white noise of unit intensity, has an instability region in the root mean square of the equilibrium state $x = 0$ in the coordinate plane of reduced noise intensity $\sigma_1^2 R^{-1}$ versus $\omega_1 R^{-1}$. It seems that for a sufficiently high noise intensity $\sigma_1^2 \gtrsim 4R$ it is necessary that $\omega_1 \gtrsim R$ for the instability to appear [135, p. 281].

7.3. The Stability of the Stationary States of a Multimode

Traveling-Wave Laser with Respect to Incoherent Perturbations

The stability of single-frequency lasing at the $\tilde{\omega}_n$ mode with respect to incoherent perturbations (12) at the $\tilde{\omega}_m$ mode is determined by the contour $\Pi = -1$ (see Section 2). In [148] the stability diagrams are determined for a single-frequency laser with different relationships among the relaxation constants at various pump powers. Such a diagram for a laser with parameters (29) is shown in Fig. 3. In a single-mode cavity ($\tilde{\omega}_n = \tilde{\omega}_m$) the single-frequency regime is stable in the small if the frequency shift of the cavity from the line center is smaller than $\sim N_{1\max}\omega_1$, where $N_{1\max}$ is the maximum number of components in the turbulence spectrum of a single-mode laser with $\tilde{\omega}_n = \omega_0$ (see Paragraph 7.2) or, more precisely,

$$x_1^2 < \frac{1}{4}\beta^{-1}\omega^2. \tag{257}$$

Two-frequency field oscillations occur in the same frequency difference region. The two-frequency oscillations are stable with respect to incoherent perturbations (12) in the region of frequency differences x_1 where three-frequency oscillations in the field are possible:

$$x_1^2 < \frac{1}{12}\beta^{-1}\omega^2. \tag{258}$$

Thus, depending on the separation between the cavity frequency $\tilde{\omega}_n$ and the line frequency ω_0, multifrequency (nonsynchronous) field oscillations lose components, and when $\frac{1}{12}\beta^{-1}\omega^2 < x_1^2 < \frac{1}{4}\beta^{-1}\omega^2$ oscillations can occur at no more than two frequencies, while when $|x_1^2 > \frac{1}{4}\beta^{-1}\omega^2$ the laser has only one stationary state, a single-frequency regime, which is, moreover, unstable. Thus, the appearance of an instability leads to nonstationary random lasing. The phase-space trajectory of the laser will wander within a certain energy region, first leaving, then returning to a single-frequency regime.

To confirm this conclusion let us examine the transition processes in a laser with the aid of the expansions (7) with slowly varying frequencies. More precisely, as in the WKB approximation we set

$$e_p(r, t) = e_p(r) \exp\left[-i \int \gamma_p \, d\tau\right], \tag{259}$$

where γ_p are slowly varying complex functions of time with

$$|A_p + B_p| \gg \omega_p |\gamma_p|; \quad |\omega_p - \omega_s| \gg |\gamma_p|, \ \gamma_1;$$

$$\sum_p \mathrm{Im}\, \gamma_p |e_p|^2 \left(\sum_p |e_p|^2\right)^{-1} \ll \gamma_1. \tag{260}$$

Under conditions (259) and (260), Eq. (5) transforms to the system of equations

$$\frac{A_p B_p}{4\omega_0^2} e_p - \frac{A_p + B_p}{2\omega_0} \gamma_p e_p = -\Omega^2 \left\{ e_p + \sum_{\omega_p = \omega_k + \omega_l + \omega_m} \frac{e_k e_l A_m \left[-i(\omega_m + \gamma_m) + \gamma_2\right] e_m}{\gamma_1 - i(\omega_l + \omega_m + \gamma_l + \gamma_m)} \right\}. \tag{261}$$

This is a generalization of Eqs. (9) used to analyze the stationary states of the field and their stability. We shall examine Eqs. (7) and (259) with two components in a single-mode cavity near the stability boundary of the single-frequency regime $x_1^2 = \frac{1}{4}\beta^{-1}\omega^2$. The results of this analysis are shown in Fig. 14. We now discuss this figure. We note, first, that points 2 and 3 represent one and the same stationary regime, single-frequency lasing in a single-mode cavity $(\tilde\omega_n = \tilde\omega_m)$. Thus, if Eq. (257) is not satisfied and the single-frequency regime is unstable with respect to arbitrarily small fluctuations, then the time dependence of the output intensity includes beating, whose amplitude grows at first and then falls to zero in accordance with the motion of the image point in Fig. 14 from point 2 to point 3. A new fluctuation causes a repetition of this process. It is clear that as a result we obtain random lasing. From Fig. 14 it is evident that in region (257) the single-frequency regime may be cut off as well, but that random lasing occurs here with more rigid conditions on the magnitude of the fluctuations. We may speak, therefore, of weak and strong excitation of random lasing with a spike repetition rate of the order of the relaxation frequency ω_1.

The stability diagram for the single-frequency regime of a single-mode laser may be constructed in (x_1^2, ω^2) coordinates for arbitrary relationships between the relaxation constants of the laser (Fig. 15). For $\gamma_3 > \gamma_1 + \gamma_2$ there is an oscillatory instability at the nutation frequency when condition (44) with the substitution

$$\omega \to \frac{1}{\sqrt{2}}\,\omega, \tag{262}$$

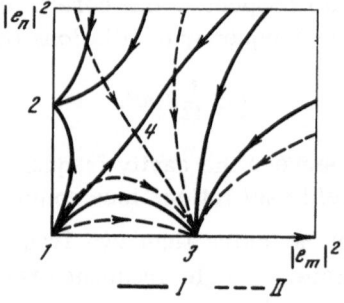

Fig. 14. The phase plane of the equation $d|e_n|^2/d|e_m|^2 =$ $\mathrm{Im}\,\gamma_n|e_n|^2/\mathrm{Im}\,\gamma_m|e_m|^2$ in the neighborhood of point 1 (see Fig. 3). I) $x_1^2 < \omega^2/4\beta$; II) $x_1^2 > \omega^2/4\beta$; the arrows on the curves specify the motion in time along the trajectories; indices 1-4 enumerate the stationary states; in (1) $|e_n| = |e_m| = 0$; points 2 and 3 denote the single-frequency regime and point 4 denotes two-frequency field oscillations.

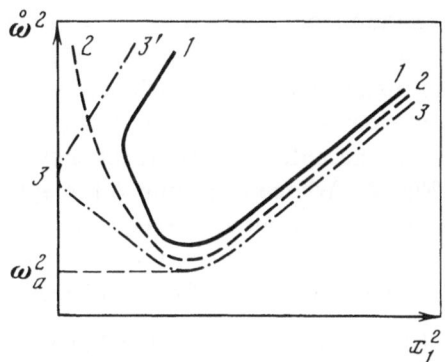

Fig. 15. The stability diagram for single-frequency lasing in a single-mode laser with respect to incoherent perturbations: Case I (curve 1): $\gamma_2 + \gamma_1 > \gamma_3$, $\omega_\alpha^2 = 2[\gamma_1(\gamma_2 + \gamma_3)]/|\gamma_2 - \gamma_3| \times [(\gamma_2\gamma_1(\gamma_2 + \gamma_3)/\gamma_3]^{1/2}$ if $|\gamma_2 - \gamma_3| \gg \gamma_1(\gamma_1|\gamma_2)^{1/2}$; case II (curve 2): $\gamma_2 + \gamma_1 = \gamma_3$; case III (curves 3, 3'): $\gamma_2 + \gamma_1 < \gamma_3$; the region D(2, 1) lies between the branches of the curves on the diagram; in the case III the ordinate and the curve 3-3' form the boundary of region D(1, 2), and the remainder of the quadrant constitutes region D(3, 0).

related to the incoherence in the perturbations,† is satisfied. The physics of this type of instability has been discussed in Section 3. It develops in region D(1, 2) of Fig. 15. In a stochastic single-mode laser a new instability appears, region D(2, 1) of Fig. 15. This is the stochastic [132] or disordered [150] instability region and may exist for a laser with arbitrary relationships among its relaxation constants. The nature of the stochastic instability is as follows: Incoherent perturbations of the field in the neighborhood of the single-frequency regime with large detuning of the cavity relax in the large, that is, they grow at first and then are damped. From the diagram (Fig. 15) it is clear that random lasing is possible when the condition $\omega_1^2 \gtrsim \gamma_1^2$ (the physics of which was discussed in detail in Paragraph 7.2) is satisfied. As the pumping level is increased the degree of randomness in the laser output increases at first and then falls. At pumping levels which satisfy condition (257) only a strong regime with random spiking is possible.

These theoretical predictions are confirmed by experiment [151, 152]. Because of the large value of γ_1 (see Table 1) the lower boundary for random lasing in semiconductor lasers is substantial, and in [151] an increase in the random regime was noted with increased pumping. In ruby γ_1 is small compared to γ_3, and the upper boundary for a weak transition to random spiking can be studied [152]. The laser in [152] worked in a random regime whose form changed with the pump level. At the threshold the spikes are very irregular in form and repetition rate. As the pump power is increased the small spikes are screened out and rare big spikes remain; thus the temporal behavior of the laser is somewhat regularized. This is in full agreement with the theoretical picture of a metamorphosis of the random regime as the pump power is changed. At sufficiently low pumping levels inequality (257) is satisfied and randomness is weakly excited in the laser. As the pump power is increased the random spiking regime becomes strong.

---

† The effect of the statistics of radiation on multiphoton processes has been discussed in connection with multiphoton absorption [149].

In a multimode laser the instability region B (of Fig. 3) appears as well as the instability region A in which disruption of the single-frequency regime leads to nonstationary random spiking. The development of an instability in the single-frequency regime in region B causes the laser to exhibit stationary random spiking since in this region multifrequency stationary field oscillations are possible, for example, the two-frequency oscillations for which the contours Π = const are plotted in Fig. 3. Where in region A the spikes are related to the relaxation, in region B they repeat at the optical nutation frequency ω and arise as in the dynamic model of a laser (Section 4). The stability diagram for a cavity in which one of the eigenfrequencies is tuned to the line center has the same form as Fig. 7 with the substitution (262).

An examination of the stability with respect to incoherent perturbations (12) of two-frequency field oscillations at eigenfrequencies located symmetrically relative to the line center ($\widetilde{\omega}_m + \widetilde{\omega}_n = 2\omega_0$) showed that stable two-frequency lasing corresponds to segments 1-4 and 5-8 of the dispersion curve of Fig. 2. Thus, if we tune the cavity (which action specifies a motion along the curve) we obtain the hysteresis loop 2-4-7-5. The sides of the lysteresis loop are of the order of the relaxation frequency ω_1. More complicated hysteresis effects associated with a strong transition to a two-frequency lasing regime in region C of Fig. 3 are possible. Actually, if we detune the cavity ($\widetilde{\omega}_m + \widetilde{\omega}_n \neq 2\omega_0$), the two-frequency regime loses its stability as the limit cycles which describe three- and two-frequency field oscillations merge. The maximum detuning for stable two-frequency oscillations in a single-mode cavity is given by Eq. (258). The single-frequency regime loses stability upon merging with the limiting cycle for two-frequency oscillations at the boundary of region (257). It may be thought that there is no place for limiting cycles with a large number of components, that is, that the N_{1max} (see Paragraph 7.2) possible stationary states of a stochastic single-mode laser form a system of cycles among which stable and unstable cycles alternate as the number of components in the spectrum of a cycle is increased. Thus strong turbulence appears and develops in the laser output and the magnitude of the fluctuations required for the turbulence spectrum to develop decreases as the number of spectral components increases. The stationary random lasing regime with the maximum possible number of spectral components may break off into a nonstationary turbulence. As a result the number of spectral components decreases and the process repeats itself. The role of the nonstationary motions becomes greater as the set of stationary states is reduced. The strong development of the turbulence and the important role of nonstationary motions distinguish the mechanism of this turbulence from that of the hydrodynamic turbulence described in [153].

7.4. Multimode Standing-Wave Laser

In a multimode laser fluctuations in the number of photons in the individual cavity modes must have a greater probability in the approximation that the total number of photons in all modes is conserved. This assertion has been confirmed experimentally in several cases [154, 155]. Under the influence of such fluctuations each mode of a stochastic laser undergoes relaxation oscillations at a low frequency ω_2 (see Sections 1 and 6). To estimate this frequency we need to know the partial intensities of the modes. They are calculated as in kinetic theory [91]. Then on replacing ω_1 by ω_2 in inequality (25) we obtain the following condition for spiking in a multimode standing-wave laser:

$$\Delta > Q' \ \beta^{-1}\left(\frac{\gamma_1}{\alpha\gamma_3}\right)^{3/2}(\Omega_{th}^2 - 1)^{-1}, \tag{263}$$

where Q' is an unimportant numerical coefficient of order unity. This inequality is derived under the assumptions that the number of working modes $N \gg 1$, $\Delta \gg \gamma_1\gamma_3^{-1}$, and $\Omega_{th}^2 - 1 \ll 1$, and it does not describe the regularization of the lasing regime at high pump levels due to the growth in amplitude of the determined component $s(t)$ in the laser output [see Paragraphs 7.1

and 7.2 and Eq. (266) below]. Equation (263) is basically the same as condition (210) for spiking with Q-switching and in a certain sense touches on the condition for regular laser kinetics derived in [29, p. 105]. According to [29], regular kinetics is observed over a time interval from the onset of lasing of less than ω_2^{-1} if $\omega_2 \gg \gamma_1$. The possibility of randomization of the lasing regime when $\omega_2 \gg \gamma_1$ (over times greater than ω_2^{-1}) is in agreement with condition (263), but the value of ω_2 in Ratner's book [29] differs from ours because he was examining a nonstationary regime in a laser with developed transverse structure. These assumptions have an effect on the estimate of the number N of generated modes.

From Eq. (263) it is apparent that for spikeless operation it is better to have a high Q cavity with a high density of modes since the total output intensity in this regime, in accordance with Eq. (26), will be less than in the spiking regime. These conclusions are confirmed experimentally [10-12, 14, 156]. Experiments with an optical delay line in the cavity so that small Λ may be obtained are of special interest [157]. In [157] the authors observed a transition to spikeless operation with an effective cavity length of the order of several tens of meters. This is in agreement with the estimate given by Eq. (263). Furthermore, according to the assumptions of this paragraph, each mode in a developed random regime (Paragraph 7.1) emits its own train of spikes. Thus if we take a time scan of the output spectrum, only one or a few modes from the entire broad spectrum will be represented in each spike. This is confirmed by the experimental data of [158]. The spikes will come in packets that repeat at a rate $\sim \omega_2$. As a crude estimate we may assume the line has a rectangular shape; then

$$\omega_2 \sim [\gamma_1 \gamma_2^{-1} \gamma_3^2 \Delta \, (\Omega_{th}^2 - 1)]^{1/2}. \tag{264}$$

We find from this estimate that the repetition rate of the packets is inversely proportional to the cavity length and directly proportional to the square root of the pumping excess above threshold. In experiments with semiconductor lasers [4, 16, 151] it has been found that the spike repetition rate depends on the laser parameters in the way indicated here, and in [151] it was noted that the spikes have a fine structure. The role of randomness in the appearance of spikes is demonstrated, for example, in [5, 159] on CO_2 lasers, An electrical discharge laser [5] worked in a random spiking regime while optical pumping with light from a $CO_2 - N_2 - He$ laser [159], which ensured more uniform laser parameters, resulted in spikeless operation.

We may consier a laser with a more complicated, two-scale mode spectrum in which there is a group of $2P_1$ transverse modes close to the axial mode frequencies and located symmetrically a distance $\Delta\gamma_3$ from one another with an interval Δ' such that

$$\gamma_1 \gamma_3^{-1} \ll \gamma_3^{-1}\Delta' \ll 1 \ll \Delta \ll \gamma_2 \gamma_3^{-1}. \tag{265}$$

It is reasonable to simplify the problem by choosing axial and transverse modes in the form of sinusoidal functions (102) and locating the frequency of one of the axial modes $\tilde{\omega}_n$ at the center of the line ω_0. We shall examine the stability of the laser in its axial modes with respect to fluctuations in the transverse modes. Let us study the case of partial randomization when the emission is coherent within the confines of each group and incoherent relative to the emission from other groups. Then the perturbations in each group develop autonomously and independently of the evolution of the perturbations in other groups. Considering only the central group of modes, in which the frequency of the axial mode coincides with the line center as in Paragraph 5.2, we find the region for growing relaxation spikes with a low frequency $\sim \omega_2$:

$$2\Delta_2^2 \frac{\gamma_1}{\gamma_2 P_1^2} \leqslant \Delta'^2 \leqslant 4 \frac{\Delta_2^2}{\gamma_2} \, [12\gamma_3^2 \Omega_{th}^2 \, |e_n^n|^2 \, L^{-1}]^{1/2}, \tag{266}$$

where $\Delta_2^2 = \gamma_2\gamma_3 L^{-1} \int_0^L \lambda'^{-1} \left(\frac{1}{\sqrt{1-\lambda'^2}} - 1\right) dr$, $\lambda' = \frac{1}{L} \sum_s |e_s^s|^2 \frac{4\gamma_3}{\gamma_1} \Omega_{th}^2 \left(\cos 2\, \frac{\widetilde{\omega}_s - \omega_0}{c}\, r\right) \times$

$\left[1 + L^{-1} \sum_s |e_s^s|^2 4\gamma_3\gamma_1^{-1}\Omega_{th}^2\right]^{-1}$, and $|e_n^n|^2$ are the partial intensities of the axial laser modes. The development of an instability yields random spikes similar to Q-switching (cf. Section 5, Paragraph 5.2).

CONCLUSION

We have developed a method for analyzing the operating regimes of a laser. The basis of the method is obtaining a closed integrodifferential equation for the field. Solutions of this equation are sought in the form of trigonometric series with real frequencies (stationary laser states), complex frequencies (in the analysis of the stability of these states), and complex frequencies which vary slowly with time (transition processes). This method, which is a variant of the Fourier method, together with the small parameter method, has been applied rigorously in dynamic and static treatments of lasers.

It has been found that the most "hazardous" form of instability in laser operation is the stochastic instability that arises when the laser parameters are modulated with large amplitude over a wide frequency spectrum. A study of the stochastic instability shows that regular kinetics is possible in cavities with a high density of high-Q modes. This conclusion, which has repeatedly been noted experimentally, is made more precise by the criterion (263) which has been confirmed in certain cases [157]. In a developed random regime a multimode laser emits packets of relaxation spikes with a (packet) repetition rate of the order of the slow relaxation frequency (264). This conclusion and Eq. (264) are confirmed in experiments with semiconductor lasers [4, 16, 151]. The output of a single-mode random laser resembles a quasiperiodic turbulence. This explains some experiments with a single-mode dysprosium laser [136].

The stochastic instability is a stronger variant of the parametric instability. In the parametric instability the individual resonances in the amplitude–frequency characteristic of the laser do not intersect and the laser emits regular spikes at the relaxation frequency [133]. The criterion (244) for going from the parametric instability into the stochastic instability is experimentally based [133].

Elimination of the stochastic and parametric instabilities by stabilizing the laser parameters yields a laser which either operates in a stable spikeless regime or emits self-Q-switched giant pulses. Because of the low stability of the limiting cycle the reasons for the instabilities and the giant pulses vary widely: coincidence of the frequencies of relaxation and nutation oscillations in the inversion in masers, burning of holes in the spatial (109) or spectral (162) inversion distributions, and dispersion in the cavity Q (138). Self-Q-switched spikes have been observed in experiments with a single-mode maser [13]. In a multimode device a giant pulse is made up of ultrashort pulses. The appearance of ultrashort pulses is due to the Cerenkov instability. The stability diagram (see Fig. 7) for self mode-locking (synchronization) agrees with the experimentally determined diagram for a He–Ne laser [107]. Our theory of mode-locking also explains the dynamic effects which occur during passive mode-locking [111, 112].

Detailed comparison of theory with experiment is difficult because of the almost complete lack of experimental stability and instability diagrams for lasers in the spikeless regime, of empirical laws for the repetition rate and duration of spikes, and of experimental data on the statistics of bursts in the stochastic regime. "The rare droplets of experimental fact are almost invisible against the broad front of theoretical developments." These words, written by L. A. Artsimovich [160] about another subject, characterize well the state of research on spiking in various types of lasers and masers. In our opinion it would be interesting to make more careful experimental investigations of several of our results, for example: the equations for the repetition rates of giant pulses in single-mode (58) and two-mode (127), (128), (134), and (135) lasers; the calculations of the regions of growing relaxation spikes in standing-wave lasers (109) and lasers with an inhomogeneous luminescence line (162); mode-locking diagrams (see Fig. 7); the conditions for the appearance of and the characteristics of low-frequency modulation in a multimode laser with an unstable cavity (210)-(212); the equations characterizing the statistics of the bursts (253)-(255) and the spectrum of the correlation function of the field (249); and the expressions for the dip in the amplitude-frequency characteristic of a laser with an inhomogeneous line (187)-(189). An experimental realization of these models would not only explain many problems in the physics of lasers and stimulate theoretical progress in this field, but would result in a laser which would operate stably in a single-frequency regime or emit short, powerful pulses (with high contrast during strong spiking). The first type of apparatus is needed for spectroscopy, for example, and the second, for thermonuclear fusion and for neuristor networks.

LITERATURE CITED

1. A. N. Oraevskii, Molecular Masers [in Russian], Nauka, Moscow (1964).
2. V. M. Fain and Ya. I. Khanin, Quantum Electronics [in Russian], Sovetskoe Radio, Moscow (1965).
3. É. M. Belenov, V. N. Morozov, and A. N. Oraevskii, Tr. FIAN, 52:237 (1970).
4. O. V. Bogdankevich, B. I. Vasil'ev, A. S. Nasibov, A. Z. Obidin, A. N. Pechenov, and M. M. Zverev, Kvant. Élektron., 1:149 (1974).
5. E. P. Velikhov, Yu. K. Zemtsov, A. S. Kovalev, I. G. Persiantsev, V. D. Pis'mennyi, and A. T. Rakhimov, Pis'ma Zh. Éksp. Teor. Fiz., 19:364 (1974).
6. D. Kato and K. Shimoda, Jpn. J. Appl. Phys., 7:548 (1968).
7. A. Heller and V. Brophy, J. Appl. Phys., 39:4086 (1968).
8. D. Andreou and V. L. Little, J. Phys. D, 6:390 (1973).
9. N. A. Borisevich, V. V. Gruzinskii, and N. M. Poltorak, Kvant. Élektron., 1:1411 (1974).
10. T. N. Zubarev and A. K. Sokolov, Dokl. Akad. Nauk SSSR, 159:539 (1964).
11. H. Wieder, J. Appl. Phys., 37:615 (1966).
12. A. M. Leontovich and V. L. Churkin, Zh. Éksp. Teor. Fiz., 59:7 (1970).
13. C. Kikuchi, J. Lambe, G. Machov, and R. W. Terhune, J. Appl. Phys., 30:1061 (1959).
14. P. Walsh and G. Kemeny, J. Appl. Phys., 34:956 (1963).
15. D. Roess, Proc. IEEE, 52:196 (1964).
16. O. V. Bogdankevich, B. A. Kovalenko, A. N. Mestvirishvili, A. S. Nasibov, A. N. Pechenov, E. A. Ryabov, and A. F. Suchkov, Zh. Éksp. Teor. Fiz., 60:132 (1971).
17. Yu. D. Golyaev, Candidates' Dissertation, Moscow State University (1973).
18. A. P. Kazantsev and V. S. Smirnov, Zh. Éksp. Teor. Fiz., 46:182 (1964).
19. A. I. Alekseev, Yu. A. Vdovin, and V. M. Galitskii, Zh. Éksp. Teor. Fiz., 46:320 (1964).
20. L. A. Ostrovskii, Zh. Éksp. Teor. Fiz., 49:1535 (1965).
21. G. N. Vinokurov, N. M. Galaktionova, V. F. Egorova, A. A. Mak, B. M. Sedov, and Ya. I. Khanin, Zh. Éksp. Teor. Fiz., 60:489 (1970).
22. R. S. Ingarden, Postepy Fiz., 25:687 (1974).
23. R. Glauber, in: Quantum Optics, Academic Press, New York (1969).
24. F. Arecchi, M. Scully, G. Haken, and W. Weidlich, Quantum Fluctuations in Laser Light [Russian translation], Mir, Moscow (1974).
25. K. Hepp and E. H. Lieb, Ann. Phys., 76:360 (1973).
26. S. T. Dembinski and A. Kossakowski, Phys. Lett., 49a:331 (1974).
27. W. E. Lamb, Phys. Rev., 134a:1429 (1964).
28. A. F. Suchkov, Tr. FIAN, 43:161 (1968).
29. A. M. Ratner, High-Divergence Lasers [in Russian], Naukova Dumka, Kiev (1970).

30. A. M. Samson, V. A. Rybakov, and N. K. Stashkevich, Zh. Prikl. Spektrosk., 10:236 (1969).
31. N. G. Basov and V. N. Morozov, Zh. Éksp. Teor. Fiz., 57:617 (1969).
32. V. S. Mashkevich, The Kinetic Theory of Lasers [in Russian], Nauka, Moscow (1971), p. 199.
33. L. V. Efimenko and V. S. Mashkevich, Ukr. Fiz. Zh., 18:756 (1973).
34. R. G. Allakhverdyan, V. N. Morozov, A. N. Oraevskii, and A. F. Suchkov, Kvant. Élektron., No. 6, p. 53 (1971).
35. S. Dmitrevsky, J. Appl. Phys., 41:1549 (1970).
36. L. D. Landau and E. M. Lifshits, Quantum Mechanics [in Russian], Fizmatgiz, Moscow (1973).
37. F. T. Arecchi, G. L. Masserini, and P. Schwendimann, La Revista del Nuovo Cimento, 1:181 (1969).
38. A. Szabo, Phys. Rev. Lett., 25:924 (1970).
39. Z. J. Kiss, Phys. Rev., 137a:1749 (1965).
40. A. L. Mikaélyan, M. L. Ter-Mikaélyan, and Yu. G. Turkov, Solid-State Lasers [in Russian], Sovetskoe Radio, Moscow (1967), p. 65.
41. N. V. Karlov, Tr. FIAN, 49:3 (1969).
42. N. G. Basov, V. V. Nikitin, and A. S. Semenov, Usp. Fiz. Nauk, 97:561 (1969).
43. L. Allen and J. Jones, Elements of the Physics of Gas Lasers [Russian translation], Nauka, Moscow (1970), p. 95.
44. N. N. Sobolev and V. V. Sokovikov, Usp. Fiz. Nauk, 91:425 (1967).
45. R. L. Abrams and P. K. Cheo, Appl. Phys. Lett., 14:47 (1969).
46. A. S. Biryukov, V. K. Konyukhov, A. I. Lukovnikov, and R. I. Serikov, Zh. Éksp. Teor. Fiz., 66:1248 (1974).
47. B. I. Stepanov and A. N. Rubinov, Usp. Fiz. Nauk, 95:45 (1968).
48. A. A. Andronov, Collected Works [in Russian], Izd. AN SSSR, Moscow (1956).
49. B. P. Demidovich, Lectures on Mathematical Stability Theory [in Russian], Nauka, Moscow (1967), p. 367.
50. I. G. Malkin, The Theory of Stability of Motion [in Russian], Nauka, Moscow (1966).
51. Yu. L. Daletskii and M. G. Krein, The Stability of Solutions of Differential Equations in Banach Space [in Russian], Nauka, Moscow (1970).
52. J. Heading, Introduction to the Method of Phase Integrals (The WKB method) [Russian translation], Mir, Moscow (1965).
53. Yu. A. Mitropol'skii, The Averaging Method in Nonlinear Mechnics [in Russian], Naukova Dumka, Kiev (1971), p. 296.
54. I. G. Malkin, Some Problems in the Theory of Nonlinear Oscillations [in Russian], Gostekhizdat, Moscow (1956).
55. N. G. Basov and A. M. Prokhorov, Usp. Fiz. Nauk, 93:572 (1967).
56. A. M. Ratner and A. M. Fisher, Izv. Vyssch. Uchebn. Zaved., 16:1510 (1973).
57. T. I. Kuznetsova and S. G. Rautian, Fiz. Tverd. Tela, 5:2105 (1963).
58. E. M. Braverman, S. M. Meerkov, and E. S. Pyatnitskii, Avtomat. Telemekh., 1:5 (1975).
59. I. I. Blekhman, The Synchronization of Dynamic Systems [in Russian], Nauka, Moscow (1971).
60. H. Statz and G. de Mars, Quantum Electronics, Columbia University. Press, New York (1960), p. 530.
61. B. L. Lifshits and V. N. Tsikunov, Ukr. Fiz. Zh., 10:1267 (1965).
62. H. Haken and H. Sauermann, Z. Phys. 176:47 (1963).
63. A. A. Galeev and R. Z. Sagdeev, in: Reviews of Plasma Physics, Vol. 7, Consultants Bureau, New York (1978).
64. L. I. Mandel'shtam, Complete Collected Works [in Russian], Izd. AN SSSR (1947), Vol. 2, p. 334.

65. L. D. Landau and E. M. Lifshits, Mechanics [in Russian], Fizmatgiz, Moscow (1958), p. 103.

66. V. A. Dement'ev and T. N. Zubarev, Dokl. Akad. Nauk SSSR, 204:66 (1972).

67. A. V. Uspenskii, Radiotekh. Élektron., 8:1165 (1963).

68. A. S. Gurtovnik, Izv. Vyssh. Uchebn. Zaved., Radiofizika 1:83 (1958).

69. A. Z. Grasyuk and A. N. Oraevskii, Radiotekh. Élektron., 9:524 (1964).

70. A. A. Andronov, E. A. Leontovich, I. I. Gordon, and A. G. Maier, The Bifurcation Theory of Dynamic Systems on a Plane [in Russian], Nauka, Moscow (1967).

71. Yu. I. Neimark, The Method of Point Images in the Theory of Nonlinear Oscillations [in Russian], Nauka, Moscow (1972).

72. N. A. Zheleztsov and L. V. Rodygin, Dokl. Akad. Nauk SSSR, 81:391 (1951).

73. V. A. Dement'ev and T. N. Zubarev, Preprint IAÉ-2045, Moscow (1968).

74. N. Levinson, Acta Math., 82:71 (1950).

75. L. S. Pontryagin, Izv. Akad. Nauk SSSR, Ser. Mat., 21:605 (1957).

76. E. F. Mishchenko, Izv. Akad. Nauk SSSR, Ser. Mat., 21:627 (1957).

77. A. A. Dorodnytsin, Prikl. Mat. Mekh., 11:313 (1947).

78. V. I. Bespalov and A. V. Gaponov, Izv. Vyssh. Uchebn. Zaved., Radiofizika 8:70 (1965).

79. N. G. Basov, V. N. Morozov, and A. N. Oraevskii, Kvant. Élektron., 1:2264 (1974).

80. A. N. Didenko, Superconducting Waveguides and Cavities [in Russian], Sovetskoe Radio, Moscow (1973).

81. H. D. Crane, Proc. IRE, 50:2048 (1962).

82. L. D. Landau and E. M. Lifshits, Electrodynamics of Continuous Media [in Russian], Fizmatgiz, Moscow (1959), p. 448.

83. Yu. I. Neimark, Dokl. Akad. Nauk SSSR, 129:736 (1959).

84. É. E. Fradkin and Z. K. Yankauskas, Opt. Spektrosk., 23:489 (1967).

85. H. Risken and H. Nummedial, J. Appl. Phys., 39:4662 (1968).

86. N. D. Milovskii and L. L. Popova, Izv. Vyssh. Uchebn. Zaved., Radiofizika, 15:19 (1072).

87. N. D. Milovskii, Izv. Vyssh. Uchebn. Zaved., Radiofizika, 14:93 (1971).

88. T. K. Lim and B. K. Garside, Phys. Lett., 43A:251 (1973).

89. N. D. Milovskii, Izv. Vyssh. Uchebn. Zaved., Radiofizika, 16:537 (1973).

90. M. M. Vainberg and V. A. Trenogin, The Theory of Branching of Solutions of Nonlinear Equations [in Russian], Nauka, Moscow (1969), p. 291.

91. C. L. Tang. H. Statz, and G. De Mars, J. Appl. Phys., 34:2289 (1963).

92. Yu. A. Anan'ev and B. M. Sedov, Zh. Éksp. Teor. Fiz., 48:779 (1965).

93. B. L. Lifshits and V. N. Tsikunov, Zh. Éksp. Teor. Fiz., 49:1843 (1965).

94. B. L. Lifshits and V. N. Tsikunov, Dokl. Akad. Nauk SSSR, 186:557 (1969).

95. V. A. Dement'ev, Preprint IAÉ-1744 (1968).

96. E. L. Klochan, L. S. Kornienko, N. V. Kravtsov, E. G. Lariontsev, and A. N. Shelaev, Radiotekh. Élektron., 19:2096 (1974).

97. A. A. Mak, V. I. Ustyugov, V. A. Fromzel', and M. M. Khaleev, Zh. Tekh. Fiz., 44:868 (1974).

98. R. Polloni and O. Svelto, IEEE J. Quant. Électron., QE-4:481 (1968).

99. L. A. Ostrovskii and E. I. Yakubovich, Izv. Vyssh. Uchebn. Zaved., Radiofizika, 8:91 (1965).

100. T. N. Zubarev and V. A. Dement'ev, Dokl. Akad. Nauk SSSR, 208:310 (1973).

101. I. G. Aramanovich, G. L. Lunts, and L. E. Elsgolc, Functions of a Complex Variable, Operational Calculus, and the Theory of Stability [in Russian], Nauka, Moscow (1968), p. 377.

102. L. S. Kornienko, N. V. Kravtsov, E. G. Lariontsev, and A. M. Prokhorov, Dokl. Akad. Nauk SSSR, 193:1280 (1970).

103. M. H. Crowell, IEEE J. Quant. Électron., QE-1:12 (1965).

104. A. N. Bondarenko, G. V. Krivoshchekov, V. M. Semibalamut, V. A. Smirnov, and M. F. Stupak, Izv. Vyssh. Uchebn. Zaved., 14:1615 (1971).

105. V. I. Malyshev, A. S. Markin, A. V. Masalov, and A. A. Sychev, Zh. Éksp. Teor. Fiz., 57:827 (1969).

106. M. A. Duguay, J. W. Hansen, and S. L. Shapiro, IEEE J. Quant. Électron., QE-6:725 (1970).

107. T. Uchida and A. Ueki, IEEE J. Électron., QE-3:17 (1967).

108. A. J. De Maria, W. H. Glenn, Jr., M. J. Brienza, and M. E. Mack, Proc. IEEE 57:2 (1969).

109. P. G. Kryukov and V. S. Letokhov, IEEE J. Quantum Électron., QE-8:766 (1972).

110. B. Ya. Zel'dovich and T. I. Kuznetsova, Usp. Fiz., Nauk, 106:47 (1972).

111. A. N. Zherikhin, V. A. Kovalenko, P. G. Kryukov, Yu. A. Matveets, S. V. Chekalin, and O. B. Shatberashvili, Kvant. Élektron., 1:377 (1974).

112. S. D. Zakharov, P. G. Kryukov, Yu. A. Matveets, S. V. Chekalin, and O. B. Shatberashvili, Kvant. Élektron., No. 5 (17), p. 52 (1973).

113. J. A. Fleck, Jr., Phys. Rev., 1B:84 (1970).

114. S. G. Rautian, Tr. FIAN, 43:3 (1968).

115. S. T. Scott, IEEE J. Quantum Electron., QE-4:237 (1968).

116. K. Uehara and K. Shimoda, Jpn. J. Appl. Phys., 10:623 (1971).

117. M. Birnbaum, P. H. Wendzikowski, and C. L. Fincher, Appl. Phys. Lett., 16:436 (1970).

118. E. I. Yakubovich, Zh. Éksp. Teor. Fiz., 55:304 (1968).

119. V. S. Idiatulin and A. V. Uspenskii, Radiotekh. Élektron., 18:580 (1973).

120. K. V. Vladimirskii, Kratk. Soobsh. Fiz., No. 10, p. 41 (1971).

121. K. V. Vladimirskii, Kratk. Soobshch. Fiz., No. 3, p. 47 (1972).

122. S. A. Akhmanov, Yu. D. Golyaev, and V. G. Dmitriev, Zh. Éksp. Teor. Fiz., 62:133 (1972).

123. Yu. D. Golyaev and S. V. Lantratov, Kvant. Élektron., 1:2197 (1974).

124. B. P. Kirsanov, A. M. Leontovich, and A. M. Mozharovskii, Kvant. Élektron., 1:2211 (1974).

125. H. G. Danielmeyer, J. Appl. Phys., 41:4014 (1970).

126. G. N. Vinokurov, Opt. Spektrosk., 31:472 (1971).

127. E. Treffetz, Math. Ann. 95:307 (1926).

128. E. S. Kovalenko and A. V. Pugovkin, Izv. Vyssh. Uchebn. Zaved., Radiofizika, 11:232 (1968).

129. E. R. Mustel' and V. N. Parygin, Light Modulation and Scanning Techniques [in Russian], Nauka, Moscow (1970).

130. H. G. Danielmeyer and F. W. Ostermayer, J. Appl. Phys., 43:2911 (1972).

131. T. R. Rout, W. L. Faust, L. S. Goldberg, and Ch. H. Lee, Appl. Phys. Lett., 25:514 (1974).

132. G. M. Zaslavskii and B. V. Chirikov, Usp. Fiz. Nauk, 105:3 (1971).

133. N. M. Galaktionova, A. A. Mak, and A. P. Khyuppenen, Zh. Éksp. Teor. Fiz., 44:1883 (1974).

134. A. P. Kazantsev, G. V. Krivoshchekov, V. M. Semibalamut, and V. S. Smirnov, Kvant. Élektron., 2:165 (1975).

135. R. Z. Khas'minskii, Stability of Systems of Differential Equations with Random Perturbations of Their Parameters [in Russian], Nauka, Mowcow (1969).

136. M. I. Dzhibladze, T. M. Murina, and A. M. Prokhorov, Dokl. Akad. Nauk SSSR, 182:1048 (1968).

137. M. I. Dzhibladze, E. M. Zolotov, T. M. Murina, A. S. Tverdokhlebov, and G. P. Shipulo, Dokl. Akad. Nauk SSSR, 195:1078 (1970).

138. D. E. McCumber, Phys. Rev. 141:306 (1966).

139. G. Batchelor, Theory of Homogeneous Turbulence, Cambridge University Press (1959).

140. J. Piest, Physica, 73:474 (1974).

141. S. V. Gantsevich, V. D. Kagan, and R. Katilyus, Zh. Eksp. Teor. Fiz., 67:1765 (1974).

142. A. G. Arutyunyan, S. A. Akhmatov, Yu. D. Golyaev, V. G. Tunkin, and A. S. Chirkin, Zh. Éksp. Teor. Fiz., 64:1511 (1973).

143. M. S. Lipsett and L. Mandel, Nature 199:553 (1963).

144. W. B. Davenport and W. L. Root, Introduction to Random Signals and Noise, McGraw-Hill (1958).

145. V. I. Tikhonov, Problems in Random Processes [InRussian], Nauka, Moscow (1970), pp. 117 and 361.

146. S. K. Isaev, L. S. Kornienko, and E. G. Lariontsev, Kvant. Élektron., No. 5(17), p. 41 (1973).

147. S. K. Isaev, L. S. Kornienko, and E. G. Lariontsev, Vestn. Mosk. Gos. Univ., Ser. III, Fiz., Astron., 15:752 (1974).

148. V. A. Dement'ev, Candidate's Dissertation, Moscow State University (1973).

149. S. Carusotto, G. Fornaca, and E. Polacco, Phys. Rev., 157:1207 (1967).

150. Yu. I. Neimark, Izv. Vyssh. Uchebn. Zaved., Radiofizika, 17:602 (1974).

151. Yu. P. Zakharov, I. N. Kompaneets, V. V. Nikitin, and A. S. Semenov, Fiz. Tekh. Poluprovodn., 3:864 (1969).

152. T. Jajima, F. Shimizu, and K. Shimoda, Proceedings of the Symposium on Optical Masers, Polytechnic Press, New York (1963), p. 111.

153. L. D. Landau and E. M. Lifshits, Continuum Mechanics [in Russian], Gostekhizdat, Moscow (1944), p. 100.

154. A. W. Smith and J. A. Armstrong, Phys. Lett., 16a:38 (1965).

155. R. V. Ambartsumyan, P. G. Kryukov, V. S. Letokhov, and Yu. A. Matveets, Zh. Éksp. Teor. Fiz., 53:1955 (1967).

156. K. Gürs, Z. Angew. Math. Phys., 16:49 (1965).

157. L. S. Kornienko, N. V. Kravtsov, E. G. Lariontsev, and N. I. Naumkin, Pis'ma Zh. Éksp. Teor. Fiz., 11:585 (1970).

158. K. G. Folin, V. V. Antsiferov, B. V. Anikeev, and V. D. Ugozhaev, Zh. Éksp. Teor. Fiz., 58:1146 (1970).

159. V. I. Balykin, A. L. Golger, Yu. R. Kolomiiskii, V. S. Letokhov, and O. A. Tumanov, Kvant. Élektron., 1:2386 (1974).

160. L. A. Artsimovich, Controlled Thermonuclear Reactions [in Russian], Fizmatgiz, Moscow (1961), p. 337.

161. T. N. Zubarev, V. M. Martynov, and Yu. A. Tarasov, Opt. Spektrosk., 34:752 (1973).

A STUDY OF INJECTION LASERS. PART II †

A. P. Bogatov, P. G. Eliseev,
and B. N. Sverdlov

The mode composition of the output of injection lasers and the ways of controlling it with selective cavities are discussed. The results of some experiments on a laser with an external dispersive cavity using one or two reflecting diffraction gratings are presented. An asymmetric interaction of the spectral modes is observed and interpreted as induced scattering of the laser light on dynamic inhomogeneities in the electron density within the active medium. The principal of isoperiodic substitution of atoms in multicomponent solid solutions is discussed and used to obtain new types of heterolasers which operate at room temperature in the infrared.

INTRODUCTION

In this paper we generalize the results of research on the physics of injection lasers and continue an earlier report [1]. Here primary attention will be devoted to the problems of controlling the properties of the output radiation from injection lasers using selective cavities and of constructing new heterojunction lasers over a wide spectral range. Chapter I is a review and deals with the mode structure of the emission from heterojunction lasers. The basis of this chapter is a lecture given by one of the authors at the International Autumn School on Semiconductor Optoelectronics (1975).

Chapter II contains the results of research on injection lasers with an external selective cavity.

In Chapter III the interaction of spectral modes, which plays a role in the formation of the spectral and spatial distribution of radiation in injection lasers, is discussed. The various nonlinear interactions of radiation with a medium due to stimulated scattering of radiation on the dynamic inhomogeneities in the electron density in the active region are examined. The possibilities for improving the coherence of the output of injection lasers are discussed.

Since publication of the first part of this article [1] important results have been obtained in development and research on injection lasers. The most important results have been the mastery of a method for making perfected heterojunctions and thus, "heterolasers." Thanks to the application of heterojunctions it has been possible to reduce the threshold current density in uncooled lasers by more than 20 times and to obtain continuous lasing at room temperature [2]. Heterolasers have replaced other types of injection lasers in those applications requiring uncooled lasers. In addition new problems have arisen, one of which is to develop heterojunctions using new compounds which cover a wide spectral range.

---

† For Part I, see Tr. FIAN, 52:3 (1970).

In Chapter IV we discuss a new principle for making perfected heterojunctions based on multicomponent solid solutions.

Research and production of new semiconductor compounds suitable for use in lasers has been done jointly with L. M. Dolginov, L. V. Druzhinina, and E. G. Shevchenko of GIDREDMET.

CHAPTER I

THE MODE STRUCTURE OF THE RADIATION FROM INJECTION LASERS

1. Introduction

Due to a number of their practical features injection lasers are the most promising light source for optoelectronic applications. These features include highly efficient conversion of electrical energy into coherent radiation, accessibility of a large frequency range by direct modulation of the output, and simplicity and compactness. With the appearance of heterogeneous junction and, primarily, injection lasers it became possible to obtain continuous lasing at room temperature and below.

For a number of reasons to be discussed below it is difficult to obtain substantial output power in a single mode from an injection laser. As the output power is increased, its degree of coherence is strongly reduced. This property is a weak point in the technology of injection lasers and has prevented their application in the recording and analyzing of holograms, in spectroscopy, and in other areas where the coherence of the radiation is important.

In general an electromagnetic field has four degrees of freedom, but in a laser cavity limitations are imposed on the field which lead to quantization of the parameters describing the field configuration. Properly speaking, by the "mode structure" of the radiation we mean the expansion of the actual electric field of the laser in terms of the cavity eigenfunctions. Theoretically it is first necessary to formulate a model of the cavity and find its eigenfunctions. The second part of the problem is to analyze the nonlinear dynamics of the oscillations in the laser which, in principle, may yield information on the mode distribution of the radiant intensity in the cavity. A discrete structure in the output spectrum which is interpretable as multimode lasing is easily observed experimentally. The modes that are excited can be identified by combined spectral and spatial measurements. In the ideal case the spectral-spatial distribution of the radiation corresponds to a single mode of oscillations characterized by the set of eigenvalues or indices for the four degrees of freedom. Under real conditions there is a deviation from coherence in the single mode regime due to a background of spontaneous emission and due to inhomogeneities and instabilities in the active medium and cavity.

Emission in a single type of cavity oscillations may be defined as single-mode or single-frequency, which in this case means the same thing. There are, however, differences between these two terms. They are often used to refer to different kinds of deviations from the ideal regime. A single-frequency regime is one with emission in a single spectral line with a smooth profile and fairly narrow width but not excluding the existence of various degeneracies or spatial modes that are slightly displaced in frequency. A single-mode regime is one with the same spatial configuration, that is, one that does not exclude a mixture of different frequencies.

In these regimes the output may be of sufficiently high quality for certain practical applications such as spectroscopy or holography. In semiconductor lasers it is possible to observe such regimes more or less reproducibly and use them in practice. Merely reducing the cavity dimensions would permit a single-frequency regime at more than ten times the threshold in GaAs lasers [3] and at up to twenty times the threshold in long-wavelength lasers using the lead chalcogenides [4]. Narrow spectral lines lines have been recorded in cw operation: less than 150 kHz half-width in GaAs lasers [5] and up to 54 kHz in $Pb_{1-x}Sn_xTe$ [6]. However, these linewidths correspond to negligible powers (10 mW or less in GaAs and at a level of 240 μW in $Pb_{1-x}Sn_xTe$).

As the desired single type of oscillations it is appropriate to choose the lowest order (longitudinal) mode of a Fabry−Perot resonator since it has the simplest configuration and ensures controlled extraction of the radiation into an external medium (i.e., useful operation of the laser). In a number of cases, however, nonlongitudinal modes are to be preferred. An example is an injection laser with distributed extraction of the radiation [7] in which the active medium has a zigzag shape between parallel mirrors in accordance with the ray trajectory in a Lummer−Gehrke interferometer. Experiment has shown that maintenance of mutual coherence at separated outlets is difficult for a number of reasons, of which the most fundamental is apparently the effect of radiation noise which builds up over the long ray trajectory.

The data given above referred to cavities with natural sides (cleavages or mechanically polished). The possibility of easily creating a high quality Fabry−Perot cavity by cleavage is one of the important advantages of semiconductor lasers. Thus, as opposed to most other types of lasers, in injection lasers it is customary to use the intrinsic cavity formed by the natural faces rather than an external cavity. In order to efficiently influence the lasing regime with the aid of external elements it is necessary to reduce the reflection from the natural surfaces. It seems that with an appropriately selective external cavity it is possible to obtain emission that is controlled in all degrees of freedom including the polarization. An intermediate case of mode selection is the composite cavity in which the natural resonator is not weakened but its selectivity is combined with that of another Fabry−Perot cavity or a dispersive element. Cavities with distributed feedback are also of interest as they make it possible to impart the spectral-selective properties directly to the active layer by periodically modulating its optical characteristics.

Here we shall examine the following questions in more detail: the optical structure and emission properties of heterolasers and the reasons for multimode lasing; the features of multimode lasing in ordinary Fabry−Perot resonators and its relation to nonstationary effects; and control of the output characteristics of heterolasers using selective cavities (composite, external, and distributed feedback).

2. The Optical Structure in Heterojunction Lasers

An important feature of heterlasers is the presence of a well-formed dielectric waveguide. It is formed by the narrow-zone emitting layer of the heterojunction and the surrounding medium with a smaller refractive index, $n = \sqrt{\varepsilon}$ (where ε is the real part of the dielectric constant). Thus, in the optical sense the active element of a heterolaser is a piece of a plane dielectric waveguide (Fig. 1). It should be noted that the waveguide effect plays an important role in homogeneous junctions as well, where it is due to variations in the dopant concentrations near the p−n transition [8]. These variations ensure a relative jump in the dielectric constant, $\delta\varepsilon/\varepsilon$, of no more than 1%. In heterogeneous structures it may be ten times greater. The dielectric waveguide permits localization of the optical flux propagating along the p−n junction within the active medium. This improves the utilization of optical amplification in the active medium and reduces diffraction losses within the cavity. A measure of the waveguide effect is the "optical constraint parameter" Γ which is defined as the fraction of the optical flux concentrated within the waveguide, that is, the high refractive index layer. The quantity Γ is related to the reduced thickness of the waveguide, D, which is defined as

$$D = kd \sqrt{\delta\varepsilon/\varepsilon}, \tag{1}$$

where $k = 2\pi/\lambda$ is the wave number within the medium. The function $\Gamma(D)$ is shown in Fig. 2

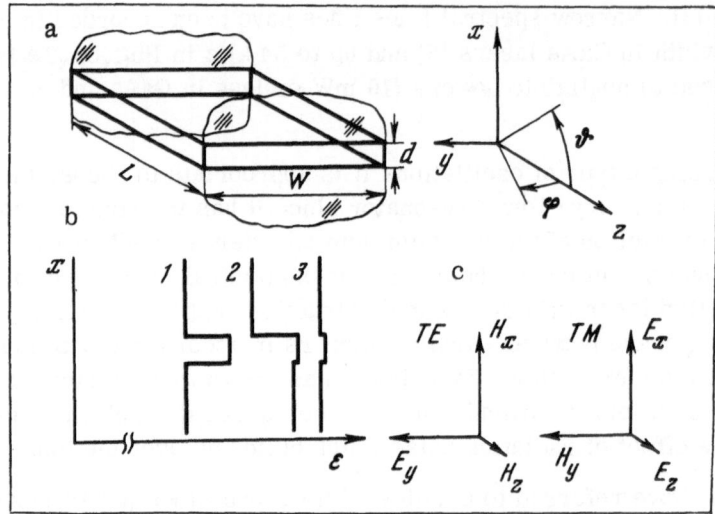

Fig. 1. The active region of an injection laser as a bounded plane dielectric waveguide: a) axes and dimensions; b) the distribution of the dielectric constant $\varepsilon(x)$ in two-sided heterojunctions (1), single-sided heterojunctions (2), and homogeneous junctions (3); c) the vectors for two types of waves in a plane waveguide for propagation along the z axis.

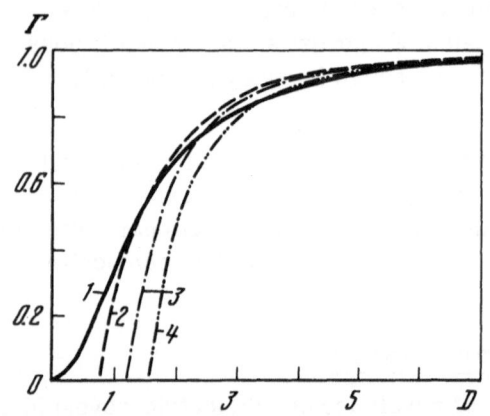

Fig. 2. The dependence of the optical constant parameter Γ on the reduced thickness of the plane dielectric waveguide D for the lowest mode in a symmetric waveguide (1) and in unsymmetric waveguides with $\eta = 2$ (2), 10 (3), and $\eta \to \infty$ (4).

(curve 1) for a symmetric waveguide, i.e., one in which both boundaries of the layer have the same absolute value of $\delta\varepsilon$ [9]. This type of waveguide is realized in two-sided heterogeneous structures, for example, those using the solid solutions $Al_xGa_{1-x}As - GaAs - Al_xGa_{1-x}As$. The drop in $\delta\varepsilon/\varepsilon$ between GaAs and $Al_{0.25}Ga_{0.75}As$ is about 10% [10]. This value is typical for heterostructures used in practice. An example of a new heterostructure is the one based on indium phosphate and a four-component solid solution which because of its composition has the same lattice period as indium phosphide, $InP - In_{1-x}Ga_xP_{1-y}As - InP$ [11]. Estimates of $\delta\varepsilon/\varepsilon$ for a laser heterojunction operating at a wavelength of about 1.1 μm at 300°K yield about 3.8%.

Solutions of the wave equation for a plane dielectric waveguide have been given previously in the literature [8, 9, 12]. We now consider some of the properties of the waveguide modes.

1. The Actual Gain for the Waveguide Modes. In an absolutely symmetric waveguide with arbitrarily small thickness there is a waveguide mode attached to the active layer. In this case, however, as the thickness of the active layer is reduced the parameter Γ falls rapidly, roughly as D^2. The actual gain g for the waveguide mode is reduced compared

to the gain g_0 for the plane waves developing in the active medium by Γ, that is,

$$g = g_0\, \Gamma, \tag{2}$$

because that part of the flux propagating outside the active medium is not amplified. If the gain g_0 increases linearly with the pump current (as is typical for relatively low pumping levels) then the lasing threshold sooner or later begins to increase as the active waveguide becomes thinner. An optimum thickness corresponding to minimal threshold current density can be found [9]. If, however, g_0 increases with the pumping level more rapidly than the square of the current density, then this growth is compensated by the drop in Γ as d is decreased, and finally the threshold current density decreases monotonically. This reduction is limited to a level which corresponds to a slowing down in the growth in g_0 compared to a quadratic dependence on the pump current density. Calculations have shown that in this way, that is, by greatly reducing the thickness of the active medium to 0.02-0.05 μm, the threshold can be reduced to below 0.5 kA/cm^2. This is confirmed experimentally.

2. The Effect of Asymmetries. If the discontinuities in the dielectric constant at the two boundaries of the waveguide are different, that is, the waveguide is asymmetric, then there is a critical thickness beginning with which a waveguide mode exists (see Fig. 2, curves 2-4). This optical thickness depends on the asymmetry coefficient η, defined as

$$\eta = \delta\varepsilon_{12}/\delta\varepsilon_{23}, \tag{3}$$

where $\delta\varepsilon_{12}$ and $\delta\varepsilon_{23}$ are the larger and smaller jumps in the dielectric constant, respectively. For a thickness such that

$$D < \tan^{-1}\sqrt{\eta - 1}, \tag{4}$$

waveguide modes (type TE) do not appear. Strictly speaking, near the critical thickness the effect of amplification is important, and, as will be shown below, large amplification can assist in the appearance of modes in the region where condition (4) holds.

An example of an asymmetric waveguide is the so-called one-sided heterostructure with the following sequence of layers: n-GaAs$-$p-GaAs$-$p-Al$_x$Ga$_{1-x}$As. It is found experimentally that in such heterostructures the threshold rises rapidly even at thickness of order 1 μm (a value which depends on the temperature). This means that in this case the critical thickness is of order 1 μm [13]. In fact, the jump in $\delta\varepsilon/\varepsilon$ at the p$-$n transition (that is, at one boundary of the active medium) is estimated to be 2-5 $\cdot$ 10^{-3}. For large η the critical reduced thickness is about $\pi/2$. The absolute thickness d corresponding to this value is

$$d = \pi/2\, k\, \sqrt{\delta\varepsilon/\varepsilon} \approx 10^{-4}\ \text{cm}, \tag{5}$$

in accordance with experiment. The temperature dependence of the critical thickness in a one-sided heterostructure is related not so much to the temperature variation of the dielectric constants of the lattices as to the temperature dependence of the threshold concentrations of free charge carriers in the active layer. As these concentrations are increased there is a reduction in ε and a reduction in the relative jumps $\delta\varepsilon/\varepsilon$, primarily at the p$-$n junction [9].

3. Transverse Modes. From the standpoint of selecting the lowest mode in a symmetric waveguide it is appropriate to choose a thickness such that D < π so no transverse modes are allowed in this range. With larger D the number of transverse modes increases linearly, roughly as $2/[(D/\pi) + 1]$, where the factor corresponds to the two different polarizations. Thus, in strong waveguides (D $\gg$ 1) the probability is high that undesired transverse modes will be excited. For $\delta\varepsilon/\varepsilon$ = 10% in a GaAs heterolaser the critical thickness for appearance of the first transverse mode is roughly 0.4 μm. It has been noted that the cavity Q for the transverse modes increases with thickness more rapidly than that for the lower modes. Here the increase in the reflectivity of the ends of the waveguide plays a role. This reflec-

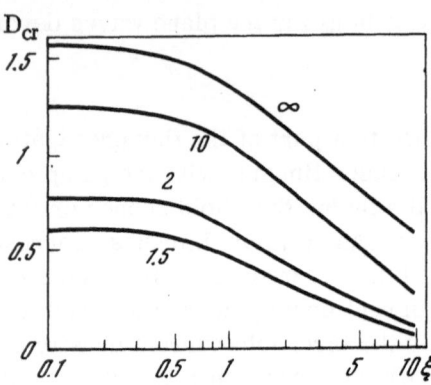

Fig. 3. The dependence of the critical reduced thickness D_{cr} of a dielectric waveguide on the parameter ξ in asymmetric waveguides. The numbers next to the curves are the asymmetry coefficients η.

tivity, which causes feedback in the cavity, differs from the reflectivity R_0 for plane waves and must be modified for each waveguide mode. For Γ close to unity the problem is simplified because the waveguide mode can be almost completely decomposed into two partial plane waves propagating at some angle to the longitudinal axis. The usual Fresnel formulas are applicable to the reflection of these partial waves. As the transverse index increases, the angle of the partial wave becomes larger and, therefore, so does its reflection coefficient (if we have TE modes in mind). Because of this, one of the higher transverse modes among the allowed modes for a given waveguide thickness [14] is most often excited in a two-sided heterostructure. Calculations of the reflection for waveguide modes have been done in [15].

4. The Effect of Amplification on the Mode Configuration. In general, localization of the electromagnetic field near the active medium may occur just because of a favorable distribution of the amplification and absorption in the transverse cross section of the cavity. This situation, however, corresponds to a high level of diffraction losses. In hetero-lasers, at least in most cases, the distribution of the refractive index (or of the real part of the dielectric constant) is of decisive importance. However, near the critical thicknesses the electromagnetic field is weakly localized and is affected by amplification in the active medium. The problem is self-consistent and may yield a nonstationary solution [16]. As the gain increases the electromagnetic field is "pulled" into the active layer, that is, the parameter Γ increases. As mentioned before, waveguide modes may then develop at thicknesses less than the critical value given by Eq. (4). In other words, the critical thickness comes to depend on the gain (or the imaginary part of the dielectric constant). An independent variable for this effect is the ratio $\xi = \mathrm{Im}\,\delta\bar{\varepsilon}/\mathrm{Re}\,\delta\bar{\varepsilon}$, where $\bar{\varepsilon}$ is the complex dielectric constant of the medium. Figure 3 shows the effect of ξ on the critical thickness in some asymmetric waveguides. It is important for $\xi > 1$.

3. The Characteristics and Physical Causes of

Multimode Excitation in Injection Lasers

As can be seen from numerous experiments, it is easy to excite multimode lasing in injection lasers. The number M_q of excited longitudinal modes determines the total spectral width (envelope) of the output since the intermode separation is usually larger for the longitudinal modes than for the transverse modes corresponding to a single longitudinal index. Thus the problem of spectral selection consists, first of all, of reducing M_q to unity. Experiments show that in ordinary cavities (without external selection) the number of modes M_q increases roughly as Y^β, where $Y = (I - I_t)/I_t$ is the relative amount by which the threshold is exceeded and β is a quantity lying between 0.2 and 0.4 [17]. The number of modes M_n excited in another degree of freedom, over the width of the cavity (that is, over the y axis; see Fig. 1), determines the broadening of the directional diagram in angle φ and may be estimated from the ratio of its width $\Delta\varphi$ to the diffraction limit λ/W (where W is the cavity width). In typical wide GaAs diodes M_n may be 10^2 or more for significant amounts above the threshold [17].

The excitation of transverse modes along the y axis has been little studied, but the function $M_n(Y)$ is qualitatively similar to $M_q(Y)$. The spectrum of the transverse modes is very dense so that they can be spectrally resolved only in narrow ("strip") cavities [18]. With large excesses over the threshold, families of different transverse modes with shifted envelopes are often observed; this, ultimately, further broadens the output spectrum. We have already touched upon the features of transverse mode excitation along the axis (that is, along the thickness of the active layer) in the previous section. In principle it is possible to obtain single-mode lasing in sufficiently thin structures. In this case a divergence close to the diffraction limit is actually observed experimentally. These observations were in essence an experimental proof of the optical constraint effect in heterostructures based on the AlGaAs system [19].

Figure 4 shows the dependence of the divergence angle $\Delta\vartheta$ in the plane perpendicular to a dielectric waveguide on its thickness d in two-sided heterostructures compared with the calculated diffraction-limited angle. A satisfactory agreement can be seen. At small d the divergence decreases because, as stated in the previous section, in a thin waveguide most of the radiation propagates in the form of a surface wave outside the active layer ($\Gamma \ll 1$) and forms a broad illumination spot on the cavity mirror. In the limit of high pumping beyond threshold it is possible to observe multimode lasing along this degree of freedom as well if the modes are resolved (that is, for sufficiently thick active layers). The number M_m of modes excited could, however, scarcely be comparable with M_q or M_n in any experiment.

The final degree of freedom, the polarization, is also subject to multimoding. In most injection lasers the preferred polarization (linear) is quite observable at threshold, but as Y increases the degree of polarization falls rapidly and, therefore, the number of modes excited in the polarization, M_p, approaches 2. It has been noted that a TE mode is highly preferred in a two-sided heterostructure with a strong waveguide and that the degree of polarization is conserved with growing Y much better than in a single-sided heterostructure and in homogeneous structures (homostructures). If a uniaxial pressure is applied to the diode perpendicular to the plane of the p−n junction (as often happens in lasers with clamp crystal holders), then it is possible to switch over to the TM mode [20]. It is known that such a pressure leads, qualitatively, to orientation of the radiating dipoles along the pressure axis, as a result of which an anisotropy in the emission probability develops in favor of the TM mode.

As opposed to a two-layer heterostructure in injection lasers with a weak waveguide, especially in single-sided GaAs heterostructures, a preferred polarization is often observed at a random angle relative to the principal angle of the waveguide. This observation may be interpreted as in-phase generation of TE and TM modes. Then the frequencies of both modes must be strictly identical. In a two-sided heterostructure this is usually not observed, and the difference is due to the fact that in a waveguide the phase velocities of TE and TM waves do not generally

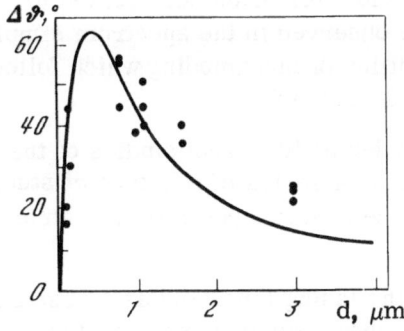

Fig. 4. The dependence of the divergence angle $\Delta\vartheta$ of the output from lasers based on two-sided heterostructures in an AlGaAs system on the thickness d of the active layer. The theoretical curve corresponds to diffraction-limited divergence for $\delta\varepsilon/\varepsilon = 0.1$; the points are experimental values.

coincide, so, in turn, a difference in the eigenfrequencies may develop for modes with other-wise identical indices. A quantitative calculation has been done in [21]. This "spectral split-ting of the polarization" has been observed experimentally in a two-sided AlGaAs heterostruc-ture in an amount of about 0.1 nm with an active waveguide 1 μm thick. Evidently, if this splitting is sufficiently large compared to the width of the cavity band, then synchronization of orthogonally polarized modes cannot occur. In a single-sided heterostructure, however, this splitting is small and does not prevent the appearance of in-phase operation.

Summarizing the above, we note that in ordinary heterolaser cavities (that is, in the ab-sence of external selective elements) a multimode regime can easily develop with a total num-ber $M = M_q M_m M_n M_p$ of excited modes so large that a single mode emits an average of 0.01 W even though the laser is generating tens of watts. Although the factors in M are sublinear in the pump power, on the whole the function M(Y) is appromixately linear with Y so that there is a tendency for the power in a single mode to saturate [17, 22]. This observation led [22] to the conclusion that there may be a fundamental limit to this quantity; however, no theoretical justi-fication has been found for this conclusion.

An explanation of the physical reasons for multimoding has presented much theoretical difficulty. Within the amplification band strong mode competition, which leads to conservation of a single mode, must occur due to rapid intracavity relaxation processes [3]. What factors can maintain this mode competition? Mode competition is weakened if the spatial overlapping of the modes is reduced, in other words, if the modes can independently use the pump power that is supplied. Thus, for example, it is easy to explain the multimode lasing pattern in nonuni-form wide diodes with a spot structure in the near zone in which independent lasing channels can exist. On the other hand, a tendency toward stable single-frequency operation has been observed [3] in small-sized lasers where the role of spatial inhomogeneities is smaller. How-ever, even in special "strip" heterostructures with a single lasing channel it is also easy to observe multimoding. Thus, it is necessary to find other causes. In a homogeneous active medium there remains the unavoidable spatial inhomogeneity in the electromagnetic field. Thus, the standing component of the longitudinal mode creates a variation in the intensity (nodes and antinodes) with period $\lambda/2$. If the inversion is efficiently used up in the antinodes of a given mode, then it is possible in principle that the inversion may build up to excess in the nodes and thereby allow another mode with shifted nodes and antinodes to acquire the necessary gain for lasing [23]. In other words, there is a possibility of several modes being excited with insufficient overlapping of their fields. Calculations have shown that because of the smoothing effect of diffusion of free current carriers this factor may lead to multimode lasing at many times the threshold [1, 23].

Another factor which must be considered is the time required for the mode competition mechanism to operate. This time corresponds roughly to the time for stationary lasing to be established and is close in order of magnitude to the effective electron lifetime in the laser regime. If we consider observations of the emission dynamics of injection lasers, it seems that, except for the case of small amounts over threshold, the lasing is spiky and essentially nonstationary. If the duration of the individual spikes is too short for mode competition to operate then it will not be surprising if numerous modes are observed in the spectrum simulta-neously. Alternatively, in a nonstationary regime the prohibition of multimoding which follows from the homogeneous properties of the emission band is no longer effective.

Most informative are high temporal resolution (of the order of 10^{-10} sec) studies of the kinetics of the spectrum and the directionality of the emission. A series of this type of studies was completed in [24-27]. This made it possible to classify the dynamic regimes in a strip laser as follows:

1. A stationary single-frequency regime is realized at the lasing threshold as a result of the narrowing of the superluminescence spectrum. Figure 5 shows some survey spectrochrono-

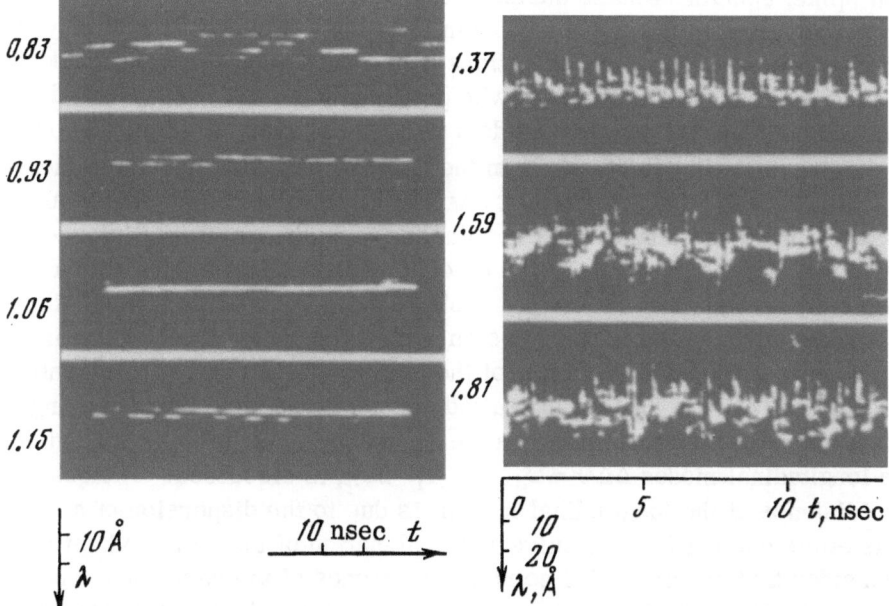

Fig. 5. Spectrochronograms of the output from strip heterolasers with pulsed excitation. The excess beyond threshold is shown on each trace. The laser characteristics are: wavelength at threshold about 870 nm, active region thickness 1.1 μm, contact width 15 μm, cavity length 320 μm, and temperature 300°K.

grams of a strip heterolaser in the superluminescence regime, where a nonstationarity is noticeable in the way the dominant mode is switched and in a stationary single-frequency regime near the lasing threshold [25].

2. A "piecewise-single-frequency" regime arises as the pump current is increased due to an increase in the frequency of switching the dominant mode. In this regime the integrated spectrum contains many modes, while they are almost never found to coexist simultaneously.

3. Regular fluctuations (self-modulation) occur with a small excess above threshold (up to 10-20 %). The frequency of the fluctuations is in the range 0.1-10 Hz and increases with the pumping level roughly as $Y^{1/2}$. Fluctuations can often be seen almost from the lasing threshold. This is typical of wide diodes where the appearance of spikes in the first lasing channel is a stimulant for all the new channels; the competition among the channels itself leads to nonstationary effects [28]. The spectral content of an individual spike, as found in [24], is practically nonreproducible from spike to spike if the automodulation depth is large. During spiking several modes actually coexist simultaneously.

4. Irregular fluctuations appear at 1.5-2 times the threshold and correspond to an extremely poorly controlled laser regime. In this regime variations in the mode frequencies of up to tens of gigahertz can be observed. Strictly speaking, the phenomenon of frequency automodulation is observed from the threshold of the fluctuating regime and at first is regular.

In all the nonstationary regimes the integrated (averaged) spectrum differs from the instantaneous spectra. Excited modes of comparable intensity are found to coexist only under nonstationary conditions, when dominant modes are being switched or in spikes. Research on the effect of external modulation has shown that at a sufficiently high frequency of hundred-percent modulation of the light the spectrum unavoidably becomes multimode [27]. The qualitative features, in particular the random nonreproducible intensity distribution over the modes in each output spike, remain the same during external modulation as during automodulation. According to the assumption in [24], the random cause of the nonreproducibility in the instantaneous spectra of the spikes is the fluctuations in the spontaneous radiant background at the moment the spike begins. V. N. Morozov, who previously pointed out the relationship between the multimode spectra and spiking, has made a theoretical analysis [29] of the spectrum during nonstationary lasing, including fluctuations in the spontaneous emission. The theoretical result is $M_q \sim Y^{1/4}$ which agrees with observation. As for the causes of the self-modulation effects in injection lasers there are a number of hypotheses including nonlinear interaction of the modes. It is known that the response of the electromagnetic field intensity to a sharp change in the pumping rate is similar to relaxation oscillations. In the case of periodic perturbations of the pump current or directly of the inversion density, resonant fluctuations in the laser intensity may develop [30] if the frequency of the perturbations equals the eigenfrequency of the relaxation oscillations ω_R. In [31] the so-called second-order self-mode-locking, which involves resonant driving of fluctuations when the frequency ω_R equals the second difference frequency of the longitudinal modes $\Delta\Delta\omega = \omega_{q+1} + \omega_{q-1} - 2\omega_q$, is discussed. The quantity $\Delta\Delta\omega$, that is, the nonequidistance of the longitudinal modes, is due to the dispersion of a semiconducting medium. This effect can hardly be a universal explanation of the self-modulation effects; however, if we consider that widely varied combination modes of the cavity may enter in the role of $\Delta\Delta\omega$ (for example, the difference frequencies of the longitudinal and transverse modes), the probability of resonant excitation of fluctuations is greatly enhanced.

There is still another factor which plays a role in multimode lasing. That is the possible direct interactions of modes to compensate their competition through the inverted population. These modes will be orthogonal by definition, so that in the linear approximation transfer of power from one mode to another does not occur. There are evidently factors which disrupt this independence of the modes, at least at sufficiently high intensities. These factors include the numerous nonlinear optical effects in which semiconductors are rich. Recently [32] a new phenomenon was identified among these effects. It might be called stimulated scattering of radiation on dynamic inhomogeneities in the electron density in the active medium of a laser. The electron density is dynamically coupled to the electromagnetic field intensity because of stimulated transitions and, on the other hand, to the refractive index because of its effect on the location of the plasma resonance and the shape of the absorption edge. This interaction must result in a growth in the intensity of the closest long wavelength modes adjacent to the mode being excited and to enhanced damping of the adjacent short-wavelength modes. The actual role of this interaction in the formation of the output spectrum is still unknown; however, the asymmetric interaction of the spectral modes predicted on the basis of this model has been observed experimentally in a heterolaser with an external "two-mode" tunable cavity [32, 33].

The excitation of multimoding is simplified in ordinary heterolaser cavities since the gain deficit for the modes adjacent to the excited mode is very small because of the large spectral width of the amplification band compared to the intermode separation. It is estimated in [17] that it is of relative magnitude 10^{-3} to 10^{-4} for the closest longitudinal mode in a cavity of typical size (0.3-0.5 mm).

4. Mode Selection Techniques

An analysis of the physical reasons for multimoding shows that to improve the coherence of the laser output the following measures are desirable: (a) reducing the threshold for the desired mode or increasing the threshold for unwanted modes (that is, essentially, mode selection), (b) the greatest possible optical homogeneity of the active medium, and (c) the suppression of self-modulation effects. Experiments seem to show that the introduction of spectral selection can increase the threshold for self-modulation fluctuations, that is, suppress spiking over some interval of Y [25]. No other methods of suppressing fluctuations without losing output power are known at present. As for the requirement of optical homogeneity of the medium, this is a customary requirement for the active medium of efficient lasers which is sought in one way or another in making heterolasers. As noted above, the contribution of spatial inhomogeneities is less in small-sized lasers. At the same time, however, the output power is reduced. The best result has been obtained in narrow GaAs lasers and corresponds to a single-frequency power level of 0.1 W [3].

We shall examine the mode selection techniques applicable to heterolasers. Internal selection may be realized by giving the active element a certain geometric form as it is made. An example is a structure with distributed feedback. External selection methods stipulate the construction of composite or external cavities including spectral or spatial filters. We shall examine these methods applied in turn to the x, y, and z axes and the polarization.

1. Spatial Selection along the x Axis. This, as noted previously, can be realized in principle by choosing a sufficiently thin active waveguide that satisfies $D < \pi$. Here there are two thickness ranges, one of which is more accessible (0.2-0.4 μm for a two-sided heterostructure in a AlGaAs$-$GaAs system) but is unsuitable because of strong diffractive divergence and the low critical power at which a heterolaser destroys itself. The other range (less than 0.2 μm) is more suitable since the directionality of the emission begins to improve while the self-destruction threshold rises and the lasing threshold is reduced (see Section 2). In experiments it has been possible to make a two-sided heterostructure with an active layer of thickness about 0.05 μm and high radiative characteristics [34]. During crystal growth, however, especially careful control of the conditions is required since in thin heterostructures the laser characteristics are extremely sensitive to the quality of the heterogeneous boundaries. In this regard it is of interest to modify the two-sided heterostructure to a form with a separating boundary [35, 36]. An example is the following heterostructure described in [36]: n^+GaAs (substrate)$-$nAl$_{0.3}$Ga$_{0.7}$As (2.5 μm)$-$nAl$_{0.1}$Ga$_{0.9}$As (0.66 μm)$-$pGaAs (0.1 μm, active medium)$-$pAl$_{0.1}$Ga$_{0.9}$As (0.76 μm)$-$pAl$_{0.3}$Ga$_{0.7}$As (0.46 μm)$-$p$^+$GaAs (1.6 μm contact layer). Here the electron boundary is provided by a jump in the fraction of aluminum arsenide of $\Delta x = 0.1$, and the optical boundary, by a jump of $\Delta x = 0.2$. Here the thickness of the waveguide is about 1.5 μm. Nevertheless, because of the central location of the active layer the most advantageous modes continue to be the lowest TE$_{0nq}$ modes. They are dominant until the threshold is considerably exceeded (an output power of several watts). The optimum location of the active medium in a thick waveguide has been examined in [37, 38]. It is shown that to maintain the lowest mode it is best to fill only two thirds of the thickness of the waveguide with the active medium. In [38] single-mode (with the index m = 0) lasing was obtained in a 2-μm-thick GaAs waveguide with a 1.3-μm-thick active medium at an output power of 0.7 W and a more than twofold excess above threshold. In [39] the lowest mode was selected with the aid of two-layer coatings which made it possible to introduce light at large angles of incidence at the end of the laser, that is, light in high-order transverse modes. An example is the layered mirror 0.0485λ_0(ZnS)$-$0.124λ_0(Al$_2$O$_3$), where λ_0 is the vacuum wavelength of the laser. Using this method together with a separating boundary it has also been possible to extract an output power of up to 0.7 W in a pulse and 35 mW average power with an off-duty factor of 10. The thickness of the waveguide was roughly 2 μm. For higher-order transverse modes the critical thickness may be increased by using special distribution profiles of ε across the perpendicular cross section of the heterostructure. Examples of this include multilayer or gradient index plane structures with minima in ε. This is an M-shaped profile in the case of a single minimum and in the case of two minima, a W-shaped profile. They are similar to the profiles developed in fiber optics for increasing the effective cross section of a single-mode fiber.

2. Spatial Selection along the y Axis. This is only partially achieved by reducing the width of the active medium [40-45]. An example of extreme narrowing is the

practically filamentary structures based on three-sided heterostructures. In these structures
the width of the active medium is close to its thickness. Wider strip heterolasers are prepared
by depositing strip contacts (using photolithography), by selective diffusion through a mask and
through the barrier layer of the semiconductor, or by proton bombardment through a mask.
In the last case the protons penetrate the semiconductor to a depth of order 1-2 μm and form
a nonconducting compensated layer on the surface. Because of this it is easy to limit the electric current in the strip which is shielded from the proton beam during the exposure [40]. In
strips that are bounded by an appropriate pumping current distribution (that is, ultimately due
to a favorable gain and absorption distribution) rather than a jump in the refractive index, the
requirements on the thickness for obtaining single-mode operation are somewhat weaker than
in heterostructure waveguides.

In strip lasers of width 10 μm it is mainly the lowest mode that is observed; at a width of
20 μm or more modes of higher order appear; and at 50 μm modes of the 10-13-th order are
observed. We note that in narrow strip lasers the threshold current density is several times
greater than in the wide diodes because of the inhomogeneous current distribution across the
p−n junction.

It is clear that a reduction in the width of a diode causes a reduction in the total output
power. To obtain higher power with a wide diode it is better to use an external cavity [46-50].
For example, synchronized single-mode operation has been reported in a battery of lasers
(five diodes) using an external cavity with diaphragms (Fig. 6) [47]. The output power was 19 W
in a pulse at slightly below room temperature. In [48] almost diffraction limited divergence
was obtained in both transverse directions of a cavity formed by external plane mirrors
matched to a diode by two lenses (see Fig. 6). All these experiments were done with homo-
structure lasers. Some data on a heterolaser with an external cavity will be presented in the
following paragraphs.

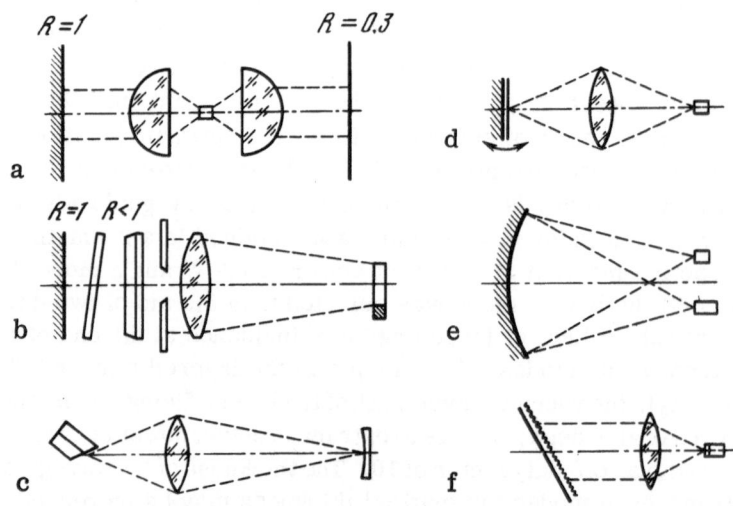

Fig. 6. Mode selection in injection lasers with the aid of external cavities: a) with an external cavity to improve the
spatial distribution of the emission [48]; b) in a laser battery
with an external cavity [47]; c) in a diode with a Brewster
angle face and an external mirror; d) with spectral selection
using a selective mirror tuned by inclining it to the cavity
axis; e) in coupled cavities of different lengths; f) in a cavity
with an external diffraction grating.

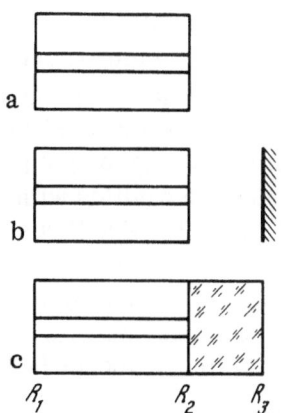

Fig. 7. Diagrams of composite Fabry−Perot cavities for injection lasers: a)   an ordinary diode; b) a diode with an additional external plane mirror; c) a diode with a transparent plane-parallel plate attached to one end.

3.   Selection along the z Axis. This refers to spectral selection of the transverse modes and has been examined theoretically in [51, 52]. Experimental work was first done with composite cavities [53–56] and with spectral filters in an external cavity [50, 55–60]. Recently heterolasers with distributed feedback have been developed to do the same thing [61–64].

The operating principle of composite Fabry−Perot cavities (Fig. 7) is that due to to the optical coupling between simple cavities of this type in a multimirror system, the Q of the longitudinal modes of the diode cavity is further modulated in frequency. Thus, for favorable tuning of the external portion of the composite cavity the Q of the desired mode may be enhanced while the Q of its neighbors may be reduced. A model analysis showed [52] that just adding a passive cavity in the form of a transparent plate of roughly the same thickness as the diode cavity length makes it possible to increase the yield in a single-frequency regime by many times. This was confirmed well in experiments with this kind of cavity [53, 54]. Figure 8 shows the spectra of a one-sided heterostructure laser with small excess pumping beyond

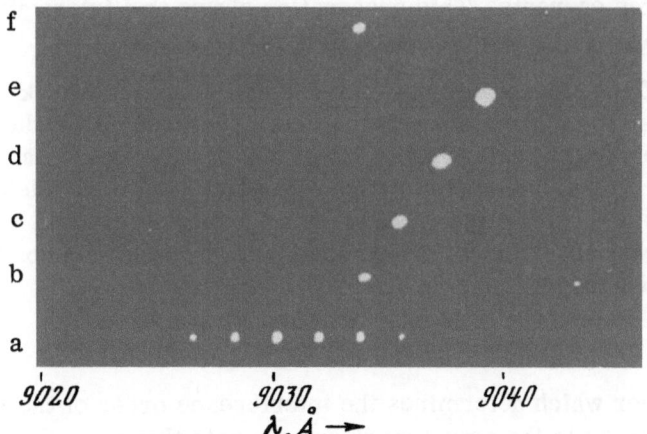

Fig. 8. Tuning an injection laser wiiι a composite cavity using piezoelectric displacement of the external mirror: a)   multimoding without a composite cavity; b–f) lasing in a composite cavity with voltages on the piezoelectric element of 0–28 V. The laser was based on a one-sided heterostructure in an AlGaAs system, the temperature was 300°K, and the threshold was exceeded by 5%.

the threshold in which the operating regime was controlled using a small plane mirror mounted a distance of about 0.2 μm from the end of the diode on a support with a piezoelectric element. The mirror was mounted parallel to the end of the diode and could be precision displaced along the cavity axis. It is clear that with this method it was possible to reproduce single-frequency operation in one of several longitudinal modes and a change in the voltage on the piezoelectric element of 3-5 V was enough to switch modes. The use of composite cavities with wide diodes made it possible to obtain a single-frequency pulsed power of about 0.5 W with a spectrum of width 0.02 nm [56], and their use with strip heterolasers yielded cw powers in a single TE_{00} mode of up to 3 mW at room temperature [25]. The limiting case of a multimirror composite cavity is a diode with an interference selective mirror. Some experiments with an external spectral selector, including an interference filter and an opaque mirror, are described in [55, 56].

The most convenient spectral selector is a reflecting diffraction grating in an autocollimation configuration [50, 57-60]. Some studies on a heterolaser in an external cavity with a diffraction grating will be described in the following paragraph. If the reflection at the ends of the diodes is reduced to zero with an antireflection coating, then the operating regime is completely controlled by the external cavity. In this case the emission is at a single frequency over the entire range of pump levels. However, several longitudinal modes of the external cavity, which are separated by an intermode distance much less than that of the transverse modes of the diode, may fall simultaneously in the spectral bandwidth of the output. In [57] a cw output power of 17 mW was obtained from a single longitudinal mode of an external cavity (of spectral width less than 200 MHz) with liquid-nitrogen cooling. Spectral selection may also be obtained when the end of the diode facing the external cavity is not antireflection coated [58, 60]. In this case the selectivity of the diode cavity itself combines with the selectivity of the external element. As with composite cavities, in this case the range in which single-frequency operation occurs is limited above by the threshold for appearance of uncontrolled intrinsic lasing of the diode. In [25] it was observed that introducing spectral selection in the cavity of a heterolaser both prevents the dominant mode's migration into the "piecewise single frequency" regime (see Section 3) and leads to the suppression of self-modulation fluctuations, naturally, in a limited range of pump currents. This observation shows that interaction of the spectral modes plays a role in the appearance of fluctuations in the output.

Another method of spectral selection is used in distributed feedback lasers [61]. Applied to injection heterolasers the methods of creating distributed feedback reduce either to a periodic modulation in the thickness of the waveguide or to use of structures that are in layers along the axis of propagation. In principle the distributed feedback effect consists of spreading the reflection and, if necessary, removing radiation from the active medium, while the period Λ of the distributed perturbation in the optical structure must satisfy the condition for resonant reflection for a chosen wavelength,

$$\Lambda = q\lambda/2, \tag{6}$$

where q is a whole number which determines the interference order of the successive reflections and λ is the wavelength in the active medium. We note that when the distributed feedback is created by modulating the refrective index or thickness of the waveguide (which is equivalent to modulating the phase retardation factor), two close wavelengths in the external medium will correspond to the resonance wavelength. This splitting practically disappears in the case of modulation of the gain coefficient while the selectivity remains at the same level in the second case. In practice, evidently, all the optical characteristics will unavoidably be modulated at once. Unlike selection with an external cavity, distributed-feedback lasers cannot be tuned spectrally. In experiments done up to now on heterolasers with distributed feedback it has been shown that single-frequency output can be obtained and the divergence can be reduced (when the emission is removed through a side surface of the waveguide). The output power and efficiency of distributed-feedback injection heterolasers are rather low. It is shown in [63] that the single-frequency output power of such a heterolaser is of the order of 10 μW with a pump current of 7 A.

    4. Polarization Selection. In two-sided heterostructure lasers the degree of polarization of one of the two basic modes TE or TM is often high even without the use of special methods. In a laser with a face cut at the Brewster angle and an external mirror [46] lasing occurred only in that polarization which is produced by feedback. Experiments on the effect of a polarizer inserted in the external cavity are described in the next chapter.

CHAPTER II

STUDY OF AN INJECTION LASER WITH A SELECTIVE CAVITY

1. A Model of a Dispersive Cavity in an Injection Laser

    Three basic ways of introducing additional spectral selectivity in an injection laser are known: (1) a composite (multimirror) Fabry−Perot cavity [53]; (2) an external dispersive cavity (with a spectrally selective element) [55, 56]; and (3) a cavity with distributed feedback [61].

    Using these techniques it is possible to create a substantial difference between the Q of the desired mode at a given wavelength and other modes lying within the amplification spectral bandwidth. The best prospects of controlling the output characteristics of the laser are obtained with an external dispersive cavity (including the possibility of fine tuning over a wide spectral interval, variations in the width of the spectral band, introduction of two or more dispersive elements, etc.). It is clear that these prospects are of great advantage in our work as well since they permit the experiments to be carried out under strictly controlled conditions.

    Here we shall consider a specific cavity scheme with an external reflecting diffraction grating (see Fig. 6f); however, the results may be generalized to several other types of dispersive cavities. A strict step-by-step analysis of the dispersive cavity of a semiconductor laser is extremely difficult; thus in order to obtain the essential results it is necessary to simplify the problem somewhat compared to the traditional cavity analysis. We shall assume that the output divergence in the plane perpendicular to the p−n junction is determined by diffraction and that transverse modes are not excited. In other words, we shall limit ourselves to laser diodes in which the active medium is a single-mode dielectric waveguide. The external elements cannot have a significant effect on the electromagnetic field distribution in such a cavity. The question is much more complicated in the case of the field distribution along the plane of the p−n junction (along the y axis). Here a large number of transverse modes are usually excited and the optical inhomogeneities in the active medium in the diode have a substantial effect on the field distribution.

    Simplifying this laser model basically reduces to assuming that the modes in the radiation field of the laser are combined incoherently; hence we can limit ourselves to analysis at the level of the radiation intensity distribution rather than the field strength. Hence the problem of the selectivity of a dispersive cavity in a semiconductor laser reduces to one which is similar to the classical problem of the apparatus function of a spectral device. Then the phase relations and information about the axial modes are lost. For a qualitative discussion, however, we shall consider the effect of the most efficiently reflecting faces and surfaces in the cavity which form an equivalent Fabry−Perot cavity.

We shall examine the effect of the selective loop of the external feedback associated with one of the ends of the laser active element. Part of the radiation filtered out by this loop returns to the end of the active element and enters the active region. We can introduce an effective reflection coefficient $R(\lambda)$ for this end which depends on the wavelength. Accordingly, the cavity losses given by $\alpha^*(\lambda)$ will also have a spectral dependence. In the next paragraph we shall examine the form of the functions $R(\lambda)$ and $\alpha^*(\lambda)$ for certain cavity parameters, and in Section 3 we shall optimize these parameters.

2. Optical Losses in a Dispersive Cavity in an Injection Laser

We shall consider the effective reflectivity $R(\lambda)$ for the intensity at the face of the laser diode turned toward the external selective cavity. On including the approximations discussed in the preceding section we have

$$R(\lambda) = R_1 + \frac{(1 - R_1)^2}{\int f^2(x, y)\, dx\, dy} \int f(-x, -y)\, \Phi(x, y, \lambda)\, dx\, dy, \qquad (7)$$

where R_1 is the reflectivity of the laser diode face itself (usually less than the natural value by means of an antireflection coating or by inclining the face at the Brewster angle), $f(x, y)$ is the distribution of output radiation across the face of the diode, and $\Phi(x, y, \lambda)$ is the distribution of radiant intensity returned to this face from the external part of the cavity. The negative sign in the argument of the function $f(-x, -y)$ under the integral sign is chosen because of the mirror reversal of the image after it passes through the external chain. For an injection laser the function $f(x, y)$ can be written in the form of the product

$$f(x, y) = \varphi(x)\, \psi(y), \qquad (8)$$

and the function $\Phi(x, y, \lambda)$ has the form

$$\Phi(x, y, \lambda) = R_g T^2 \int dx' \int dy'\, F(x, y, \lambda, x'\, y')\, f(x', y'), \qquad (9)$$

where R_g is the reflective efficiency of the grating, T is the transmission of the lens, and $F(x, y, \lambda, x', y')$ is the "instrument function" which can also be written as the product

$$F(x, y, \lambda, x', y') = F_x(x, \lambda, x')\, F_y(y, y'). \qquad (10)$$

This expression is possible to the extent that the spectral dispersion is only along the x axis. In addition, because the diode and lens can be adjusted axially it is enough to limit consideration of the optical aberrations of the system to the "scattering spot" corresponding to spherical aberration. We take both functions (F_x and F_y) in the Gaussian approximation,

$$F_x(x, \lambda, x') = \frac{1}{\sqrt{\pi\sigma}} \exp\left[-\frac{(x - x' - C\delta\lambda)^2}{\sigma}\right], \qquad (11)$$

where $\sqrt{\sigma}$ is the diameter of the scattering spot, $C = dx/d\lambda$ is the linear dispersion of the external cavity, and $\delta\lambda = \lambda - \lambda_0$ is the wavelength difference relative to the central wavelength of the cavity λ_0, which satisfies the alignment condition for the cavity, $\lambda_0 = (2d/m)\sin\Theta$ (where d is the grating period, m is the interference order, and θ is the angle between the cavity axis and the normal to the plane of the grating). Similarly, for $F_y(y, y')$ we have

$$F_y(y, y') = \frac{1}{\sqrt{\pi\sigma}} \exp\left[-\frac{(y - y' - \xi)^2}{\sigma}\right], \qquad (11a)$$

where ξ is a quantity which characterizes the cavity alignment along the y axis. For example, if the image of the point y = 0 falls at the point y = 0, then ξ = 0. In all other cases the coordinate of the image of y = 0 is ξ. Substituting Eqs. (11) and (11a) in Eqs. (9) and (7), we obtain

$$R(\lambda) = R_1 + \frac{(1-R_1)^2 R_p T^2}{\sqrt{\pi\sigma} \int \varphi^2(x)\, dx} \int\int \varphi(-x)\, \varphi(x') \exp\left[-\frac{(x-x'-C\delta\lambda)^2}{\sigma}\right] dx\, dx' \times$$

$$\times \frac{1}{\sqrt{\pi\sigma} \int \psi^2(y)\, dy} \int\int \psi(-y)\, \psi(y') \exp\left[-\frac{(y-y'-\xi)^2}{\sigma}\right] dy\, dy'. \tag{12}$$

We note that for homogeneous lasing along the p−n junction in a laser diode a characteristic range of variation of this function is usually more than tens of microns while the scattering spot may be reduced to a few microns by using a high-quality lens. This permits replacement of the Gaussian along the y axis by a delta function $\delta(y - y' - \xi)$. Then we obtain

$$R(\lambda) = R_1 + \frac{(1-R_1)^2 R_g T^2 \int \psi(y)\, \psi(\xi-y)\, dy}{\pi\sigma \int \varphi^2(x)\, dx \int \psi^2(y)\, dy} \int\int \varphi(-x)\, \varphi(x') \exp\left[-\frac{(x-x'-C\delta\lambda)^2}{\sigma}\right] dx\, dx'. \tag{13}$$

The specific form of the function $\psi(y)$ essentially affects only the magnitude of the constant factor in front of the integral in Eq. (13) whereas the form of the function $\varphi(x)$ determines the form of the spectral dependence of $R(\lambda)$. We shall use a Gaussian approximation

$$\varphi(x) = \exp(-x^2/\sigma_1), \tag{14}$$

where $\sqrt{\sigma_1}$ is the size in the x direction of the illuminated spot at the end of the diode, that is, the thickness of the layer in the laser diode in which the bulk of the radiant flux is concentrated. Using Eq. (14) we obtain

$$R(\lambda) = R_1 + \frac{(1-R_1)^2 R_g T^2 \int \psi(y)\, \psi(\xi-y)\, dy}{\int \psi^2(y)\, dy} \sqrt{\frac{2\sigma_1}{2\sigma_1+\sigma}} \exp\left(-\frac{C^2\delta\lambda^2}{2\sigma_1+\sigma}\right). \tag{15}$$

We rewrite this equation in the form

$$R(\lambda) = R_1 + R_c \exp\left[-\left(\frac{\lambda-\lambda_0}{\Delta\lambda}\right)^2\right], \tag{16}$$

where R_c is the preexponential factor in Eq. (15) and

$$\Delta\lambda = \sqrt{\frac{2\sigma_1+\sigma}{C^2}} = \left(\frac{\partial\lambda}{\partial\theta}\right)\frac{1}{\mathscr{F}} \sqrt{2\sigma_1+\sigma}, \tag{17}$$

where $\mathscr{F}$ is the focal distance of the lens and $d\lambda/\partial\theta$ is the reciprocal angular dispersion of the external cavity. In Eq. (16) R_c plays the role of an effective reflectivity at wavelength λ_0 to which the cavity is tuned. The quantity $\Delta\lambda$ characterizes the spectral selectivity of the cavity and depends on the linear dispersion $\mathscr{F}/(\partial\lambda/\partial\theta)$, the quality of the lens, and the thickness of the region within the diode in which the optical flux is concentrated. As a result we see that our cavity is equivalent to a cavity with a selective mirror $R(\lambda)$ described by Eq. (16) and a non-selective second mirror R_0 formed by the external end of the laser diode. The optical loss coefficient in such a cavity is given by

$$\alpha^\bullet(\lambda) = \alpha_0 - \frac{1}{2L}\ln R_0 - \frac{1}{2L}\ln R(\lambda), \tag{18}$$

where α_0 is the distributed loss coefficient in the cavity and L is the length of the laser diode cavity. Including Eq. (16) we obtain

$$\alpha^*(\lambda) = \alpha_c - \frac{1}{2L} \ln \left\{ R_1 + R_c \exp \left[-\left(\frac{\lambda - \lambda_0}{\Delta \lambda} \right)^2 \right] \right\}, \tag{19}$$

where $\alpha_c = \alpha_0 - (1/2L) \ln R_0$.

In this approximation it is easy to obtain formulas for different kinds of selective cavities. Let the external cavity consist of a spectral filter

$$T(\lambda) = T_f \exp \left[-\left(\frac{\lambda - \lambda_0}{\Delta \lambda_f} \right)^2 \right] \tag{20a}$$

and a mirror R_3. Then R_c in Eq. (19) is changed and becomes equal to

$$R_c = \frac{(1 - R_1)^2 R_3 T^2 T_f^2 \int \psi(y) \psi(\xi - y)\, dy}{\int \psi^2(y)\, dy} \sqrt{\frac{2\sigma_1}{2\sigma_1 + \sigma}}, \tag{20b}$$

while $\Delta \lambda$ now becomes

$$\Delta \lambda = \frac{1}{\sqrt{2}} \Delta \lambda_f. \tag{20c}$$

In another variant a Fabry–Perot etalon and a diffraction grating are used in the cavity. In this case the transmission of the cavity is equal to

$$T_{FP} = \left(\frac{q}{1 - \rho} \right)^2 \frac{1}{1 + \eta^2 \sin^2(2\pi l/\lambda)}, \tag{21}$$

where $\eta^2 = 4\rho/(1 - \rho)^2$, ρ and q are the reflection and transmission coefficients of these mirrors, and l is the distance between them (for simplicity we assume the refractive index of the material between the mirrors to be unity). This sort of cavity must be built so that the selectivity band of the grating, $\Delta \lambda$ [see Eq. (17)], is less than the dispersion range $\Delta \lambda_{FP}$ of the Fabry–Perot etalon,

$$(\partial \lambda / \partial \theta)(1/\mathscr{F}) \sqrt{2\sigma_1 + \sigma} \leqslant \Delta \lambda_{FP} \approx \lambda^2/2l. \tag{22}$$

Then the selectivity of the etalon will be fully used. This has the advantage of narrowing the spectral bands of the cavity by a factor of $\pi \eta$. For example, when $\rho = 0.9$ this is about 50 times. In Eq. (19) the following substitution must be made:

$$R_c = \frac{(1 - R_1)^2 R_p T^2 [q/(1 - \rho)]^4}{\int \psi^2(y)\, dy} \sqrt{\frac{2\sigma_1}{2\sigma_1 + \sigma_c}} \int \psi(y) \psi(\xi - y)\, dy, \tag{23}$$

$$\Delta \lambda_s = \lambda^2/\pi \eta l. \tag{24}$$

3. Optimizing the Design of a Dispersive Cavity

We shall examine the properties of a cavity with the inner end of the laser diode well antireflection-coated, i.e., with $R_1 \ll R_c$. Then

$$\alpha \approx \alpha_0 - \frac{1}{2L} \ln R_c + \frac{1}{2L} \left(\frac{\lambda - \lambda_0}{\Delta \lambda} \right)^2. \tag{25}$$

It should be noted that it is correct to neglect R_1 if lasing does not occur when the external elements of the cavity are removed (that is, there is no lasing up to the threshold in "internal" modes of the diode). The longitudinal modes of the internal cavity formed by the outer face R_0 and the diffraction grating have K resonances with

$$K = 2\Delta\lambda\mathscr{L}/\lambda^2 \tag{26}$$

in the selection band of the grating, where $\mathscr{L}$ is the effective optical path length of the cavity,

$$\mathscr{L} = \int\left(n - \frac{\partial n}{\partial\lambda}\lambda\right)dx, \tag{27}$$

with the integral taken over all the cavity length including the laser diode. In the other limiting case $R_1 \gg R_c$ (composite cavity) Eq. (19) can be written in the form

$$\alpha = \alpha_0 - \frac{1}{2L}\ln R_1 - \frac{1}{2L}\frac{R_c}{R_1}\exp\left[-\left(\frac{\lambda - \lambda_0}{\Delta\lambda}\right)^2\right]. \tag{28}$$

A comparison of Eqs. (25) and (28) shows that the maximum increase in losses in undesired neighboring modes compared with the selected modes $\Delta\alpha$ is

$$\Delta\alpha = \begin{cases} (1/2L)\ln(R_c/R_1) & \text{for} \quad R_1 \ll R_c, \tag{29} \\ (1/2L)(R_c/R_1) & \text{for} \quad R_1 \gg R_c. \tag{30} \end{cases}$$

The value of selection in the second case is quite limited compared with the first case if we recall the difference in the losses which may develop in two neighboring longitudinal modes of the laser diode with optimal adjustment and tuning of the cavity. It is clear that to obtain narrowband output at the highest power for a given spectral width (and limited device size) it is appropriate to reduce R_1 as much as possible, that is, to realize a purely external cavity with dispersive losses described by Eq. (25). This simultaneously expands the spectral range over which the laser can be tuned by turning the diffraction grating and helps to smooth out the dependence of the laser power on the tuning.

To increase R_c it is necessary to use lenses that are antireflection-coated at the laser wavelength, use a highly effective diffraction grating, and choose homogeneous laser diodes. When R_1/R_c is small the increase in selectivity of the cavity must be achieved by reducing $\Delta\lambda_3$ and K. We note that in a lens corrected for spherical aberration the scattering spot is determined by diffraction,

$$\mathfrak{s} \approx (\mathscr{F}\lambda/\mathscr{D})^2, \tag{31}$$

where $\mathscr{D}$ is the diameter of the lens aperture and $\mathscr{F}$ is its focal distance. This limit must be taken into account in optimizing the cavity design. We shall assume that the relative aperture of the lens $S = \mathscr{D}/\mathscr{F}$, and the blaze angle of the diffraction grating are given. These two quantities cannot be increased arbitrarily for technical reasons. For $S = 1:2$ we shall have $\sigma \approx 4\lambda^2$. The reciprocal angular dispersion $d\lambda/d\theta$ of a grating operating in an autocollimating configuration with angle θ is

$$d\lambda/d\theta = \lambda/\tan\theta = \lambda/\tan\arcsin\frac{m\lambda}{2d}, \tag{32}$$

where m is the interference order and d is the grating period. Substituting Eq. (32) in Eq. (16)

we obtain

$$\Delta\lambda = \lambda^2/\mathcal{F}S \tan\theta, \tag{33}$$

from which it is clear that the selectivity is improved as $\mathcal{F}$ is increased (for fixed S)   This is explained by the increased number of illuminated grooves on the grating. For $\mathcal{F} = 5$ cm, S = 1 : 2, $\theta = 60°$, d = 10 μm, and m = 20 we obtain $\lambda = 0.9$ μm and $\Delta\lambda = 0.02$ nm. Then the grating size is 6 × 2.5 cm. Increasing $\mathcal{F}$ above this value leads to technical difficulties and a bigger cavity. Thus to increase the selectivity further it is more appropriate to place a Fabry–Perot etalon in the cavity. It is necessary, however, that condition (22) be satisfied and that the radius of the interference ring on which the etalon is turned be much greater than the width of the active region, in other words, that the fragment of the interference ring projected onto the end of the diode appear to be a straight band across the diode width. From Eq. (22) it follows that we must have $l \leqslant \mathcal{F}$. In the following we shall assume that $l = \mathcal{F}$ and in this case the greatest spectral resolution of the system is realized. If we work in the third interference ring, then the angle of inclination φ of the etalon to the cavity axis is $\varphi = \sqrt{3\lambda/\mathcal{F}}$ and the radius of the ring is $r = \sqrt{3\lambda\mathcal{F}}$. If the width of the diode (active region) is W, then according to the above, it must satisfy the inequality

$$W < \sqrt{8\lambda\sqrt{3\lambda\mathcal{F}}}. \tag{34}$$

The spectral band $\Delta\lambda_s$ is determined by $\lambda^2/\pi\eta\mathcal{F}$ but it cannot be less than diffraction permits, i.e.,

$$\Delta\lambda_s > \frac{2\lambda^2}{\mathcal{F}} \sqrt{\frac{3\lambda}{\mathcal{F}}}. \tag{35}$$

Therefore, the cavity selectivity can be increased as η is increased until a reflection coefficient of $\rho_{max} \approx 1 - 10 \sqrt{\lambda/\mathcal{F}}$ is achieved in the etalon. For $\mathcal{F} = 5$ cm and $\lambda = 0.9$ μm it appears that it is useless to try to increase ρ above 95%. Where a cavity with a grating may have $\Delta\lambda = \lambda^2/\mathcal{F}$, adding a Fabry–Perot etalon that has been optimized for this case can, in principle, bring this value to $\Delta\lambda = (2\lambda^2/\mathcal{F}) \sqrt{3\lambda/\mathcal{F}}$ without significantly increasing the dimensions of the device. For $\mathcal{F} = 5$ cm a calculation for a cavity with a grating yields $\Delta\lambda = 2 \cdot 10^{-2}$ nm while for a cavity with a grating and an etalon $\Delta\lambda = 2 \cdot 10^{-4}$ nm, which is apparently the limit attainable in a cavity smaller than a few $\mathcal{F}$.

To estimate K we take Eq. (33) for $\Delta\lambda$ and take the minimum possible cavity length to be $\mathcal{F} + \frac{\mathcal{F}}{2} \tan\theta$. Then, according to Eq. (27) we obtain

$$K \approx \frac{4 + 2\tan\theta}{\tan\theta}, \tag{36}$$

that is, the number of resonances is independent of $\mathcal{F}$ and depends only on the angle θ; K cannot be made less than 3, and for $\theta = 60°$ we have K $\approx$ 4. If, however, we use an additional Fabry–Perot etalon then K can be reduced to unity, which would correspond to the single-mode regime.

Returning to the case $R_1 > R_c$, we note that this variant is suitable for mode selection when operating at small amounts above the threshold, in particular, in cw lasers operating at high temperatures (since large excesses above threshold do not occur in them). In certain lasers based on two-sided heterostructures the end cannot be antireflection-coated well, so it is necessary to operate with $R_1 \geq R_c$.

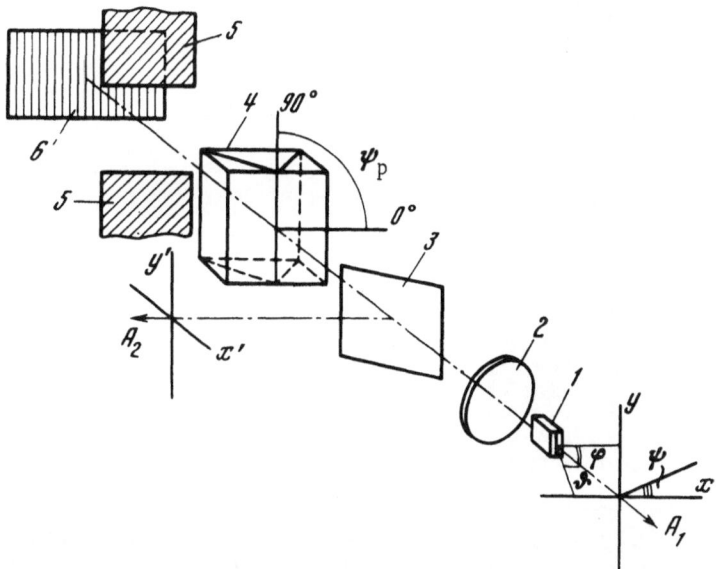

Fig. 9. A diagram of an external cavity with mode selection according to spectrum, directionality, and polarization: 1) laser diode; 2) collimator lens; 3) beam splitter; 4) prism polarizer; 5) slit diaphragm; 6) reflecting diffraction grating.

4. An Experimental Study of an Injection Laser

with a Selective Cavity

The best instrument for studying the limits of controllability of the light from a semiconductor laser is the external cavity scheme shown in Fig. 9. Here spectral selection with the aid of a reflecting diffraction grating, spatial selection along the y axis using a slit diaphragm, and polarization control with a prism polarizer are provided. An external cavity containing these selectors is matched to one end of the diode with a lens. Most of the experiments involved diodes of which the end facing the external cavity has an antireflection coating of silicon monoxide with thickness $\lambda/4$. Diodes with an end cleaved at an angle of 9–16° to the plane of the p−n junction were also studied.

The external part of the cavity filters the light which returns to the diode, purifying it of undesired modes. This light is observed at the output A_1 (see Fig. 9) after a single pass through the cavity medium. The beam splitter at the beginning of the external part of the cavity makes it possible to observe the light at output A_2 after two passes through the active element. Comparing the compositions of these two beams makes it possible to trace the degradation of the light interacting with the active element (spectral broadening, the appearance of side lobes in the directional pattern, depolarization). Some of the laser light goes into spurious scatter from the diffraction grating. When an antireflection−coated single-sided heterostructure diode is used as an active element (with an active layer of thickness about 2 μm), it is possible to obtain pulsed power of 4–6 W in a bandwidth of less than 0.1 nm at 300°K. The tuning range of the laser wavelength is 10–20 nm near the central wavelength of 905 nm. Experiments of this sort have been reported in [59, 60].

The observation of spectral selection in an external cavity makes it possible to conclude that the output spectrum at a level close to the maximum diode power can be narrowed by 20–50 times and that its final width is determined by the width of the selective feedback bandpass. There is no loss in the output power or efficiency of the laser in this case.

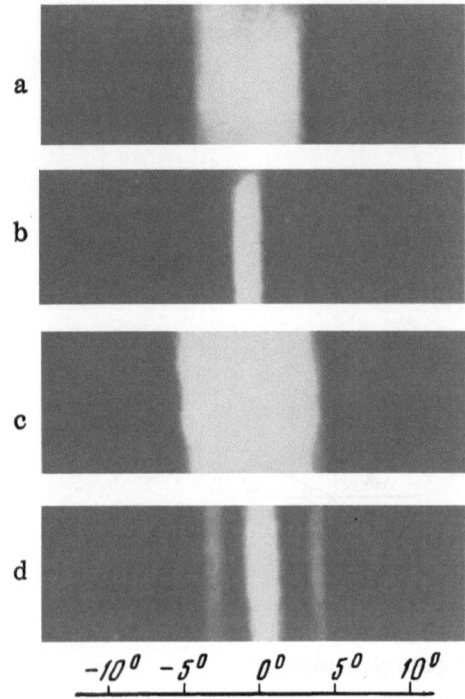

Fig. 10. A picture of the far zone of the injection laser with narrowing by a slit diaphragm in an external cavity (see Fig. 9); the slit width is 40 mm (a, c) and 2.5 mm (b, d); amount beyond threshold 1.5 (a, b) and 2.0 (c, d).

A slit diaphragm in front of the grating makes it possible to narrow the directional pattern (Fig. 10) almost to the diffraction limit (in this experiment to 0.8°); however, beginning with some value of the divergence angle (about 2.5°) the output power decreases along with the slit width. Thus it is possible to narrow the directional pattern by up to 10 times, but the brightness is increased by only 3-4 times. It has been noticed that at high powers side lobes appear in the directional diagram at roughly the same wavelength as the center lobe and that they are not removed even if the lobes symmetric to them are quenched at the diaphragm. The mechanism for this effect is not clear.

Controlling the polarization showed that when the polarizer maintains the natural polarization of the diode, the degree of polarization can be increased by 1.3-1.6 times at high output power, and in successful experiments it is 0.95-0.97 at the output $A_{\uparrow}$. Nevertheless, the orthogonally polarized component increases by roughly 50-100 times in a single pass. The preferred plane of polarization of a heterolaser may be smoothly rotated by rotating the polarizer; however, in general the angle of rotation of this plane and the polarizer angle are different. The output polarization seemingly is pulled to the principal planes of the waveguide. This behavior is illustrated in Fig. 11, which shows the variation in the degree of polarization as the polarizer is rotated. The polarization is minimal when the polarizer is in an intermediate position between the principal polarizations of the waveguide. Less carefully controlled samples were observed that had less polarization and larger deviations from the polarizer angle. Figure 12 shows an example in which the degree of polarization for favorable tuning is lower than in Fig. 11, and, in addition, the laser does not work at one of the two orthogonal polarizations.

From these measurements we can conclude that using selection with the aid of an external cavity generally increases the radiation density in the spectrum, directionality, and polarization by 80 to 320 times for an output power of 1 W or more (at lower powers the effect is stronger). This factor represents a gain over the power which we can obtain by filtering the radiation from an uncontrolled heterolaser outside the cavity.

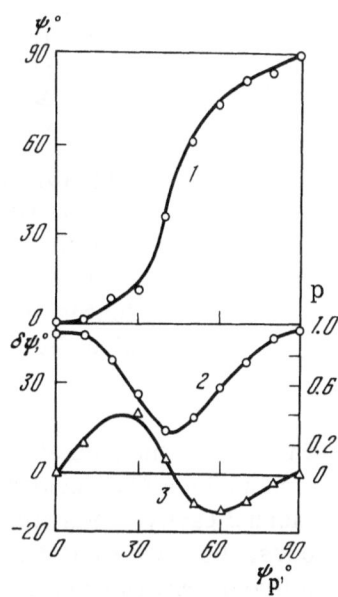

Fig. 11. Controlling the output polarization of a single-sided heterostructure injection laser based on an AlGaAs system using an external cavity (see Fig. 9): 1) the effect of the rotation angle of the polarizer, ψ_p, on the angle of preferred output polarization ψ; 2) the degree of polarization; 3) the difference $\delta\psi = \psi_p - \psi$. $\psi = 0$ corresponds to a TM wave and $\psi = 90°$ to a TE wave.

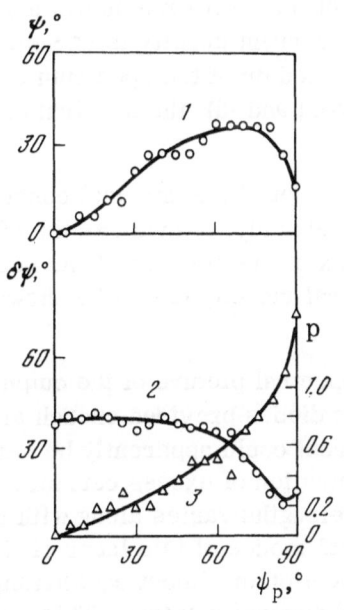

Fig. 12. The same as in Fig. 11 for another, inferior sample laser diode. Note the greater depolarization of the radiation in the active medium, the lower degree of polarization, and features that are manifested when the laser is tuned to the TE wave.

These results show that the coherent properties of the output of injection heterolasers can be greatly improved by using mode selection techniques similar to those traditionally used in other kinds of lasers. In addition, there are internal selection techniques (such as distributed feedback) which make it possible to avoid substantially increasing the overall dimensions of the device. The development of these special selective techniques for heterolasers has just begun.

5. Experiments on Intracavity Refractometry and Spectroscopy of an Injection Laser with a Selective Cavity

The tunable narrow-band injection laser described in the previous paragraph opens up interesting opportunities for research on the optical properties of various media, primarily

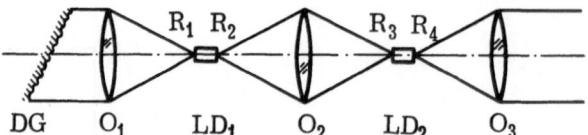

Fig. 13. A diagram of coupled cavities with two active elements LD_1 and LD_2 and a diffraction grating DG for spectroscopic experiments. O_1, O_2, and O_3 are the matching lenses, and the reflection coefficients are given by $R_1R_3 \approx 0.01$ and $R_2R_4 = 0.32$.

the study of the active medium of identical injection lasers. In the next chapter we shall describe experiments on a "two-mode" laser with two independently tunable diffraction gratings inside an external cavity. This type of selective cavity allows us to study the interaction of spectral modes. Here we shall describe phenomena in coupled cavities with two active elements. The linear arrangement used in the experiments is shown in Fig. 13. The following effects were studied: (1) the change in the refractive index of the active medium of a semiconductor laser in the working wavelength band as the pump current density is varied; (2) the evolution of the laser radiation as the pump current is increased or of the spectrum of a laser amplifier operating with an input of single-frequency radiation; and (3) the spectral bandwidth for damping when lasing is established.

These studies will make it possible to describe in more detail the physical conditions for generation of coherent radiation in semiconductors and, particularly, to evaluate the effect of the free-charge-carrier concentration on the refractive index of the medium. This nonlinear factor has an effect on the interaction dynamics of the spectral modes, as will be shown in the next chapter.

In the first experiment we observed the complicated spectral picture of the output from the coupled cavities without a diffraction grating. One of the diodes provided enough amplification to maintain lasing while, in the second, the pump current could apparently be varied outside the saturation regime. Since in this case the concentration of excess carriers varies with the current, and the refractive index at the working wavelengths varies along with the excess carrier concentration, the frequencies of the longitudinal modes of this laser diode were somewhat displaced. The overall laser spectrum was a collection of almost equidistant longitudinal modes whose intensities varied so their envelope had several maxima. This was a result of interference of modes in cavities with different optical path lengths. The change in the shape of the envelope with the current in the unsaturated diode was evidence of a change in the optical path length of its cavity. In this experiment the diode LD_1 had a cavity of length $L_1 = 675$ μm and LD_2 had $L_2 = 525$ μm. Both diodes had a width of 175 μm. In LD_1 the pumping was fixed (current pulses lasting 0.1 μsec and with amplitude 50 A). In LD_2 the synchronous current pulses had a variable amplitude. It was found that the maxima in the mode envelopes shifted toward long wavelengths as the current was increased. This displacement was 1.4 nm when the current density j was changed from 24 to 40 kA/cm². The quantity $d\lambda/dj$ evaluated in this experiment was $-2.5 \cdot 10^{-12}$ cm³/ A. If we assume that the lifetime of the electrons in diode LD_2 is 1 nsec, then an estimate of $\Delta n/\Delta N_e$, where n is the refractive index and N_e is the electron density, yields values of the order of $-4 \cdot 10^{-21}$ cm³ at a wavelength near 906 nm, in satisfactory agreement with theory. It was shown earlier [56] that it is possible to obtain mode selection and single-frequency lasing in coupled cavities of different optical path lengths.

In the second experiment we observed the emission of a laser amplifier LD_2 with a nonselective cavity and narrow-band ("single-frequency") input from a laser with a selective

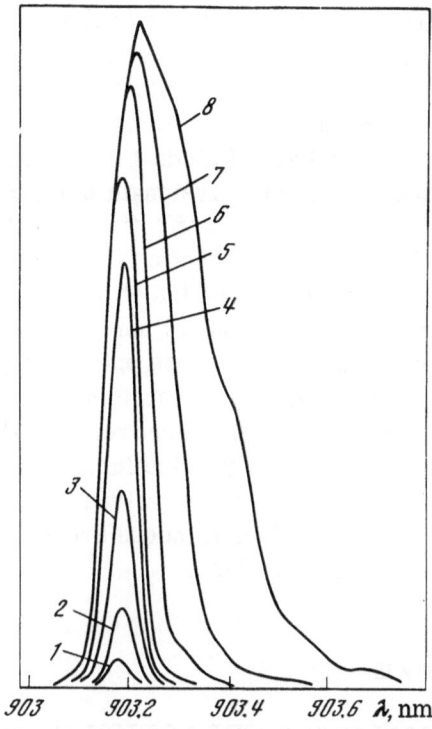

Fig. 14. The output spectrum of a laser amplifier for different pump currents through the diode LD$_2$ (see Fig. 13) and a fixed operating regime in the narrow-band (width about 0.05 nm at 903.19 nm) driver laser LD$_1$. The current in diode LD$_2$ is: 1) 32; 2) 36; 3) 42; 4) 44; 5) 48; 6) 52; 7) 56; 8) 61 A.

cavity and active element LD$_1$. We have studied the location and shape of the spectral band passed by the laser amplifier for a given input spectrum line. The operating regime of this laser amplifier corresponded to injection of monochromatic light from an external source when the cavities of the driver and amplifier lasers are weakly coupled.

Figure 14 shows the evolution of the laser amplifier spectrum as the pump current through it is increased. The linewidth is found to increase by several times mainly due to the development of the long-wavelength wing of the line. This asymmetric broadening of the amplified line is in qualitative agreement with the model of stimulated scattering of laser radiation on dynamic inhomogeneities in the electron density which will be discussed in detail in the next chapter.

In the third experiment we studied the spectral quenching of the radiation as narrow-band lasing is initiated. To do this we measured the luminescence intensity at various points of the

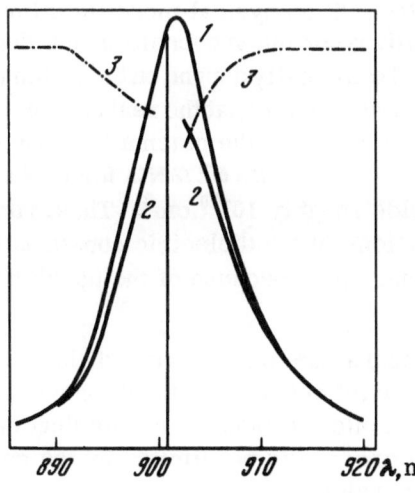

Fig. 15. The output spectrum of diode LD$_2$: 1) in the superluminescence regime with diode LD$_1$ turned off (see Fig. 13); 2) when narrow-band radiation near 900.8 nm is injected from laser LD$_1$; 3) relative change in the spectral density of the output when narrow-band radiation is injected.

spectral bandwidth in the two regimes for a single pump current in the laser diode LD_2 being studied. In one regime the coupling with the laser oscillator cavity LD_1 is interrupted and the output from LD_2 is a superluminescence band. In the second regime the input to the second diode is the narrow-band radiation from the laser oscillator tuned roughly to the center of the luminescence band of diode LD_2. Then, of course, a considerable increase in the intensity at the laser oscillator wavelength is observed, while at other wavelengths the output of LD_2 may be attenuated due to an effective reduction in the excess electron—hole pair concentration as the intense electromagnetic field builds up. Figure 15 shows an example of such a study plotted as the spectral distribution of the relative quenching. For this experiment the width of the quenching band is about 10 nm (about 14 meV). This quantity cannot be identified with the homogeneous width because optical amplification played a significant role in the formation of the initial spectrum. Because of this the quenching bandwidth defined in terms of the half-maximum is obviously narrower than the homogeneously broadened bandwidth. Nevertheless, it is very important that the luminescence is not quenched far from the center of the band, especially on the short-wavelength side. This observation is direct evidence in favor of the involvement of inhomogeneous broadening in the formation of the luminescence spectral band of a semiconductor laser.

CHAPTER III

ANOMALOUS INTERACTION OF MODES IN A SEMICONDUCTOR LASER

1. The Physical Nature of the Nonlinear Interaction of Modes in a Semiconductor

Semiconductor materials are media with a substantial optical nonlinearity. Self-modulation in the intensity and frequency in semiconductor lasers can be shown to be an example of the effect of optical nonlinearity on the dynamics of a laser. The nonlinear part of the complex dielectric constant $\tilde{\varepsilon}$ has already been taken into account in an analysis of spiking in semiconductor lasers [65, 66].

Here we examine a new nonlinear effect which leads to an anomalous interaction of modes in the laser spectrum. This effect differs from the usual mode competition due to homogeneous broadening, and is asymmetric since the longer wavelength mode of two neighboring modes will predominate. This interaction occurs over small spectral separations (less than 10^{11} Hz), which is less than the homogeneous broadening (10^{12}-10^{13} Hz). The physical model is stimulated scattering of the laser radiation in the active medium itself on dynamic inhomogeneities in $\tilde{\varepsilon}$ due to variations in the electron density. These variations in the electron density depend on the radiant intensity which determines the rate of stimulated emission. It is well known that the real part of $\tilde{\varepsilon}$ in semiconductors depends on the electron density N_e due to the fact that the plasma frequency and the location of the absorption edge depend on N_e. An estimate of Re $(\partial\tilde{\varepsilon}/\partial N_e)$ for GaAs at its intrinsic laser wavelength near the absorption edge yields roughly 10^{-20} cm^3. Thus, variations in the electron density N_e create corresponding variations in the dielectric constant (in the real part because of the above effects and in the imaginary part because of the population inversion).

The asymmetric interaction of the spectral modes occurs due to scattering of the radiation by dynamic variations in the electron density. These variations are a kind of "recombination" waves which appear due to discrepancies in the pump-emission balance rather than due to real plasma oscillations. Because of this the scattering process is not limited by the conservation of momentum since these "waves" do not have a dispersion law.

In Section 2 the experimental results which indicate the existence of anomalous mode interactions are presented. In Section 3 the distortion of the spectral gain profile near the laser frequency ω_0 is examined theoretically. The variation in the gain is described by an odd function of the frequency difference $\Omega = \omega_0 - \omega$, which causes the asymmetry in the mode interaction.

2. An Experimental Study of Spectral Mode Interaction Effects

The experiments were performed at 300°K with single-sided heterostructure AlGaAs−GaAs lasers with an antireflection coating and an external cavity containing spectrally selective elements (Fig. 16). Lasing was obtained with linear and ring cavities including a reflecting diffraction grating and with a cavity containing two independently tunable diffraction gratings. Cavities with a single grating [57-60] make it possible to obtain a narrow-band output tunable over some spectral interval. The highest output power (up to 5 W) was obtained in a width of less than 0.1 nm while at the lowest power level (0.5 W) the bandwidth was less than 0.01 nm. This kind of spectral band is treated as a "spectral mode" in the following even when it contains several longitudinal modes of the external cavity. The reason is that their spatial configurations within the active medium coincide closely while their spectral widths are much less than the intermode distance of the diode cavity. At high pump powers it is sometimes possible to observe internal modes simultaneously with the narrow-band external mode. It was noticed that a strong external spectral mode suppresses weak internal modes more effectively on the short-wavelength side than on the long-wavelength side.

In order to study this form of mode interaction a cavity with two gratings was made. Lasing was obtained in two modes. The mode interaction seemed to depend on the spectral distance between the modes and on the location of one relative to the other. Two experiments were done. In one experiment the intensity of the tunable mode was measured as a function of the wavelength for fixed pump current. The results are shown in Fig. 17. The quantity Φ_λ was measured under conditions such that only one tunable mode was excited (the second grating was covered by a black screen). If a fixed mode at λ_0 is excited simultaneously with the tunable mode, then the function Φ_λ changes somewhat, the tuning range is reduced from 4 to 3 nm, and a discontinuity develops at λ_0. We note the nonmonotonic variation of Φ_λ to the left of λ_0.

The tunable mode is effectively quenched when it approaches the fixed mode from the short-wavelength side.

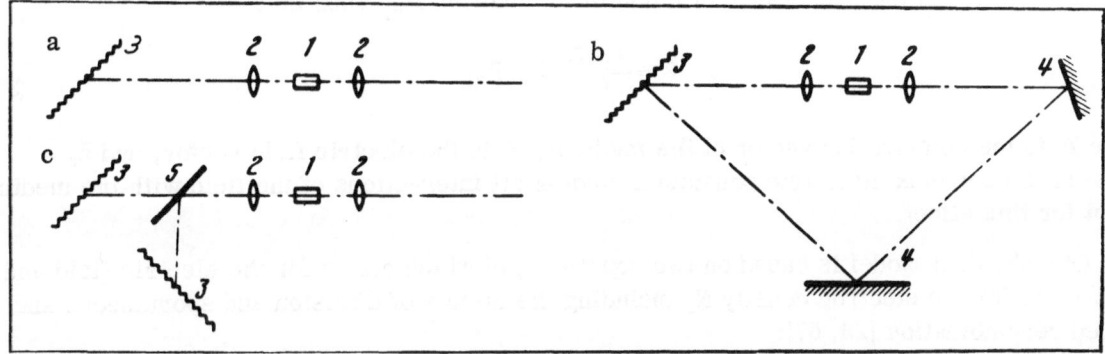

Fig. 16. Diagrams of the external cavities with diffraction gratings: a) a linear autocollimating scheme; b) ring-laser scheme; c) a scheme with two independently tuned gratings for studying the interaction of spectral modes. 1) Laser diode; 2) lens; 3) reflecting diffraction grating; 4) mirror; 5) beam splitter.

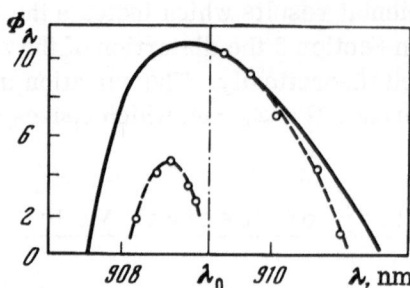

Fig. 17. The spectral dependence of the intensity of the tunable mode Φ_λ in a laser with an external cavity of the design shown in Fig. 16c when the second grating is covered (smooth curve) and when the second grating is opened for narrow-band lasing (dashed curve). The asymmetry of the mode interaction is distinct.

The second experiment consists of measuring the intensity of two spectral modes separated by 0.3 nm as a function of the pump current. It was found that the intensity of the short-wavelength mode saturates at a high current while the intensity of the long-wavelength mode increases as the current rises.

These results show that the interaction of nearby (separated by $\leq 10^{11}$ Hz) modes is asymmetric in favor of the long-wavelength mode if the laser intensity is sufficiently high ($\geq 10^4$ W/cm^2). The electron concentration in these experiments was more than 10^{18} cm^{-3}, which corresponds to a maximum pump current density of up to $4 \cdot 10^4$ A/cm^2.

3. An Analysis of the Mechanism for the Anomalous

Interaction of the Spectral Modes

The asymmetric distortion of the gain profile may be due to nonlinear induced scattering of the laser light at frequency ω_0 on dynamic inhomogeneities in the electron density. The analysis will be limited to the case of a weak signal at frequency $\omega_1 = \omega_0 - \Omega$. We assume that the relaxation time of the polarization of the medium is small compared to $1/\Omega$ and, therefore, that the polarization amplitude of the medium follows the field amplitude. The effect we are studying consists of the appearance of a nonlinear addition $\mathbf{P}_{nl}$ such that

$$\mathbf{P} = \frac{1 - \tilde{\varepsilon}_0}{4\pi} \mathscr{E} + \mathbf{P}_{nl} , \tag{37}$$

where $\mathbf{P}$ is the polarization vector of the medium, $\mathscr{E}$ is the electric field vector, and $\tilde{\varepsilon}_0 = \varepsilon + i\varepsilon'$ is the complex dielectric constant including all interactions of the field with the medium except for this effect.

Our physical model is based on two equations, of which one is for the electric field and the other is for the electron density N_e including the effects of diffusion and spontaneous and induced recombination [23, 67]:

$$(\tilde{\varepsilon}_0/c^2)\,(\partial^2\mathscr{E}/\partial t^2) - \nabla^2\mathscr{E} = -(4\pi/c^2)\,(\partial^2\mathbf{P}_{nl}/\partial t^2), \tag{38}$$

$$\partial N_e/\partial t - D\nabla^2 N_e + N_e/\tau + aN_e\,|\,\mathscr{E}\,|^2 = G, \tag{39}$$

where D is the diffusion coefficient, τ is the lifetime of the electrons in the absence of the electromagnetic field, G is the pumping rate, and α is a proportionality coefficient for the rate of induced recombination.

We shall write the electromagnetic field as a series in terms of the modes of an ideal cavity [67]:

$$\mathscr{E}(\mathbf{r}, t) = \sum_j [\mathbf{E}_j(t) + \text{c.c.}]\,\varphi_j(\mathbf{r}). \tag{40}$$

The orthonormal function $\varphi_j(\mathbf{r})$ is the wave vector. The analysis will be limited to the case in which a strong field exists in the single mode

$$\nabla^2\varphi_j(\mathbf{r}) + \mathbf{k}_j^2\varphi_j(\mathbf{r}) = 0, \tag{41}$$

where $\mathbf{k}_j$ is the wave vector. The analysis will be limited to the case in which a strong field exists in the single mode,

$$\mathscr{E}(\mathbf{r}, t) = [\mathbf{E}_0(t) + \text{c.c.}]\varphi_0(\mathbf{r}), \tag{42}$$

while the oscillations in the other modes are very weak (that is, a single-mode regime is realized), i.e.,

$$|\mathbf{E}_0(t)| \gg |\mathbf{E}_j(t)|, \quad j \neq 0. \tag{43}$$

We have to study the behavior of the field in one of the nonlaser modes $\mathscr{E}_1(\mathbf{r}, t)$ in the presence of the strong field $\mathscr{E}_0(\mathbf{r}, t)$. We substitute Eq. (40) in Eq. (38), multiply both parts of the resulting equation by $\varphi_j(\mathbf{r})$, and then integrate over the cavity volume V. This operation yields the equation

$$\partial^2\mathbf{E}_1/\partial t^2 + \tilde{\omega}_1^2\mathbf{E}_1 = -(4\pi/\varepsilon_0)\int_V (\partial^2\mathbf{P}_{n1}/\partial t^2)\,\varphi_1(\mathbf{r})\,dV, \tag{44}$$

where $\tilde{\omega}_1 = \omega_1 - i\gamma_1$, and γ_1 is the damping decrement for the oscillations in the corresponding mode. To compute the right side of Eq. (44) it is necessary to find the form of the inhomogeneity in the nonlinear polarization, which may be written in a linearized form as

$$P_{n1} = (\mathscr{E}_0/4\pi)\delta\tilde{\boldsymbol{\varepsilon}} = (\mathscr{E}_0/4\pi)(\partial\tilde{\boldsymbol{\varepsilon}}/\partial N_e)\delta N_e, \tag{45}$$

where $\delta\tilde{\varepsilon}$ is the variation in the complex dielectric constant and δN_e is the variation in the electron density. Both these variations are a consequence of superimposing the fields $\mathscr{E}_0$ and $\mathscr{E}_1$, and they oscillate at frequency Ω. Such variations in the electron density have no effect on the oscillations at frequency ω_0 as long as $\mathbf{E}_1$ is sufficiently small and Ω is large compared to the output spectral width. The threshold level for optical amplification is due to the average value of the electron density in time since the instantaneous deviations δN_e are quite ordinary for a multimode laser.

These deviations generate a dynamic phase lattice

$$\delta\tilde{\boldsymbol{\varepsilon}}(\mathbf{r}, t) = (\partial\tilde{\varepsilon}/\partial N_e)\delta N_e$$

which scatters the laser light at ω_0 in the same direction with a Stokes shift Ω. This nonlinear contribution to the polarization of the medium at frequency $\omega_1 = \omega_0 - \Omega$ is represented by the right side of Eq. (44). The oscillations in $\mathbf{E}_1$ will have a growing amplitude if this contribution can compensate the damping on the left side of the equation. As for backscattering, it is less important since the variations in N_e over small distances are effectively smoothed out by rapid electron diffusion. We shall deal with variations δN_e averaged over regions of extent of the order of $\lambda_0/2\sqrt{\varepsilon}$; This approximation corresponds to neglecting backscattering. To obtain $\delta N_e(\mathbf{r}, t)$ it is necessary to return to Eq. (39). In Section 4 it is found that the right side of Eq. (44) can be approximated by a combination of averaged values of the electron density $\overline{N_{e0}}$, the recombination probability Γ, and the intensity $|\mathbf{E}_0|^2$. Now the equation for E_1 takes the form

$$\partial^2 \mathbf{E}_1/\partial t^2 + \tilde{\omega}_1^2 \mathbf{E}_1 = -\frac{\omega_1^2}{\varepsilon_0}(\partial\tilde{\varepsilon}/\partial N_e)\mathbf{E}_1(\alpha\overline{N}_{e0}\mathscr{K}\,|\,\mathbf{E}_0\,|^2)/(i\Omega + \Gamma), \tag{46}$$

where $\mathscr{K}$ is the "overlap" coefficient introduced in Section 4. This equation has unstable solutions when the inequality

$$(\omega_1/2)\alpha\overline{N_{e0}}|\,\mathbf{E}_0\,|^2\mathscr{K}\,|\,\mathrm{Im}\,[(1/\tilde{\varepsilon}_0)(\partial\tilde{\varepsilon}/\partial N_e)(i\Omega + \Gamma)] > \gamma_1 \tag{47}$$

is satisfied. The following transformations may be used:

$$\mathrm{Im}\left(\frac{1}{\tilde{\varepsilon}^0}\frac{\partial\tilde{\varepsilon}}{\partial N_e}\frac{1}{i\Omega + \Gamma}\right) \approx -\frac{1}{\varepsilon}\left(\frac{\Omega}{\Omega^2 + \Gamma^2}\mathrm{Re}\frac{\partial\tilde{\varepsilon}}{\partial N_e} - \frac{\Gamma}{\Omega^2 + \Gamma^2}\mathrm{Im}\frac{\partial\tilde{\varepsilon}}{\partial N_e}\right) =$$

$$= \frac{1}{\varepsilon}\mathrm{Re}\frac{\partial\tilde{\varepsilon}}{\partial N_e}\frac{\Omega - \Gamma\xi}{\Omega^2 + \Gamma^2} = -\frac{1}{\varepsilon\Gamma}\mathrm{Re}\frac{\partial\tilde{\varepsilon}}{\partial N_e}F(\Omega, \xi), \tag{48}$$

where $\xi \equiv \left(\mathrm{Im}\dfrac{\partial\tilde{\varepsilon}}{\partial N_e}\right)\Big/\left(\mathrm{Re}\dfrac{\partial\tilde{\varepsilon}}{\partial N_e}\right)$ and the function F is introduced in the form

$$F(\Omega, \xi) = \frac{(\Omega - \Gamma\xi)\,\Gamma}{\Omega^2 + \Gamma^2}. \tag{49}$$

It is clear from this result that the interaction of the spectral modes calculated here represents an additional gain (absorption) in a weak mode and this gain is due to a parametric effect of a strong mode. The spectral shape of this additional gain is determined by the function $F(\Omega, \xi)$ shown in Fig. 18 for the special case $\xi = 0.3$ (approximate estimate). This additional gain

$$\delta g = -(\omega_1/c\sqrt{\varepsilon})\left(\mathrm{Re}\frac{\partial\tilde{\varepsilon}}{\partial N_e}\right)\alpha\overline{N}_{e0}|\,E_0\,|^2\mathscr{K}F\,(\Omega)/\Gamma \tag{50}$$

is the superposition of two effects, specifically: scattering of laser light on variations in $\mathrm{Re}\,\tilde{\varepsilon}$ with an odd profile $\Omega/(\Omega^2 + \Gamma^2)$ and similar scattering on variations in $\mathrm{Im}\,\tilde{\varepsilon}$ with an even profile $\xi\Gamma/(\Omega^2 + \Gamma^2)$. The first of these effects is important for explaining the asymmetric mode interaction; δg is positive for $\Omega > \Gamma\xi$ and negative from the short wavelength side, where $\Omega < \Gamma\xi$. The second type of scattering has a negative contribution but still causes some displacement

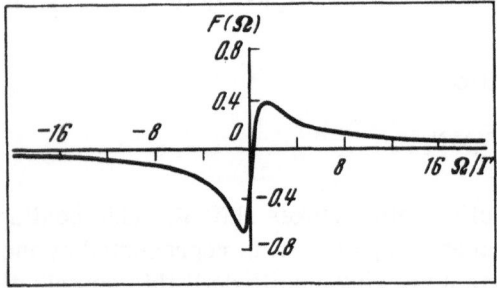

Fig. 18. The function $F(\Omega)$ for $\xi = 0.3$, showing the spectral distribution of the additional gain (absorption) due to induced scattering of the laser light in the active medium on dynamic electron density waves.

of the asymmetric curve from the point $\Omega = 0$ without changing its qualitative character. The extrema of δg lie at the following points:

$$\Omega_{1,2} = \Gamma\,(\xi \pm \sqrt{1 + \xi^2}).$$ (51)

The maximum value of δg saturates for $\alpha\,|E_0|^2 \gg 1/\tau$ and shifts toward long wavelengths if the intensity $|E_0|^2$ increases. This maximum value can be expressed in the form

$$\delta g_{max} \approx -\frac{\bar{N}_{e0}\omega_1}{4c\,\sqrt{\varepsilon}}\left(\mathrm{Re}\,\frac{\partial\tilde{\varepsilon}}{\partial N_e}\right)\frac{\sqrt{1+\xi^2}}{1+\xi+\sqrt{1+\xi^2}},$$ (52)

where the estimate of $\mathscr{K}$ obtained in Section 4 is used. According to Eq. (52) the limiting value of the additional gain for a GaAs laser depends on the average electron density and is roughly

$$\delta g_{max} \approx (2 \times 10^{16}\,\text{cm}^2)\overline{N_{e0}}.$$ (53)

This equation is valid for estimating the additional gain due to induced scattering only as to order of magnitude. In view of the finite width of the lasing band at ω_0 such a value of the additional gain may not be observed. At larger Ω the additional gain falls as $1/\Omega$. This value is about 10^{-1} cm^{-1} for the neighboring transverse mode of a typical diode cavity and is quite comparable to the gain deficit. Thus, this mechanism can lead to unstable single-frequency lasing at a sufficiently high optical flux density.

We assume that the experimental results presented here are evidence of a new type of mode interaction. It cannot be explained as due to formation of a simple dip in the gain profile near the laser frequency since it is strongly asymmetric. In addition, the asymmetry in the gain profile typical of semiconductor lasers is not enough to fit the experimental data. From Fig. 17 it is clear that the efficiency of the asymmetric interaction of the spectral modes is high for small spectral separations between the modes and decreases at larger separations. Recently it was found that the delay in lasing relative to the beginning of the current pulse may be spectrally dependent. However, near the optimum spectral point there is a broad interval where both this delay and the threshold current value are constant. All our observations are in this interval. By additional experiments we verified directly that there is no significant difference in the time behavior of the two interacting modes.

The effect of the intensity on the nature of the interaction is similar to that which is known for other induced scattering mechanisms (see, for example, the similar dependence of induced Brillouin backscatter [68]). The short-wavelength component which serves as a pump wave typically tends to saturate.

The experimental conditions described here are favorable for observation of the anomalous mode interaction since they make it possible to avoid masking effects due to both the spatial inhomogeneity of the diode and the broadening of the laser light. It is also important that the measurements were made at a high electron density level (our experiments correspond to a density of 10^{18} cm^{-3} or above). We could observe asymmetric mode competition, which depends on the spectral separation that agrees qualitatively with the theoretical model of induced scattering of the laser light on dynamic inhomogeneities in the electron density. This new mechanism applies solely to inverted media in which the real part of the dielectric constant depends on the population inversion.

We assume that this mechanism can operate in a wider range of experimental conditions, including those where it is difficult to observe it directly. We recall that in ordinary semiconductor lasers there are many spectral modes with roughly the same amplification above threshold. A small perturbation in the gain profile due to anomalous mode interaction may cause significant changes in the laser output such as the shifting of the dominant mode and disruption of single-frequency operation observed in [24-27]. With this model it is possible to explain how inserting a spectrally selective element in the laser cavity leads to suppression of spontaneous fluctuations [25]. Unfortunately, in view of its complexity the very interesting problem of the dynamics of a multimode laser including this new mechanism does not possess an analytic solution. We limited our theoretical discussion to the case of a small signal in the second mode for this reason. The effect of other types of induced scattering in semiconductor lasers, specifically, due to temperature variations ("temperature" scattering) and ejection of electrons from high field regions, was evaluated in a similar fashion. The contribution of these processes was less important than that of the mechanism described above.

4.   Calculating the Nonlinear Polarization

The electron density $N_e(\mathbf{r}, t)$ can be written in the form

$$N_e(\mathbf{r}, t) = N_{e0}(\mathbf{r}) + \delta N_e(\mathbf{r}, t), \tag{54}$$

where $N_{e0}(\mathbf{r})$ is the stationary part of the local electron density and $\delta N_e(\mathbf{r}, t)$ is the time-dependent part which averages out to zero. Substituting Eq. (54) in Eq. (39) results in its breaking up into two equations in which we retain only the linear terms in $\delta N_e(\mathbf{r}, t)$ and $\mathscr{E}_1$. The first equation is for the spatial variation $N_{e0}(\mathbf{r})$,

$$-D\nabla^2 N_{e0}(\mathbf{r}) + [N_{e0}(\mathbf{r})/\tau] + aN_{e0}(\mathbf{r})|\mathscr{E}_0(\mathbf{r}, t)|^2 = G, \tag{55}$$

and the second is for $\delta N_e(\mathbf{r}, t)$,

$$(\partial \delta N_e/\partial t) + (\delta N_e/\tau) - D\nabla^2 \delta N_e + a\delta N_e|\mathscr{E}_0|^2 = -aN_{e0}\mathscr{E}_0^*\mathscr{E}_1 + \text{c.c.}, \tag{56}$$

where the asterisk denotes the complex conjugate. A valid simplification of Eq. (56) is possible if δN_e is averaged over small regions along the cavity axis z. The periodic variations δN_e produced by the standing wave can be neglected due to rapid diffusion of electrons. Thus, there is no need to evaluate the short-wavelength variations in $|\mathscr{E}_0|^2$ and $\mathscr{E}_0^*\mathscr{E}_1$. The smoothing effect of diffusion will be efficient if the following expression holds:

$$4D\varepsilon/\lambda^2 > \Omega. \tag{57}$$

On the other hand, the long-wavelength variations δN_e are not subject to the influence of electron diffusion since

$$c/\sqrt{\varepsilon D\tau} > \Omega. \tag{58}$$

Under these assumptions we can drop the diffusion term and take the averages

$$\overline{|\varphi_0|^2} = (2\sqrt{\varepsilon}/\lambda) \int_z^{z+\lambda/2\sqrt{\varepsilon}} |\varphi_0|^2 \, dz', \tag{59}$$

$$\overline{\varphi_0\varphi_1} = (2\sqrt{\varepsilon}/\lambda) \int_z^{z+\lambda/2\sqrt{\varepsilon}} \varphi_0\varphi_1 \, dz'. \tag{60}$$

Now Eq. (56) can be rewritten in the form

$$\partial \delta N_e/\partial t + \delta N_e/\tau + a\delta N_e|\mathscr{E}_0|^2\overline{|\varphi_0|^2} = -aN_{e0}E_0^*E_1\overline{\varphi_0\varphi_1} + \text{c.c.} \tag{61}$$

The solution of this equation is written in the form

$$\delta N_e = - \frac{\alpha N_{e0} E_0^* E_1 \overline{\varphi_0 \varphi_1}}{i\Omega + (1/\tau) + \alpha \, |E_0|^2 \overline{|\varphi_0|^2}} + \text{c.c.} \tag{62}$$

Now we use Eqs. (45) and (62) to write Eq. (44) for E_1 in the form

$$\partial^2 E_1/\partial t^2 + \widetilde{\omega}_1^2 E_1 = \frac{1}{\widetilde{\varepsilon}} \frac{\partial \widetilde{\varepsilon}}{\partial N_e} \int \frac{\alpha N_{e0} \varphi_0(r) \, \varphi_1(r) \, \overline{\varphi_0 \varphi_1} \left[\partial^2 \left(E_0 E_0^* E_1 \right)/\partial t^2 \right] dV}{i\Omega + (1/\tau) + \alpha \, |E_0|^2 \overline{|\varphi_0|^2}} =$$

$$= -\frac{\omega_1^2}{\widetilde{\varepsilon}} \frac{\partial \widetilde{\varepsilon}}{\partial N_e} \, |E_0|^2 E_1 \int \frac{\alpha N_{e0} \varphi_0 \varphi_1 \overline{\varphi_0 \varphi_1} \, dV}{i\Omega + (1/\tau) + \alpha \, |E_0|^2 \overline{|\varphi_0|^2}} \, . \tag{63}$$

Further analysis requires accurate calculations of the distribution N_{e0} and the mode configurations. If the cavity Q is sufficiently high, these distributions are not too important in evaluating this effect. We now take the approximation

$$\int \frac{\alpha N_{e0} \varphi_0 \varphi_1 \overline{\varphi_0 \varphi_1} \, dV}{i\Omega + (1/\tau) + \alpha \, |E_0|^2 \overline{|\varphi_0|^2}} \approx \frac{\alpha \overline{N_{e0}} \mathscr{K}}{i\Omega + \Gamma} \, , \tag{64}$$

where $\overline{N_{e0}}$ and Γ are the average values of the stationary electron density and the effective recombination probability, respectively, and $\mathscr{K}$ is the "overlap" coefficient of the inversion and combined mode configuration. As an example, in the plane-wave approximation in an ideal cavity this coefficient is about 0.5/V, where V is the cavity volume.

CHAPTER IV

A STUDY OF INJECTION LASERS USING NEW SEMICONDUCTOR MATERIALS

1. Isoperiodic Substitution as a Principle for Creating New Materials for Heterolasers

In this chapter we consider the prospects for using multicomponent solid solutions in injection lasers. As the experience of recent years has shown, the highest output characteristics are obtained with heterolasers [2]. Until recently, however, the number of pairs of semiconducting materials suitable for laser heterostructures was limited to a system of gallium arsenide−aluminum arsenide solid solutions. The advantage of this system is that because of the closeness of the covalent radii of gallium and aluminum their interaction does not cause a significant change in the period of the crystal lattice. This in turn makes it possible to produce improved heterojunctions free of noncorrespondence defects. There is a natural interest in other pairs of compounds based on the interaction of gallium and aluminum such as GaN−AlN and GaSb−AlSb. In addition, further expansion of the class of such materials is limited because of the lack of other pairs of elements in the III and V groups of the periodic chart with the same covalent radius (see Table 1). In addition the lattice period in a GaSb−AlSb or even in a GaAs−AlAs system does not completely coincide, so residual voltages or noncorrespondence defects result.

Because of the use of multicomponent solid solutions new prospects appear. We shall show that in such systems with four or more components there are discontinuous series of compositions with a constant lattice period. This greatly expands the likelihood of making improved heterostructures for new spectral ranges. This same principle can be extended to

TABLE 1. Lattice Periods and Thermal Expansion Coefficients
of Some III-V Compounds with a Sphalerite Type Lattice

Compound	Lattice period a at 300°K, nm	Thermal expansion coefficient $\alpha \cdot 10^6$	Compound	Lattice period a at 300°K, nm	Thermal expansion coefficient $\alpha \cdot 10^6$
AlP	0.54625	—	GaSb	0.60954	6.9
AlAs	0.56675	5.2	InP	0.58687	4.5
AlSb	0.61355	4.9	InAs	0.60584	5.3
GaP	0.54511	3.5	InSb	0.64788	5.5
GaAs	0.56535	5.8			

other types of compounds, in particular, to II-VI and IV-VI compounds which are used in lasers. The principle of isoperiodic (that is, conserving the lattice period) substitution consists of simultaneously changing the molar fraction of two or more elements in a way that their effect on the lattice period is mutually compensated. Because of the increased number of degrees of freedom which must be controlled in preparing these systems, the technology is much more complex than for the AlGaAs system; however, it is fully realizable.

The first example of a heterolaser prepared by isoperiodic substitution of two components without using mutual substitution of gallium and aluminum was reported by us in [11] for the case of the four-component system GaInPAs. Later it was possible to create an uncooled heterolaser at a wavelength of roughly 1.1 μm with a record low threshold current density for this range of 10 kA/cm^2. At present a number of similar heterolasers have been built. They will be reviewed in Section 3. The basic problems in choosing materials are, first, choosing isoperiodic pairs for the heterogeneous structures and, second, choosing suitable substrates. Binary compounds are preferred for the latter since they can be prepared in the form of bulk ingots. The best solution is to use available binary substrates. If this is not successful then it is necessary to anticipate the fabrication of heterostructures by making so-called gradient layers in which the lattice period gradually changes from the period of the substrate to the period of the heterostructure.

2. An Interpolation Technique for Predicting the Properties of a Multicomponent Solid Solution

Planning for improved heterostructures requires preliminary estimates of the basic characteristics of new solid solutions and their composition dependence.

Let us consider the hypothetical solid solution

$$A^1_{x_1} A^2_{x_2} \ldots A^i_{x_i} B^1_{y_1} B^2_{y_2} \ldots B^j_{y_j}, \tag{65}$$

where A^i and B^j are elements of, for example, the III and V groups, respectively, of the chart. The molar fractions x_i and y_j satisfy the stoichiometry condition

$$\sum_i x_i = 1, \quad \sum_j y_j = 1. \tag{66}$$

Our problem is to predict the physical properties of solution (65) from the known properties of the elements, binary compounds, and certain of their mixtures. As unknown parameters we shall consider the lattice period, the coefficient of thermal expansion, and the forbidden-band width. The lattice period may be evaluated approximately by using the method of covalent radii.

In a binary compound which crystallizes with a sphalerite-type lattice there is a simple relationship between the edge of the elementary cell (a cube) a_{ij} and the covalent radii of the component elements, r_i and r_j:

$$a_{ij} = \frac{4}{\sqrt{3}} (r_i + r_j), \tag{67}$$

so interpolation according to this method reduces to the calculation

$$a(x_1, x_2, \ldots, x_i, \; y_1, y_2, \ldots, y_j) = \frac{4}{\sqrt{3}} \left(\sum_i r_i x_i + \sum_j r_j y_j \right). \tag{68}$$

The choice of the isoperiodic compositions corresponds to solving the equation

$$a(x_1, \ldots, x_i, y_1, \ldots, y_j) = C, \tag{69}$$

where the constant on the right side is the desired lattice period. The most interesting case of Eq. (68) corresponds to $C = a_{ij}$, i.e., equality to the period of one of the binary compounds, since it is best to use binary compounds as a substrate.

It is known that the covalent radii are approximately conserved in various compounds. For example, as can be seen from the table, mutual substitution of aluminum and gallium in the arsenide causes a change in the lattice by 0.004 nm. This slight variation in the covalent radii can be taken into account by interpolation among the binary compounds. If we assume that in the composite crystal the atoms are distributed completely randomly, then the composition (65) can be written in the form of an expansion in the binary compositions

$$(A^1 B^1)_{x_{11}} (A^1 B^2)_{x_{12}} \cdots (A^i B^j)_{x_{ij}}, \tag{70}$$

where the molar fractions of the binary components x_{ij} obey the condition

$$\sum_i \sum_j x_{ij} = 1, \tag{71}$$

and the quantities x_{ij} are found using the simple rule

$$x_{ij} = x_i y_j. \tag{72}$$

The interpolation formula among the binary compounds can be written in the form

$$a(x_{11}, x_{12}, \ldots, x_{ij}) = a_{11} x_{11} + a_{12} x_{12} + \ldots + a_{ij} x_{ij} = \sum_i \sum_j a_{ij} x_{ij}, \tag{73}$$

where a_{ij} is the lattice period of the binary compound. The interpolation formula (73) is approximately valid only for those material parameters which are not substantially affected by the substitution. For example, a similar equation for the thermal expansion coefficient

$$\alpha(x_{ij}) \approx \sum_i \sum_j \alpha_{ij} x_{ij}, \tag{74}$$

where α_{ij} are the thermal expansion coefficients of the binary compounds, has to be verified experimentally. Certain parameters absolutely cannot be interpolated linearly. They include, for example, the thermal conductivity coefficient, the freezing point of the liquid mixture, and others. In particular, the effect is large for the coefficients of thermal conductivity: For the two-component mixture $GaP_{0.5}As_{0.5}$ a drop in the thermal conductivity by a factor of 3-5 compared to the binary compounds GaP and GaAs is typical.

Linear interpolation over the binary compounds can also be applied to the parameters of the energy spectrum; however, the error in the results may be unsatisfactory. A preliminary equation for mixtures based on linear interpolation over the binary states has a form analogous to Eqs. (73) and (74):

$$E(x_{ij}) = \sum_i \sum_j E_{ij} x_{ij}, \tag{75}$$

where E_{ij} is the energy parameter of the band structure of a binary semiconductor.

The deviation from linear interpolation in the width of the interband gap is large for some compositions. For example, in the mixture InAs–InSb the forbidden-band width E_g falls to 0.1 eV, which is 2.5 times less than the interpolation value. For such mixtures the quadratic approximation is usually used:

$$E(x) = E(0) + [E(1) - E(0)]x + 4\varepsilon_x x(1-x), \tag{76}$$

where $x_1 = x$ and $x_2 = (1-x)$, $E(0)$ and $E(1)$ are the energy parameters of the binary comounds, and ε_x is the bending in the curve $E(x)$ at the point $x = 0.5$ (the "depth" of the curve), i.e., the difference between the physical value $E(0.5)$ and the linear interpolation value. A similar approximation is acceptable for the lattice period as well if it is desired to take into account its deviation from the Vegard law (i.e., from linearity). The ε_x must be determined for each two-component mixture. As for the band parameters it may be noted that they have been experimentally determined for the majority of the two-component mixtures.

There is a third variant of interpolation over two-component mixtures – stepwise interpolation. Thus, for the four-component composition

$$A^1_x A^2_{1-x} B^1_y B^2_{1-y} \tag{77}$$

or its equivalent

$$(A^1 B^1)_{xy} (A^1 B^2)_{x(1-y)} (A^2 B^1)_{(1-x)y} (A^2 B^2)_{(1-x)(1-y)} \tag{78}$$

the following interpolation formula is valid:

$$\begin{aligned}E(x,y) = {}& E_{00} + (E_{10} - E_{00} + 4\varepsilon_{x0})x - 4\varepsilon_{x0}x^2 + (E_{01} - E_{00} + 4\varepsilon_{0y})y - \\ & - 4\varepsilon_{0y}y^2 + [E_{00} + E_{11} - E_{10} - E_{01} + 4\varepsilon_{x1} + 4\varepsilon_{1y} - 4\varepsilon_{x0} - 4\varepsilon_{0y}]xy + \\ & + (4\varepsilon_{x0} - 4\varepsilon_{x1})x^2 y + (4\varepsilon_{0y} - 4\varepsilon_{1y})xy^2. \end{aligned} \tag{79}$$

Here we have neglected the mixing effect typical of a four-component mixture. A check for $x = y = 0.5$ should make the approximation more precise. If the actual value $E(0.5; 0.5)$ differs from the interpolation value (79) by an amount ε_{xy} then this effect may be included in Eq. (79) by means of the term

$$16\varepsilon_{xy} x(1-x)y(1-y), \tag{80}$$

which vanishes on all sides of the rectangle $0 \le x \le 1$, $0 \le y \le 1$.

3. Examples of Injection Heterolasers Based on Isoperiodic Substitution in Multicomponent Solid Solutions

The use of four-component solid solutions in heterolasers was begun in [69], where the active layer had the composition $Al_x Ga_{1-x} P_y As_{1-y}$ and the wide-band layers were distinguished by a high concentration of aluminum. Thus, isoperiodic substitution of aluminum and gallium was used here. We note that with this four-component solution it was not possible to go beyond the spectral range attained with other, simpler systems, in particular, the $Al_xGa_{1-x}As$ system.

The principle of simultaneous substitution of two elements for preserving the lattice period in a four-component system was first applied in [11] where the active material had the composition $Ga_xIn_{1-x}P_{1-y}As_y$ isoperiodic with indium phosphide which served as the material for the wide-band layers. Another example of a heterolaser based on isoperiodic substitution is a new laser with gallium antimonide as an active medium and with the composition $Al_xGa_{1-x}Sb_{1-y}As_y$ in the wide-band layers. A small arsenic impurity (about 2%) in this mixture compensates the noncorrespondence in the lattice caused by adding aluminum. (We note that the difference between the lattice periods of aluminum and gallium antimonides is greater than in the arsenides of the same elements at 0.66%.) The heterostructure is grown on gallium antimonide. The pattern for both heterostructures is shown in Fig. 19.

Let us examine these two systems in more detail. Our goal is to choose a composition for the solid solution $Ga_xIn_{1-x}P_{1-y}As_y$ for a heterolaser operating in the range 1.05-1.1 μm (that is, with a forbidden-band width about 1.2 eV in the active layer). Interpolation of the lattice period leads to the following equation for compositions isoperiodic with indium phosphide:

$$5.8687 = 5.8687\ (1-x)(1-y) + 5.4511\ x\ (1-y) + 5.6535\ xy + 6.0584\ (1-x)\ y, \qquad (81)$$

which yields

$$y = \frac{0.4176x}{0.1897 + 0.0127x}. \qquad (82)$$

This equation must be substituted in the interpolation formula for the forbidden-band width set equal to the desired value of 1.2 eV. The isoperiodic curve corresponding to Eq. (82) is given in Fig. 20. The preliminary values of x and y for this forbidden-band width are 0.09 and 0.20, respectively.

Structures with an active layer of $Ga_xIn_{1-x}P_{1-y}As_y$ with $0.18 \le x \le 0.2$ and $0.09 \le y \le 0.10$ were made experimentally. Coherent output was obtained in the spectral range 1.05-1.15 eV both at 77°K and at room temperature. In the best sample the threshold current density at 300°K was 10.5 kA/cm^2. This is roughly 8 times smaller than the previously achieved thresh-

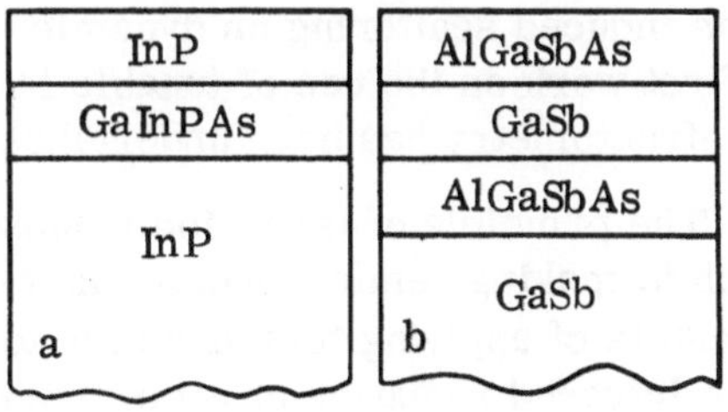

InP	AlGaSbAs
GaInPAs	GaSb
InP	AlGaSbAs
	GaSb
a	b

Fig. 19. The arrangement of the epitaxial layers in laser heterostructures with $Ga_xIn_{1-x}P_{1-x}As_y$ (a) and GaSb (b) as active media.

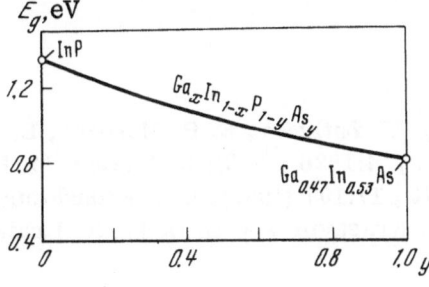

Fig. 20. The computed dependence of the forbidden-band width E_g along the isoperiodic curves in the solid solution $Ga_xIn_{1-x}P_{1-x}As_y$ for a lattice period equal to that of InP.

old current density in homolasers at the same wavelength range. The development of low-threshold lasers at wavelengths of 1.05-1.15 μm is of great practical interest for two reasons. First, this type of laser can be used to simulate high-power neodymium lasers (1.06 μm) and, second, they can be used in glass-fiber optical communication since their emission falls in one of the best transparency windows of optical glasses and quartz (with low damping and low spectral dispersion). It may be hoped that in lasers with a binary material it might be possible to obtain lower thermal resistance in wide-band layers; moreover, indium phosphide has a higher thermal conductivity than gallium arsenide. Finally, from the standpoint of optimizing the lifetime of lasers it is desirable to use a lower photon energy from an acceptable wavelength range. In this regard lasers at 1.05-1.15 μm seem most promising.

In a $Al_x Ga_{1-x} Sb_{1-y} As_y$ system the condition for isoperiodicity with gallium antimonide has the form

$$y = \frac{0.0401x}{0.4419 + 0.0261x} .$$ (83)

Since for heterostructures with gallium antimonide as an active medium ($E_g \approx 0.72$ eV) it is sufficient to choose a composition with a wide forbidden band ($E_g \geq 0.72$ eV + 4 kT), we may specify a sufficiently high molar fraction x and then choose the fraction y. For x = 0.28 a calculation yields y = 0.025. The important role of the arsenic impurity is illustrated by the following data. The best result with a two-sided heterostructure without arsenic, AlGaSb−GaSb−AlGaSb, corresponds to lasing at 77°K with a threshold current density of 2-3 kA/cm^2. At room temperature lasing was not observed. The best result for the two-sided heterostructure AlGaSbAs−GaSb−AlGaSbAs corresponds to lasing at 77°K with a threshold current density of 0.22 kA/cm^2 (that is, almost an order of magnitude less) and lasing at 300°K with a threshold current density of 9.5 kA/cm^2.

CONCLUSION

This study of the spectral and spatial distribution of the output of injection lasers with an external cavity has demonstrated the wide prospects for controlling the composition of the light with the aid of selective elements and for obtaining high quality single-frequency output. It has been shown that spectral selection occurs practically without loss of pump power. It was shown experimentally that several little-studied nonlinear effects characteristic of semiconductors affect the spectral and spatial distribution of the light. One of these was identified as induced scattering on dynamic electron density waves in the active region of the laser. The first work on the use of tunable lasers with an external cavity for intracavity spectroscopy and refractometry has been undertaken.

The principle of isoperiodic substitution in multicomponent solid solutions is a new approach to making semiconductor lasers over a wide spectral range. Here we have presented the results of applying this principle to new four-component systems capable of producing coherent infrared radiation at room temperature.

LITERATURE CITED

1.   P. G. Eliseev, Tr. FIAN, 52:3 (1970).
2.   Zh. I. Alferov, V. M. Andreev, D. Z. Garbuzov, Yu. V. Zhilyaev, E. P. Morozov, E. L. Portnoi, and V. G. Trofim, Fiz. Tekh. Poluprovodn., 4:1826 (1970); I. Hayashi, M. B. Panish, R. W. Foy, and S. Sumski, Appl. Phys. Lett., 17:109 (1970); I. K. Bronshtein, L. M. Dolginov, L. V. Druzhinina, P. G. Eliseev, I. V. Krasavin, and L. D. Libov, Kratk. Soobshch. GIREDMET, Ser. V, No. 21 (1970).

3.   P. G. Eliseev and V. P. Strakhov, Pis'ma Zh. Éksp. Teor. Fiz., 16:606 (1972); P. G. Eliseev and N. N. Shuikin, Kvant. Élektron., 3(15), p. 5 (1973).

4.   A. R. Calawa, J. Lumin., 7:477 (1973).

5.   W. E. Ahearn and J. M. Crowe, IEEE J. Quant. Electron., QE-2:597 (1966).

6.   E. D. Hinkley and C. Freed, Phys. Rev. Lett., 23:277 (1969).

7.   L. A. Rivlin, Kvant. Élektron., No. 3, p. 34 (1971).

8.   W. W. Anderson, IEEE J. Quant. Electron., QE-1:228 (1965).

9.   P. G. Eliseev, FIAN Preprint No. 33, Moscow (1970).

10. H. C. Casey, D. Sell, and M. B. Panish, Appl. Phys. Lett., 24:63 (1974).

11. A. P. Bogatov, L. M. Dolginov, L. V. Druzhinina, P. G. Eliseev, B. N. Sverdlov, and E. G. Shevchenko, Kvant. Élektron., 1:2294 (1974).

12. V. F. Vzyatyshev, Dielectric Waveguides [in Russian], Sovetskoe Radio, Moscow (1970).

13. P. G. Eliseev, Kvant. Élektron., 6(12), p. 3 (1972).

14. I. Hayashi, M. B. Panish, and R. K. Reinhart, J. Appl. Phys., 42:1929 (1971).

15. T. Ikegami, IEEE J. Quant. Electron., QE-8:470 (1972).

16. R. G. Allakhverdyan, A. N. Oraevskii, and A. F. Suchkov, Fiz. Tekh. Poluprovodn., 4:341 (1970).

17. P. G. Eliseev and Chan Min Thai, Kvant. Élektron., 1:1138 (1974).

18. T. H. Zachos and J. E. Ripper, IEEE J. Quant. Electron., QE-5:29 (1969).

19. L. M. Dolginov, P. G. Eliseev, L. D. Libov, I. Z. Pinsker, E. L. Portnoi, G. G. Kharisov, and E. G. Shevchenko, Kratk. Soobshch. Fiz., No. 12, p. 63 (1970).

20. P. G. Eliseev, A. I. Krasil'nikov, A. V. Khaidarov, and G. G. Kharisov, Kvant. Élektron., 1:196 (1974).

21. A. P. Bogatov, L. M. Dolginov, L. V. Druzhinina, P. G. Eliseev, and L. D. Libov, Fiz. Tekh. Paluprovodn., 6:43 (1972).

22. M. S. Sommers, Jr., J. Appl. Phys., 44:1263 (1973).

23. H. Statz, C. L. Tang, and J. M. Lavine, J. Appl. Phys., 35:2581 (1964).

24. A. P. Bogatov, P. G. Eliseev, V. I. Panteleev, and E. G. Shevchenko, Kvant. Élektron., No. 5, p. 93 (1971).

25. A. P. Bogatov, P. G. Eliseev, M. A. Manko, L. P. Ivanov, A. S. Logginov, and K. Ya. Senatorov, IEEE J. Quant. Electron., QE-9:392 (1973).

26. A. P. Bogatov, P. G. Eliseev, L. P. Ivanov, A. S. Logginov, and K. Ya. Senatorov, Kvant. Élektron., p. 14 (1973).

27. P. G. Eliseev, L. P. Ivanov, A. S. Logginov, and K. Ya. Senatorov, Kvant. Élektron., 1:151 (1974).

28. K. Kobayashi, K. Yonezu, F. Saito, and Y. Nannichi, Appl. Phys. Lett., 19:323 (1971).

29. V. N. Morozov, Kvant. Élektron., 1:634 (1974); R. G. Allakhverdyan, V. N. Morozov, and A. F. Suchkov, Kratk. Soobshch. Fiz., No. 6, p. 3 (1972).

30. N. G. Basov and V. N. Morozov, Zh. Éksp. Teor. Fiz., 57:617 (1969).

31. T. L. Paoli and J. E. Ripper, Phys. Rev. Lett., 22:1085 (1969).

32. A. P. Bogatov, P. G. Eliseev, and B. N. Sverdlov, IEEE J. Quant. Electron., QE-11:510 (1975).

33. A. P. Bogatov, P. G. Eliseev, and B. N. Sverdlov, Kvant. Élektron., 1:2286 (1974).

34. P. A. Kirkby and G. H. B. Thompson, Appl. Phys. Lett., 22:638 (1973).

35. G. H. B. Thompson and P. A. Kirkby, Electron. Lett., 9:295 (1973).

36. H. C. Casey, M. B. Panish, W. O. Schlosser, and T. L. Paoli, J. Appl. Phys., 45:322 (1974).

37. J. K. Butler, J. Appl. Phys., 42:4447 (1971).

38. T. L. Paoli, B. W. Hakki, and B. I. Miller, J. Appl. Phys., 44:1276 (1973).

39. B. W. Hakki and C. J. Hwang, J. Appl. Phys., 45:2168 (1974).

40. L. A. D'Asaro, J. Lumin., 7:310 (1973).

41. W. Susaki, H. Namizaki, H. Kan, and A. Ito, J. Appl. Phys., 44:2893 (1973).

42. P. G. Eliseev, Kratk. Soobshch. Fiz., No. 3, p. 9 (1974).

43. M. Takusagawa, S. Ohsaka, N. Takagi, H. Ishikawa, and H. Takanashi, Proc. IEEE, 61:1758 (1975).

44. K. Yonezu, I. Sakuma, K. Kobayashi, T. Kamejima, M. Ueno, and Y. Nannichi, Jpn. J. Appl. Phys., 12:1585 (1973).

45. V. M. Andreev, V. I. Borodulin, V. P. Konyaev, G. T. Pak, A. I. Petrov, E. L. Portnoi, and V. I. Shveikin, Fiz. Tekh. Poluprovodn., 6:1739 (1972).

46. E. Mohn, T. F. Broom, C. Deutsch, and J. Hats, Phys. Lett., 24A:561 (1967).

47. J. M. Crowe and W. E. Ahearn, IEEE J. Quant. Electron., QE-4:169 (1968).

48. E. M. Phillip-Rutz and H. D. Edmonds, Appl. Opt., 8:1859 (1969).

49. H. Bachert and S. Raab, Monatsber. Dtsch. Akad. Wiss. Berlin, 10:911 (1968).

50. H. D. Edmonds and A. W. Smith, IEEE J. Quant. Electron., QE-6:356 (1970).

51. P. G. Eliseev, Yu. M. Popov, and N. N. Shuikin, Zh. Éksp. Teor. Fiz., 56:1412 (1969).

52. Yu. M. Popov and N. N. Shuikin, Fiz. Tekh. Poluprovodn., 4:45 (1970).

53. P. G. Eliseev, I. Ismailov, M. A. Man'ko, and V. P. Strakhov, Pis'ma Zh. Éksp. Teor. Fiz., 9:594 (1969).

54. P. G. Eliseev and M. A. Man'ko, Kratk. Soobshch. Fiz., No. 4, p. 47 (1970).

55. D. Ackermann, H. Bacher, P. G. Eliseev, A. Keiper, M. A. Manko, and S. Raab, Laser und Ihre Anwendungen, Intern. Tagung, ZOS, Berlin (1970), Teil 12, S. 743.

56. D. Ackermann, P. G. Eliseev, A. Keiper, M. A. Man'ko, and S. Raab, Kvant. Élektron., No. 1, p. 85 (1971).

57. R. Ludeke and E. P. Harris, Appl. Phys. Lett., 20:499 (1972).

58. D. Ackermann, G. Eliseev, M. A. Man'ko, S. Raab, Chan Min Thai, A. V. Khaidarov, and N. N. Shikin, Kratk. Soobshch. Fiz., No. 6, p. 9 (1973).

59. J. A. Rossi, S. R. Chinn, and H. Heckscher, Appl. Phys. Lett., 23:25 (1973).

60. D. Ackermann, A. P. Bogatov, P. G. Eliseev, S. Raab, and B. N. Sverdlov, Kvant. Élektron., 1:1145 (1974).

61. R. F. Kazarinov and R. A. Surus, Fiz. Tekh. Poluprovodn., 6:1359 (1972).

62. Zh. I. Alferov, S. A. Gurevich, R. F. Kazarinov, M. N. Muzerov, E. L. Portnoi, and R. P. Seisyan, Fiz. Tekh. Poluprovodn., 8:832 (1974).

63. D. R. Scifres, R. D. Burnham, and W. Streifer, Appl. Phys. Lett., 25:203 (1974).

64. D. B. Anderson, R. R. August, and J. E. Coker, Appl. Opt., 13:2742 (1974).

65. R. G. Allakhverdyan, A. N. Oraevskii, and A. F. Suchkov, Fiz. Tekh. Poluprovodn., 4:341 (1970).

66. R. G. Allakhverdyan, V. N. Morozov, A. N. Oraevskii, and A. F. Suchkov, Kvant. Éledtron., No. 6, p. 53 (1971).

67. É. M. Belenov, V. N. Morozov, and A. N. Oraevskii, Tr. FIAN, 53:119 (1970).

68. M. Maier, W. Rother, and W. Kaiser, Phys. Lett., 23:83 (1966).

69. R. D. Burnham, N. Holonyak, Jr., H. W. Korb, H. M. Macksey, D. R. Scifres, J. B. Woodhouse, and Zh. I. Alferov, Appl. Phys. Lett., 19:25 (1971).

ACTIVE MEDIA, DESIGNS, AND PLANS FOR POWERFUL RAMAN LASERS

A. Z. Grasyuk, V. F. Efimkov, I. G. Zubarev, A. V. Kotov, and V. G. Smirnov

The possibility of using liquid nitrogen and liquid oxygen as active media in high-power Raman lasers [lasers using induced combination (Raman) scattering] is established. Some results from a study of nonlinear light losses which cause a reduction in the transparency of liquid nitrogen and liquid oxygen are presented. Ways of reducing the nonlinear losses are discussed, including filtration systems, overpressures, etc. Cryogenic cuvettes for liquid N_2 and O_2 are described in which the active media have high optical homogeneity over lengths of up to 1 m. Systems for ensuring high spatial uniformity of pump light over the length of the active medium are discussed, specifically a focusing prism screen combined with a light guide of square cross section. Various practical schemes for pulsed Raman lasers are presented.

INTRODUCTION

For the past several years work has been done at the quantum electronics laboratory of the Lebedev Physics Institute on the construction and investigation of Raman lasers, or lasers employing stimulated combination (Raman) scattering. Interest in this type of coherent radiation source has grown recently in connection with research on the selective interaction of radiation with matter. This research includes laser stimulation of chemical reactions [1-3], laser isotope separation [4-6], and laser spectroscopy [7]. Powerful lasers with tunable wavelengths are required to solve these and other basic and applied problems. Raman lasers open up new possibilities since they efficiently transform the pump radiation into another (usually at a shorter wavelength) wavelength range. In a number of cases this transformation is accompanied by an improvement in the characteristics of the radiation; that is, its divergence is reduced, its intensity is increased, and so on. In [8, 9] one of us has summarized a group of our papers [10-16] on the physical bases of Raman lasers. The present paper contains the results of some studies of active media, designs, schemes, and possible applications of powerful pulsed Raman lasers with tunable wavelengths. The purpose of these studies was to develop Raman lasers as a new type of device suitable for solving basic and applied problems.

CHAPTER I

THE PROPERTIES AND FEATURES OF ACTIVE MEDIA

1. Liquid Nitrogen and Liquid Oxygen

The decisive stage in building powerful Raman lasers was the search for an active medium since the characteristics of any laser are primarily determined by the properties of the ac-

TABLE 1.   Characteristics of Liquid Nitrogen and Liquid Oxygen
at 77°K

Parameters	Nitrogen	Oxygen	References
Emission spectrum	Transparent from the IR to the UV	Absorbs near frequencies ν_h (cm^{-1}); $\nu_n = 7930 + n \cdot 1552$; $n = 0, 1, 2, 3, \dots$	
Gain coefficient, cm/MW			
$\lambda_p = 0.6943$ μm[†]	$1.6 \cdot 10^{-2}$	$1.6 \cdot 10^{-2}$	[19]
$\lambda_p = 1.06$ μm	10^{-2}		[14]
Frequency shift of the Stokes component, cm^{-1}	2326	1552	[18—20]
Optical durability, GW/cm^2	1	1	[18, 20, 21]
Critical self-focusing power. MW	0.5: 7	0.3: 5	[19], [23]
Width of the spontaneous Raman scattering line, cm^{-1}	0.067	0.11	[20]
Vapor pressure, atm	1	0.2	
Density, g/cm^3	0.8	1.2	[24]
Specific heat, J/g · deg	2.1	1.67	[24]

[†] λ_p is the wavelength of the exciting (pump) radiation.

tive medium. In this case there was a special difficulty since the active medium had to satisfy extremely rigid requirements, the principal ones of which are as follows:

(1) Transparency in the infrared
(2) A high gain coefficient for induced Raman scattering
(3) Optimal frequency shift of the Stokes component: on the one hand, the shift must be small enough to ensure a high energy conversion efficiency at 100% quantum yield, and on the other, large enough to ensure frequency decoupling between the Raman laser and the pump source
(4) Optical durability: the medium must withstand intensities of greater than 100 MW/cm^2 without significant changes in its properties
(5) A high critical self-focusing power
(6) Optical homogeneity

These requirements made the search for an appropriate active medium extremely difficult. However, even in 1964, while working on the construction of optically pumped semiconductor lasers, we found that liquid nitrogen satisfies these requirements to a great extent and, in particular, converts ruby laser light into the Stokes component very efficiently. We used this component to excite a GaAs semiconductor laser [17, 18]. This practical experience enabled us to build a Raman laser in which liquid nitrogen was the active medium [10]. Later, Stoicheff showed [19] that liquid oxygen is very close to liquid nitrogen in its nonlinear optical properties. Thus, liquid nitrogen, and later liquid oxygen [15] as well, became our principal active media. Their principal characteristics are shown in Table 1. Let us now consider some of these properties. Liquid nitrogen is transparent over a wide spectral range, from the infrared (IR) to the ultraviolet (UV). Liquid oxygen has a number of absorption lines, of which the lowest frequency ones are at wavelengths of 1.26 and 1.06 μm [25]. In other words, liquid O_2 cannot be excited by light from a neodymium laser. At shorter wavelengths the absorption lines occur roughly every 1550 cm^{-1}. One of them coincides with the wavelength of a He—Ne laser ($\lambda = 0.6328$ μm).

The critical self-focusing power in nitrogen and oxygen is tens of times greater than in other materials [23].

The high gain coefficient for stimulated (induced) Raman scattering in liquid nitrogen and oxygen makes stimulated Raman scattering predominate over all other nonlinear optical effects. Only for pulse durations $\tau \gtrsim 500$ nsec can other kinds of stimulated scattering compete with induced Raman scattering (for example, stimulated Brillouin scattering and stimulated temperature scattering [15]).

The large frequency shift in the Stokes component, ν_c, makes it possible to excite Raman lasers without the special decouplers required in stimulated Brillouin lasers [12]. At the same time this shift makes it possible to realize highly efficient conversion. For example, for pumping with a neodymium laser ($\nu_p = 9440$ cm^{-1}) the limiting efficiency given by the Manley–Rowe theory is $\eta = [(\nu_p^{Nd} - \nu_c)/\nu_p^{Nd}] \cdot 100\% = 75\%$. The optical homogeneity of the liquid nitrogen or oxygen is determined by the design of the cryostat and is better the more precisely the temperature of the working volume is controlled.

2. Nonlinear Light Losses in Liquid Nitrogen

and Liquid Oxygen

By nonlinear light losses in liquid nitrogen and oxygen we mean the reduction in transparency of the active medium observed in our experiments as the intensity and duration of the laser pulse are increased [26]. The transparency T% is defined in this case as

$$T = [(\Sigma N_{iS} + N_P^{out})/N_P^{in}] \, 100\,\%, \tag{1}$$

where N_P^{in} and N_P^{out} are the number of pump photons entering the active medium and leaving it, and N_{iS} is the number of photons of the i-th Stokes component leaving the medium.

This reduction in the transparency was not due to resonance absorption. The transparency reduction was qualitatively the same and occurred in both liquid nitrogen and liquid oxygen when the cryogenic liquids were acted upon by light from different lasers in the free-lasing mode ($\lambda_p = 0.6943$ μm, $\lambda_p = 1.06$ μm, $\lambda_p = 1.315$ μm).

The dependence of the transparency of liquid nitrogen and oxygen on various parameters of the exciting radiation was examined on the apparatus shown in Fig. 1. Laser light was

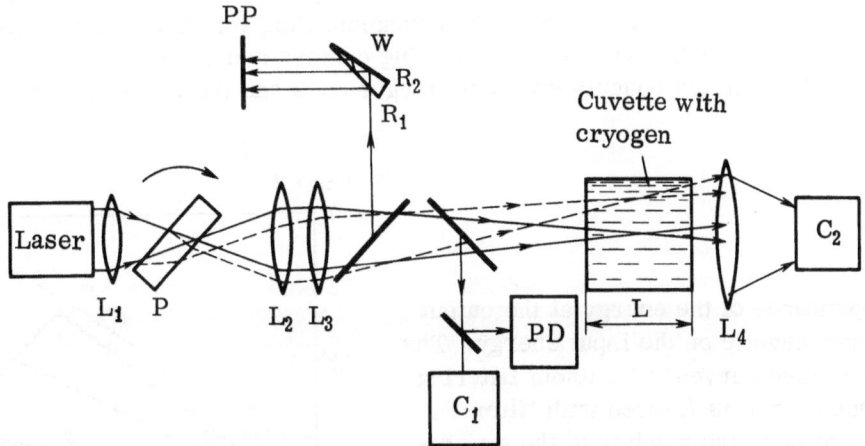

Fig. 1. Block diagram of the experimental apparatus used to study the mechanism of the nonlinear losses of laser light in cryogens. W is a wedge made up of mirrors with reflectivities $R_1 = 50\%$ and $R_2 = 100\%$; L = 50 cm is the length of the active medium.

focused by lens L_3 (focal length $F = 1$ m) into the center of a cuvette containing the active medium. The input (pump) energy was measured with calorimeter C_1 and the energy passing through the medium was measured with calorimeter C_2. A photodiode (PD) was used to measure the duration of the pump pulse. The intensity distribution at the focus of lens L_3 was measured with the mirrored wedge W [27] (serving as a stepwise attenuator) and a photographic plate PP located at the focal plane of lens L_3. Depending on the purpose of the experiment, one of three elements could be placed within the telescope $L_1 - L_2$. These elements were a 3-mm Plexiglas plate, a rotating metal disc with apertures, or a rotating glass plate P (shown in the figure). This made it possible to obtain three interaction regimes of the laser light with the cryogenic fluid.

1. By changing the position of the plastic plate with respect to the focal plane of lens L_1 it was possible to control the duration of the interaction between the radiation and the material. The closer the plate was situated to the focal plane the earlier it was destroyed and scattered the remainder of the energy. The threshold value of the energy density at which destruction occurred was about 40 J/cm^2.

2. The rotating disc with apertures was used to break up the free lasing pulse into five pulses lasting $\tau \simeq 200$ μsec each and following one another by 150 μsec.

3. The rotating glass plate was used to shift the focal spot (of diameter 3 mm) by roughly 10 mm during the free lasing period (τ_p). This made it possible to reduce the interaction time of the radiation by about a factor of 3 in the focal plane where the intensity is greatest.

Figure 2 shows the interaction of light from a neodymium laser with industrial-grade liquid nitrogen without additional filtration (smooth and dashed curves). It is clear that the dependence of the output energy on the input energy is irreversible. The shape of the loop depends on how the input energy was changed. The smooth curve is for experiments in which the energy input to the cuvette was gradually increased until the output energy fell sharply. This value of the input energy corresponds to the first two experiments in the other series, denoted by 1' and 2' on the dashed curve.

After experiment 5 of the first series and experiment 1' of the second, strong scattering of an alignment He−Ne laser beam was observed in the liquid nitrogen. After 24 h the transparency of the liquid nitrogen was recovered.

This dependence of the transparency of liquid nitrogen on the input energy is explained by the presence of fairly heavy particles in the cryogenic fluid which are pulverized by the laser light. This increases the number of scattering centers and reduces the transparency. With time the small particles coagulate to form larger ones and the transparency returns.

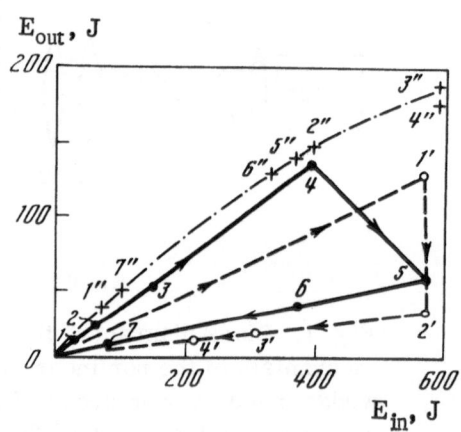

Fig. 2. Dependence of the energy at the output of the nitrogen cuvette on the input energy. The smooth and dashed curves are without filtering; the dot-dashed curve is filtered with filter F_p; 1, 2, 3, etc. refer to the number of the experiment; L = 50 cm, λ = 1.06 μm, τ = 1.4 μsec.

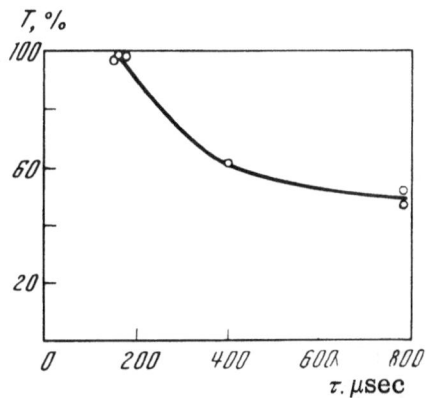

Fig. 3. Dependence of the transparency of liquid nitrogen on the duration of a free-lasing pulse from a neodymium laser. L = 50 cm, λ = λ .06 μm, I = 1.4 MW/cm^2, S_F = 0.12 cm^2.

The dot-dashed curve of Fig. 2 is for experiments with liquid nitrogen purified in a simple way. Only a preliminary cleaning filter F_p was used here.† Clearly, the light losses were reduced and the irreversible reduction in the transparency ceased. However, the nonlinear dependence of the output energy on the input energy remained.

The most probable reason for these light losses is scattering on vapor bubbles which develop about absorption centers. A similar effect was observed in [28, 29] where it is shown that a substantial part of the light flux may be scattered by such bubbles. Our studies confirmed this.

Figure 3 shows the dependence of the transparency of liquid nitrogen on the duration of the pump pulse for a constant intensity‡ I_p = 1.4 MW/cm^2. It is clear that if the pulse interacts with the matter for τ > 0.2 μsec, there will also be a reduction in the transparency.

The variations in the transparency of liquid nitrogen with the radiant intensity I_p for different pumping methods are compared in Fig. 4. The sharpest reduction in the transparency corresponds to the case in which the radiation acted on the material for a time of the order of τ = 1.4 μsec. The curves corresponding to a light pulse of duration τ = 0.2 μsec and a series of similar pulses following one another by 0.15 μsec are almost the same. This means that in this case the time for the liquid nitrogen to regain its transparency is roughly equal to the duration of each pulse.

Pumping so that the focal spot is displaced perpendicular to the propagation direction of the laser light during the duration of the pulse (Fig. 4) makes it possible to transmit an energy in excess of the usual amount by roughly the number of times the displacement of the focal spot exceeds its diameter.

This pumping method reduces the requirements on cleaning the medium and may be used to excite a Raman laser with a series of pulses. For example, a similar method was used later on [30] to excite an acetone Raman laser with periodically repeating pulses. This allowed us to eliminate the thermal defocusing which occurs in the active medium during ordinary pumping.

---

† In later experiments, if it is not specially mentioned, liquid nitrogen purified with a filter F_p made in the form of a 50-cm-long tube filled with pressed metal powder is used.

‡ Here we mean the maximum intensity in the focal spot of the pump without taking the off-duty factor of the free-lasing spikes into account; that is, $I_p = E_p /S_F \tau$, where E_p is the pump energy, τ is the pulse duration, and S_F is the area of the focal spot at the half-maximum intensity level. The maximum values of I_p did not exceed 10-15 MW/cm^2. Further increases caused destruction of the cuvette windows due to increased feedback as induced temperature and Brillouin scattering appear [15].

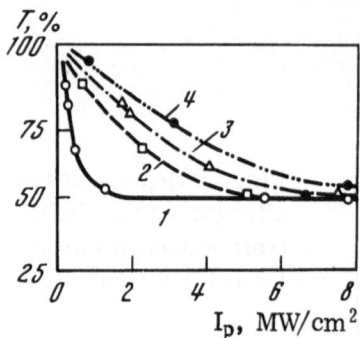

Fig. 4. Dependence of the transparency of liquid
nitrogen on the intensity of a neodymium laser
for various interactions with the material (λ =
1.06 μm, L = 50 cm). 1) A free-lasing pulse τ =
1.4 μsec; 2) free-lasing pulse broken up into five
pulses of duration τ = 0.2 μsec each separated
by 0.15 μsec; 3) free-lasing pulse with τ = 0.2
μsec; 4) free-lasing pulse with τ = 1.2 μsec (with
a rotating glass plate).

3. Nonlinear Losses of Q-switched Laser Light

We also used a Q-switched ruby laser as a pumping source for studying nonlinear losses.
The light pulse duration did not exceed $\tau_p \sim$ 50 nsec. This excluded the possibility of induced
temperature or Brillouin scattering and made it possible to raise the pump intensity to the
levels needed for efficient transformation. The block diagram of Fig. 1 applies in this case
but the telescope L_1, L_2 is absent here.

Figure 5 shows the dependence of the transparency of liquid N_2 and O_2 on the pump intensity. The transparency of the cryogenic fluids decreased when the intensity was increased as
before. However, this dependence was not so distinct for liquid oxygen and the losses were
mainly determined by resonant absorption. In fact, both media were cooled by liquid nitrogen
and at a temperature T = 77°K. In addition, the working chamber was at a pressure slightly
in excess of atmospheric with p $\sim$ 0.1-0.3 atm. As a consequence the oxygen was in an aggre-
gate state roughly 9° from the boiling point, which ensured a weaker dependence of the transparency on the pump intensity.

The results of this experiment were yet another indirect proof that the reduction in the
transparency is due to scattering on vapor bubbles.

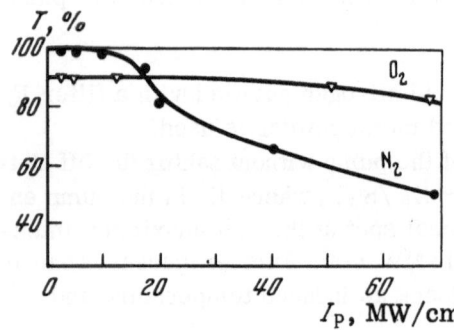

Fig. 5. Dependence of the transparency of liquid
N_2 and O_2 on the intensity of a Q-switched ruby
laser. T = 77°K; p = 0.1-0.3 atm; λ = 0.69 μm;
L = 25 cm.

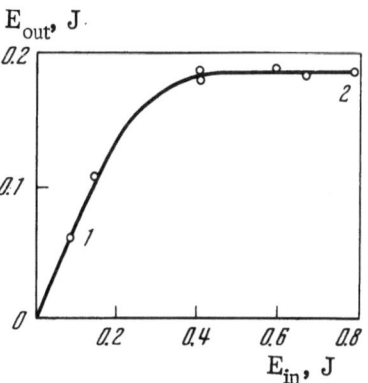

Fig. 6. Energy from a Q-switched ruby laser at the output of a cuvette with unfiltered nitrogen as a function of the input energy. The curve is theoretical, and the points, experimental: $\lambda = 0.69\ \mu m$; $\tau = 25$ nsec, $S_F = 1.2\ cm^2$; $L = 50$ cm.

The expression for the intensity of light passing through the perturbed liquid has the form [29]

$$I(x, t) = \frac{I(0, t)\, \sigma_{abs}^{1/2}\, \exp(-\sigma_{abs}\, nx)}{\{\gamma I^2(0, t)\, [1 - \exp(-2\sigma_{abs}\, nx)]\}^{1/2}}, \qquad (2)$$

where σ_{abs} is the absorption cross section, n is the concentration of impurity particles, and γ is the scattering characteristic.

Figure 6 shows a theoretical curve plotted from Eq. (2) passing through experimental points 1 and 2. It is clear that Eq. (2) describes the transparency of the medium as a function of the intensity quite well.

4. Methods of Reducing the Nonlinear Losses
in Liquid Nitrogen and Liquid Oxygen

4.1. Filtering

The Petryanov and metal-ceramic filters with pore sizes of 20–50 μm that we tested yielded no significant results: the first because of the friability of their structure, and the second because of the excessive pore size. A solid filter element that did not introduce impurities in the purified medium and had a pore size of the order of a fraction of a micron was needed.

To do this we have developed a filtering apparatus (Fig. 7) made up of a demountable Dewar (D) containing coarse (F_1) and fine (F_2) cleaning filters (Fig. 8a). The coarse cleaning filter F_1 is made in the form of a tube,[†] filled with compressed copper powder. The fine cleaning filter is made up of a chamber (1) and a quartz filter element (2) compressed by a spring (3) isolated from the chamber by a bellows (4). Between the bottom washer and the bottom of the chamber there is an indium gasket (5) to prevent the fluid's flowing past the filter element.

The filter element in the fine cleaning filter (Fig. 8b) had pores of diameter d ~ 0.2 μm and was made in the shape of a drinking glass from quartz washers and a disc. The pore size in such a filter is determined by the surface quality of the quartz washers. We used a set of 40 washers with surfaces finished to the 12th class and the following dimensions: outer diameter 60 mm, inner diameter 10 mm, and thickness 2 mm.

---

*Filter F_1 was designed the same way as the preliminary cleaning filter F_p and differed from it only in having a smaller length.

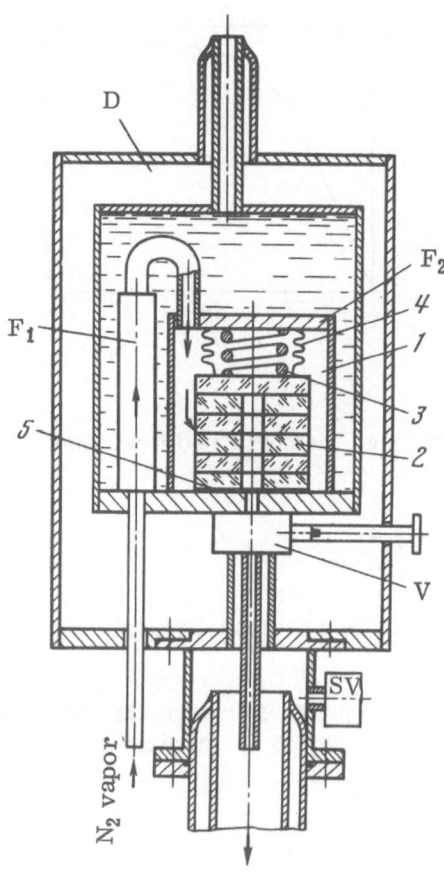

Fig. 7. Design of the filtering apparatus. D, demountable Dewar; V, valve; F_1, coarse filter; SV, safety valve; F_2, fine filter. 1) Chamber; 2) quartz filtering element; 3) spring; 4) bellows; 5) indium washer.

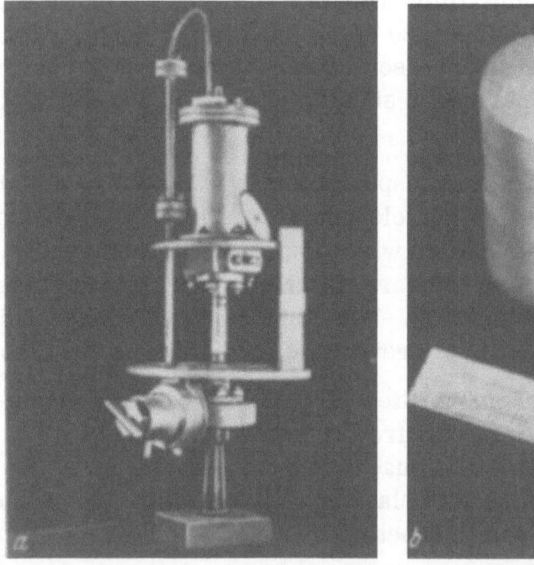
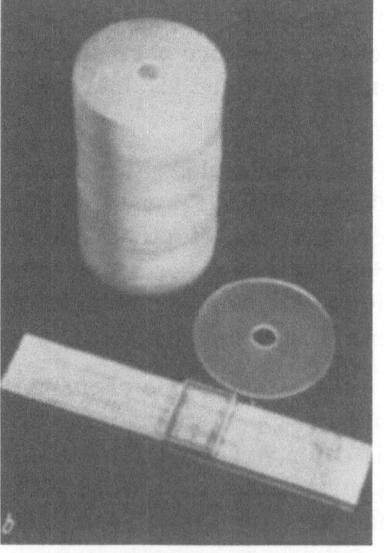

Fig. 8. The filter: a) Coarse and fine cleaning filters; b) quartz filtering element of filter F_2.

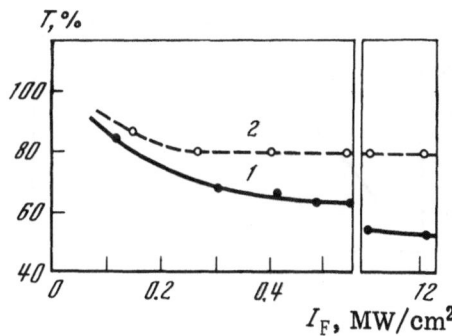

Fig. 9. Transparency of liquid nitrogen as a function of the intensity of a neodymium laser in the free-lasing regime: 1) filtering with filters F_p and F_1; 2) with filters F_p, F_1, and F_2.

In the quartz filter F_2 the cryogenic fluid (N_2, O_2) is driven between the compressed quartz washers at a pressure p ~ 2 atm. Then the liquid nitrogen (oxygen) is purified of undissolved impurities (CO_2, C_2H_2, etc.) and mechanical particles with sizes d $\geqslant$ 0.2-0.5 μm.

One feature of this design is that the valve V through which the cryogenic fluid enters the cuvette is cooled to liquid-nitrogen temperature. Otherwise, a gas with a speed v_g ~ 100 m/sec would flow through a room-temperature valve during filtration and cause the working volume of the cuvette to be filled at a rate v ~ 0.5 liter/h. Then, as we found, the strong flow would polish the valve walls and bring mechanical impurities into the purified material.

The results of filtering liquid nitrogen by the scheme described below (Fig. 20) are shown in Fig. 9. The liquid nitrogen was illuminated by light from a neodymium laser operating in the free-lasing regime. It is clear that at a focal intensity I_F ~ 12 MW/cm² purification with a quartz filter (F_2) made it possible to increase the transparency by more than a factor of 1.5.

4.2. Reducing Nonlinear Radiative Losses by Supercooling

and Excess Pressure†

As our results on irradiating liquid nitrogen and oxygen (Fig. 5) show, the transparency of the medium may be increased by lowering the temperature of the fluid and creating an overpressure in the working chamber. Then vaporization will take place at higher intensities, which is equivalent to an increase in the transparency for a given light flux.

The overpressure was produced by gaseous helium which is initially filtered in the same way as liquid nitrogen. The temperature of the liquid nitrogen was reduced by pumping its vapor from the temperature-controlled chamber. However, it is difficult to keep liquid nitrogen in a supercooled state in this manner for a long time. In practice, we cooled only liquid oxygen, keeping its temperature constant with liquid nitrogen, and only nitrogen was subjected to an overpressure.

Figure 10 shows the dependence of the transparency of liquid nitrogen on the overpressure when a neodymium laser operating in the free-lasing regime acted on it. As the overpressure is increased the transparency increases for a given amount of filtering. For a focal intensity I_F ~ 1 MW/cm², the optimum overpressure is p ~ 0.8 atm.

4.3. Protracted Settling of the Active Medium

During our work we noticed that the transparency of liquid nitrogen increases with time. There are at least two reasons for this:

---

† The breakdown strength of cryogenic cables is increased in similar fashion [31].

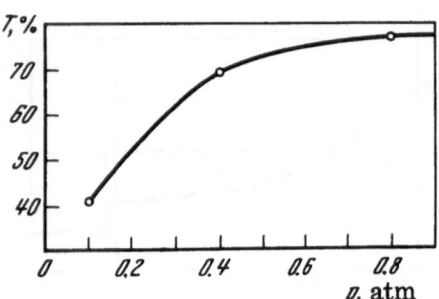

Fig. 10. Typical dependence of the transparency of liquid nitrogen on the pressure above atmospheric.

(a) The coagulation of small particles into larger ones which then settle on the bottom of the working chamber in the form of white flocs. They can be observed visually for several tens of hours after the cuvette is filled. It is necessary to keep the thermostat chamber filled to prevent mixing of the medium due to convective flows.

(b) The solubility of such impurities as CO_2, C_2H_2, and so on, in liquid nitrogen [32].

Both in the first and second cases the number of absorption centers around which vapor bubbles can develop is decreased.

Using a device for automatically controlling the fluid level in the cryostats (see the Appendix), we were able to keep the thermostat chamber filled for a long time (up to several months). This made it possible to examine the time dependence of the transparency of liquid nitrogen.

Some curves which characterize this dependence are shown in Fig. 11.

The liquid nitrogen was irradiated with light from a Q-switched ruby laser ($\tau_p \sim 30$ nsec) at different times after the cuvette was filled. After three days the transparency reached 85%.

These methods made it possible to achieve (to within the measurement error, which did not exceed 5%) 100% transparency of liquid nitrogen for a light flux of intensity $I_p \lesssim 20$ MW/cm^2 ($\tau_p \sim 30$ nsec). As will be shown below, these intensities are sufficient for effective conversion of the laser light at a quantum efficiency close to the limiting value.

Figure 12 shows the dependence of the energy leaving the cuvette and of the transparency of liquid nitrogen on the pump energy and intensity. The experiments were done the day after the cuvette was filled. Gaseous helium was used to produce an overpressure p ~ 2 atm. The active medium was excited in two ways: a parallel beam (data points denoted by a cross) and focusing in the center of the cuvette with a lens. At intensities such that the conversion efficiency is maximal ($I_p \sim 30$ MW/cm^2) the transparency is 90%.

Fig. 11. Dependence of the transparency of liquid nitrogen on the intensity of a Q-switched ruby laser ($\lambda = 0.69$ μm, L = 50 cm, $\tau = 30$ nsec, $S_F = 0.1$ cm^2): 1) on the day the cuvette was filled; 2) after 50 h; 3) after 70 h.

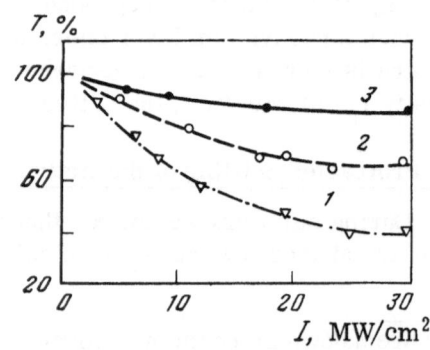

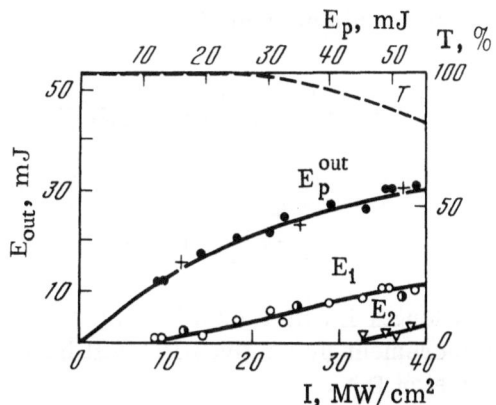

Fig. 12. Nonlinear losses and stimulated Raman scattering in liquid nitrogen ($\lambda = 0.69 \ \mu m$, $L = 50$ cm, $S_F = 0.07 \ cm^2$, $p = 2$ atm); E_1 is the energy of the first Stokes component ($\lambda_1 = 0.83 \ \mu m$); E_2 is the energy of the second Stokes component ($\lambda_1 = 1.03 \ \mu m$); E_P^{out} is the transmitted pump energy ($\lambda_p = 0.694 \ \mu m$). The cavity was filled using a preliminary filter F_p and the filtering apparatus F_1, F_2, and the 2 atm overpressure in the working chamber was produced by helium.

When the active medium was excited with the aid of a lens the intensity I_p was defined as

$$I_p = \frac{P_p}{\frac{\pi}{4} D_0 d_F} = \sqrt{I_0 I_F},$$

where P_p is the pump power, D_0 and d_F are the diameters of the light spots in the plane of the cuvette window and in the focal plane, and I_0 and I_F are the corresponding intensities.

Since in the following we shall take the pump intensity to be given by $I_p = (I_0/I_F)^{1/2}$, we now demonstrate the validity of this equality for the case in which the excited region can be represented as two equal conical frustums (Fig. 13).

From Fig. 13 it is clear that the diameter of the pump beam at different cross sections from the window to the focal plane is given by

$$D(x) = (D_0 - d_F)\left(I - \frac{2x}{L}\right) + d_F.$$

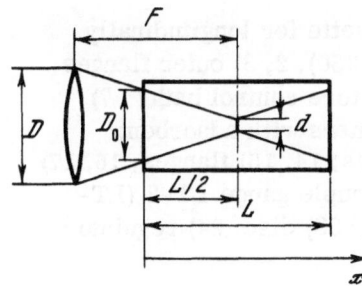

Fig. 13. Illustrating the derivation of the expression for the pump intensity I_p when the medium is excited by conical light beams.

Then the gain increment in a medium excited this way will be

$$b = 2gP_P \int_0^{L/2} \frac{dx}{\pi/4 \, D^2(x)} = \frac{gP_PL}{(\pi/4)D_0d_F}.$$

Thus,

$$I_P = \frac{P_P}{(\pi/4)D_0d_F} = \sqrt{I_0 I_F}.$$

Thus, if we excite a medium of length L with a light beam that has been focused by a lens, then when $D_{eff} = (D_0d_F)^{1/2}$ the gain increment (for a given P_P) will have the same value as for excitation by a parallel beam of diameter D_{eff}.

CHAPTER II

THE DESIGN OF HIGH-POWER RAMAN LASERS

1. Optical Cryogenic Cuvettes

An optical cryogenic cuvette for a high-power Raman laser must satisfy the following requirements:

(1) A sufficiently long working chamber (0.25-1 m)
(2) Isolation of the active medium from the atmosphere
(3) A highly homogeneous active medium to ensure alignment of the optical elements through the cuvette with an accuracy of at least 10^{-4} rad

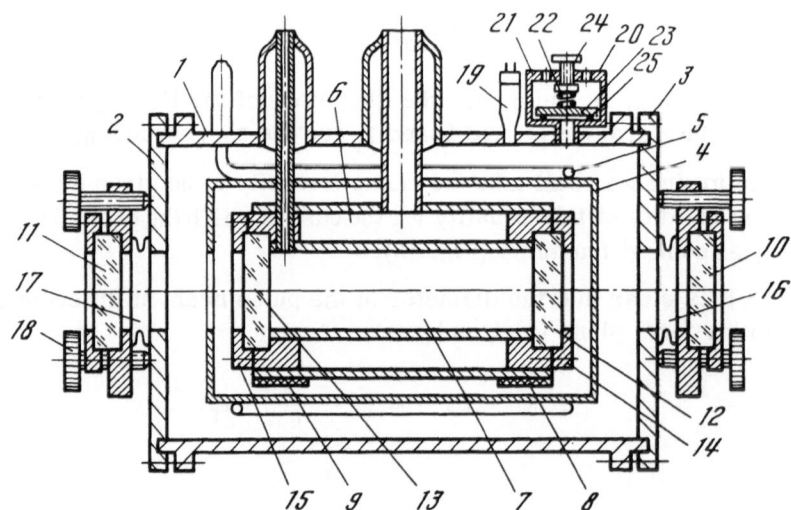

Fig. 14. Design of an optical cryogenic cuvette for longitudinally pumped Raman lasers: 1) Vessel ($\emptyset 110 \times 350$); 2, 3) outer flanges; 4) cooled shield; 5) coil of pipe; 6) temperature control bath; 7) working chamber ($\emptyset 28 \times 250$); 8, 9) containers with adsorbent; 10, 11) outer windows; 12, 13) inner windows; 14, 15) flanges; 16, 17) bellows; 18) adjusting screws; 19) thermocouple gauge LT-2 (LT-4m); 20) safety valves; 21) case; 22) spring; 23) disc; 24) regulator screw; 25) rubber O-ring.

(4) Adjustable external windows

(5) Capability of safe use for a long time

Based on known designs for cryogenic devices [32, 33], we developed special cuvettes for Raman lasers using liquid N_2 and O_2. These cuvettes were designed with these requirements foremost in mind.

Figure 14 shows the design of one of the cuvettes (Fig. 15).

Activated carbon was used as an adsorbent to maintain the vacuum when the working chamber was filled with liquid nitrogen, and silaca gel was used when it was filled with liquid oxygen. The external windows and flanges 2 and 3 were sealed with rubber gaskets, and the inner windows, with 1-mm-thick indium wire [32, 33].

Here we must mention one circumstance of practical importance. In order to seal the inner windows it was best that the flanges 14 and 15 be made of Dural for K-8 glass windows, and of copper for quartz windows.

The case was made of stainless steel. The shield, coil of pipe, and the temperature-control and working chambers were made of copper. The cuvette was pumped to a pressure of about $5 \cdot 10^{-3}$ mm Hg with an absorption pump. The rate at which the cooling fluid was pumped from the temperature-control bath with an uncooled shield was $v \sim 80$ g/h.

The optical homogeneity achieved in the active medium inside the working chamber of the cuvette is illustrated in Fig. 16, which shows images of the target and crosshair of an AKT-400 autocollimator. Both objects were placed in the focal plane of the F = 40 cm lens of the autocollimator and photographed through a cuvette with an active medium of length L = 250 mm. Evidently the images of the objects are clear enough to permit alignment of the optical elements to within the thickness of the lines (i.e., with an angular accuracy of at least 5-$8 \cdot 10^{-5}$ rad).

Figures 17 and 18 show a cuvette intended for studies of transversely pumped lasers. The cuvette is made in the form of a vessel with a cover through which come the filling pipes and from which are suspended the temperature-control baths, the coupling tubes, the working

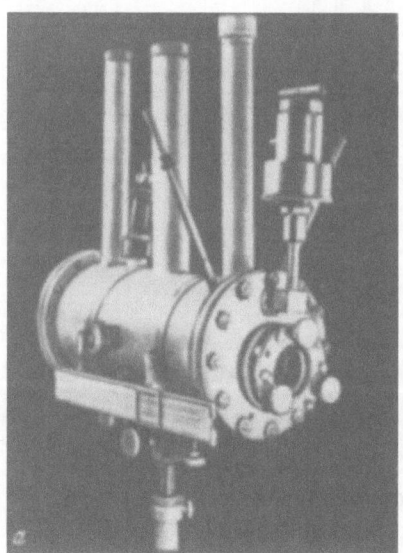

Fig. 15. Optical cryogenic cuvettes for longitudinally pumped Raman lasers: a) diameter 28 mm, length 250 mm; b) diameter 68 mm, length 500 mm.

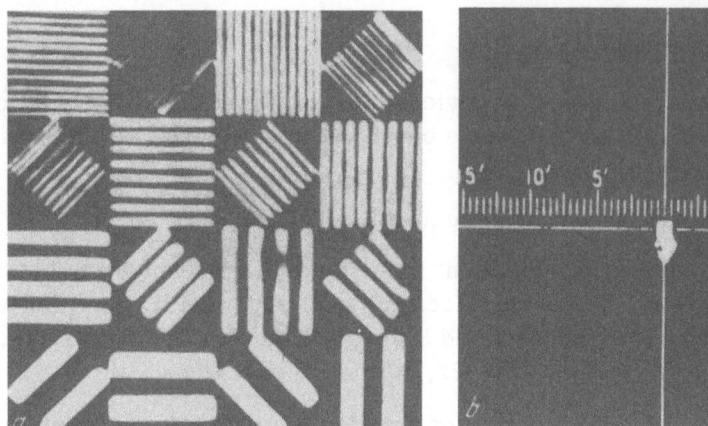

Fig. 16. A viewing target with a maximum of n = 12 lines per milli-
meter (a) and the autocollimator crosshair (b). One division on the
crosshair scale corresponds to 30 sec and the thickness of the line is
10 sec ($5 \cdot 10^{-5}$ rad). The central square has been damaged by laser
light.

chamber, and the shield. Containers of adsorbent are attached to the lower temperature-
control bath. The outer and inner windows were mounted and the inner windows were aligned
as in the previous case.

Filling the cuvette while filtering the liquid nitrogen (oxygen) should ensure a minimum
amount of impurities in the active medium. The extinction coefficient must not exceed $\alpha \sim 10^{-3}$
cm^{-1}; that is, it has to be at least of the same order of magnitude as α for the best optical
glasses.

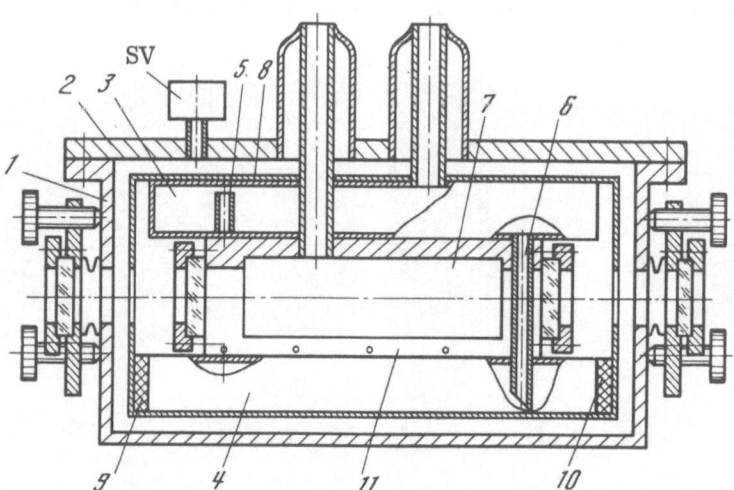

Fig. 17. The design of an optical cryogenic cuvette for transversely pumped Raman lasers: 1) vessel ($150 \times 200 \times 330$);
2) cover; 3, 4) temperature-control baths; 5, 6) pipes for coolant filling; 7) side window; 8) shield; 9, 10) containers of adsorbent; 11) working chamber ($20 \times 20 \times 200$). SV is a safety
valve.

Fig. 18. The optical cryogenic cuvette for transversely pumped Raman lasers. The length of the working chamber is 200 mm and its cross section is 20 × 20 mm.

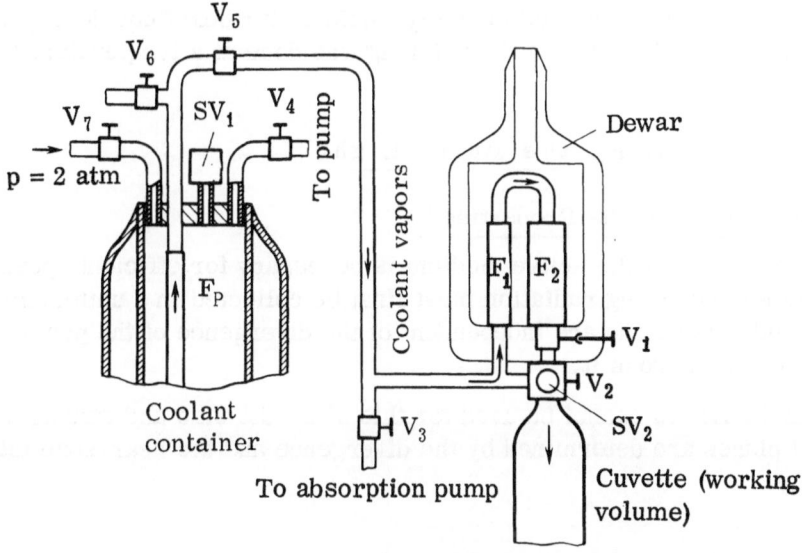

Fig. 19. Block diagram illustrating the filling of the working chamber of the cuvette with liquid nitrogen. V_1-V_7, valves; F_P, F_1, and F_2, preliminary, coarse, and fine cleaning filters; SV, safety valves.

The process of filling the cuvette is illustrated in Fig. 19. At the entrance to the cuvette chamber there is a valve V_2 through which the working chamber of the cuvette is pumped. Above it lies the filter apparatus which consists of a demountable Dewar, the coarse and fine cleaning filters,† and valve V_1. The inlet of filter F_1 is linked to an adsorption pump through valve V_3 and to a preliminary cleaning filter (mounted in a coolant container) through valve V_5. The outlet of filter F_2 is connected to the working chamber through V_1. At the entrances of the container and the working chamber there are safety regulator valves which operate at an overpressure of $p = 0.5-3$ atm.

Filling takes place in the following sequence. The section of pipe up to the closed valve V_5, the filters F_1 and F_2, and the working chamber of the cuvette are pumped through the open valves V_1-V_3. Valves V_2 and V_3 are closed and the Dewar and the temperature-control bath of the cuvette (not shown in the figure) are filled with liquid nitrogen. The latter takes at least 30-40 min to fill, so the inner glass windows are cooled evenly. The coolant vapors are pumped out of the container through valve V_4 over 1-2 min. This causes the liquid nitrogen to cool rapidly, and the impurities dissolved in it (for example, CO_2) are precipitated. This ensures more efficient filtering of the coolant by filter F_P. Then valve V_4 is closed and an overpressure p of about 2 atm is produced by the container through valve V_7. For roughly 5 min the coolant vapors are released to the atmosphere through valve V_6 to wash filter F_P and the unpumped part of the pipe. Then V_6 is closed, V_5 is opened, and the coolant is distilled through filters F_P, F_1, and F_2 into the working chamber of the cuvette. After the working chamber is filled V_1 and V_7 are closed and the Dewar is warmed. Then the coolant from filters F_1 and F_2 is distilled into the container and the excess pressure is released through the safety valve SV_1. About an hour after the temperature reaches equilibrium over the entire volume of the working chamber the cuvette is ready for operation. During operation coolant (N_2) vapor is sent through a coil of pipe 5 on the shield 6 (see Fig. 14) to improve the temperature control of the active medium and thereby the optical homogeneity of the medium.

Calorimetric measurements showed that the extinction coefficient in purified liquid nitrogen was $\alpha \sim 5 \cdot 10^{-4}$ cm^{-1}; that is, to within 5%, the light loss in a medium $L = 100$ cm long was not detectable. Similar results were obtained with liquid oxygen.

The same kind of filters have been used to purify liquid sulfur hexafluoride SF_6 [34] and compressed hydrogen [35]. In the latter case the filtering was done at a temperature $T = 77°K$ to freeze out impurities.

2. Systems for Concentrating the Pump Light

2.1. The Focusing Prism Screen and Its Properties ‡

Spatially uniform excitation of the active medium is necessary for efficient operation of a Raman laser. To do this the exciting radiation must first be collected in a uniformly illuminated spot whose shape and dimensions are independent of the divergence of the pump beam and the distribution of intensity across it.

Spherical lenses and mirrors cannot be used for this since the size and structure of the light spots in their focal planes are determined by the divergence and the near-zone intensity distribution, respectively.

---

† See Chapter I.

‡ Homogeneity of the light field along the length of the active medium is achieved by a square-cross-section light guide whose input lies in the focal plane of the prism screen (see paragraph 2).

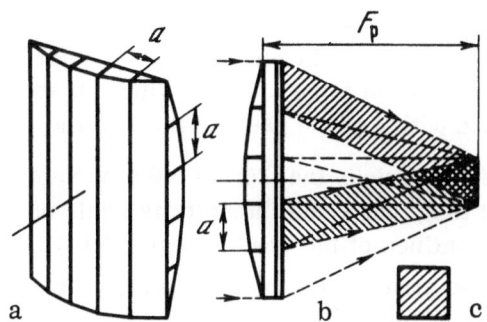

Fig. 20. The design (a), ray paths (b), and focal spot (c) of the focusing prism screen.

To obtain a uniform intensity distribution over the focal spot cross section we used the prism screen described below [36]. This screen structure is made up of two crossed sets of wedges with an edge width a. The profile of each set is part of a polygon inscribed in a circle. The two sets added together form a set of prisms, a prism screen (Fig. 20a).

We now estimate the light intensity in the focal plane of the screen and demonstrate that for certain conditions the size of the focal spot is independent of the directional pattern of the light to be focused. The light incident on the screen is broken up into a group of beams each of which passes through the same place in the focal plane (Fig. 20b) in which a square focal spot is formed (Fig. 20c). The intensity distribution in the spot is practically independent of the inhomogeneities in the light falling on the screen since the random inhomogeneities in the beams are averaged when they are combined. Since the optical effect of all the cells in the prism screen is the same, we shall examine one of them (we neglect the inclination of the beams to the principal optical axis of the screen).

From a cell with side a (Fig. 21) let light of power P with a nonzero directional diagram $f(\varphi, \psi)$ go into a circle $\varphi^2 + \psi^2 \leqslant (\vartheta_m/2)^2$, where $a > \vartheta_m F_p$ (F_p is the focal length of the screen) and $\vartheta_m \gg \lambda/a$ (λ is the wavelength of the incident light). The function $f(\varphi, \psi)$ is symmetric with respect to rotations about the OO_1 axis. The incident radiation can be represented as a superposition of homogeneous parallel beams of square cross section (elementary beams), each of which is characterized by the direction φ and ψ of its symmetry axis, power $\Delta P = f(\varphi, \psi)\,\Delta\varphi\Delta\psi$, and intensity $\Delta I = \frac{1}{a^2}\Delta P$. Evidently $P = \iint f(\varphi, \psi)\,d\varphi d\psi$ and $\varphi^2 + \psi^2 \leqslant (\vartheta_m/2)^2$.

The set of all axes of symmetry of the elementary beams forms a cone with its vertex in the center of the cell and a vertex angle of ϑ_m. The cross section of this cone in the focal plane of the screen is a circle $x^2 + y^2 \leqslant \left(\frac{\vartheta_m F_p}{2}\right)^2$, which is the geometric locus of the centers of all the elementary beams.

Fig. 21. Diagram for calculating the intensity at the focus of the prism screen. ABCD is a cell with side a: O_1O is the axis of the directional diagram of the light leaving the cell; O_1M and O_1O are the symmetry axes of the elementary cells; XY is the focal plane of the screen.

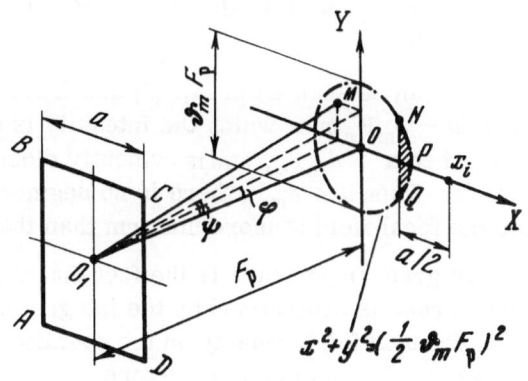

In order to find the intensity I(x, y) of light at a point x, y in the focal plane, it is necessary to sum the intensities of those elementary beams which pass through the given point; I(x, y) is a surface that is symmetric with respect to a rotation of 90°. Let us find, for example, the form of the axial intersection of this surface with a plane passing through the x axis.

For points on the x axis obeying $x \geqslant \dfrac{a + \vartheta_m F_p}{2}$, it is obvious that I(x, 0) = 0 since not even one elementary beam passes through these points. At the point x_i the intensity is made up of those elementary beams whose centers lie within the confines of the shaded region NPQ (Fig. 21) determined by the circle $x^2 + y^2 = \left(\dfrac{1}{2}\vartheta_m F_p\right)^2$ and the straight line x = x_i − (a/2). Transforming to the angular variables

$$\varphi = \frac{x}{F_p}, \qquad \psi = \frac{y}{F_p}, \tag{3}$$

and integrating, we find

$$I(x_i, 0) = \frac{1}{a^2} \iint\limits_{\sigma} f(\varphi, \psi)\, d\varphi\, d\psi, \tag{4}$$

where the region of integration, σ, is expressed in angular variables in accordance with the coordinate transformation (3). In the plane of the variables φ, ψ this region is also a segment formed by the circle $\varphi^2 + \psi^2 = \left(\dfrac{\vartheta_m}{2}\right)^2$ and the line $\varphi = \dfrac{1}{F_p}\left(x_i - \dfrac{a}{2}\right)$.

For $x \leqslant \dfrac{a - \vartheta_m F_p}{2}$ the integral (4) is taken over the entire circle $\varphi^2 + \psi^2 \leqslant \left(\dfrac{\vartheta_m}{2}\right)^2$. In this region of values of x the intensity I(x, 0) is constant and equal to its maximum value

$$I(x, 0)\Big|_{x \leqslant \frac{a - \vartheta_m F_p}{2}} = I_{max}.$$

For x = a/2 the integral is taken over the half circle $\varphi^2 + \psi^2 \leqslant (\vartheta_m/2)^2$ and, evidently, $I\left(\dfrac{a}{2}, 0\right) = \dfrac{1}{2} I_{max}$.

Omitting the subscript on x, we obtain an expression for the intensity in the focal plane:

$$I(x, 0) = \begin{cases} 0 & \text{for } |x| > \dfrac{a + \vartheta_m F_p}{2}, \\[2mm] I_{max} & \text{for } |x| \leqslant \dfrac{a - \vartheta_m F_p}{2}, \\[2mm] \dfrac{1}{2} I_{max} & \text{for } |x| = \dfrac{a}{2}. \end{cases} \tag{5}$$

Figure 22 shows the results of calculations with Eq. (4) and some experimental data obtained by passing neodymium laser light through the screen (Fig. 23). Figure 22a shows (on the same scale) the intensity distribution in the focal plane of a spherical lens with the same focal length.

As can be seen from Eq. (5) and from Fig. 22b and c, the function I(x, 0) has a segment of size $a - \vartheta_m F_p$ over which the intensity is constant [I(x, 0) = I_{max}]. Such a segment always exists when $a > \vartheta_m F_p$, and is evidently closer to the size of the cell, a, the stronger the inequality. When $a \leq \vartheta_m F_p$ there is no segment in which I(x, 0) = const; however, even in this case the focal field is more uniform than that with a spherical lens (mirror).

Of great importance is the fact that the intensity at an arbitrary point of the focal plane of the screen is determined by the integral of the directional diagram $f(\varphi, \psi)$. This means that I(x, 0) depends fairly weakly on the specific form of $f(\varphi, \psi)$, which is especially important when $f(\varphi, \psi)$ has a complicated structure.

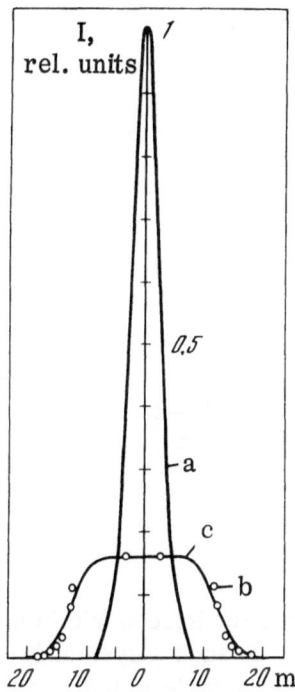

Fig. 22. Intensity distribution in the focal plane of a spherical lens (a) and a prism screen (b, c) with the same focus; a) spherical lens (experiment) with $F_L = 2$ m; b) prism screen (experiment) with $d = 25$ mm and $F_p = 2$ m; c) prism screen [calculated according to Eq. (4) using the directional diagram obtained from curve a].

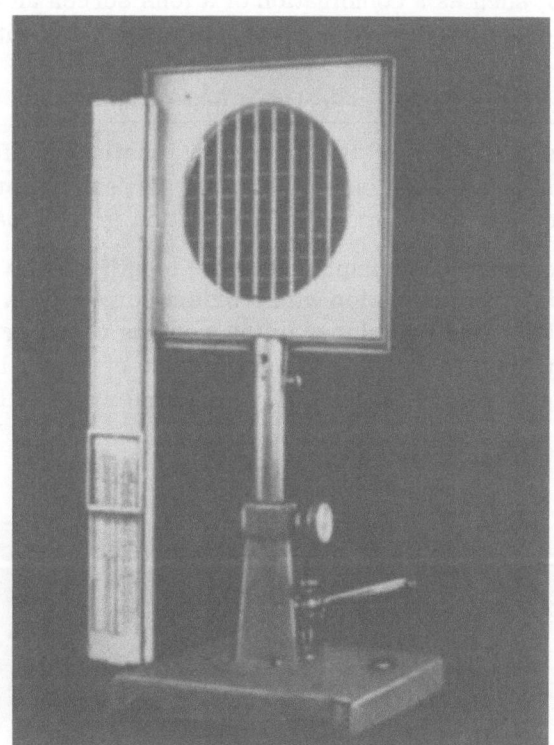

Fig. 23. A focusing prism screen.

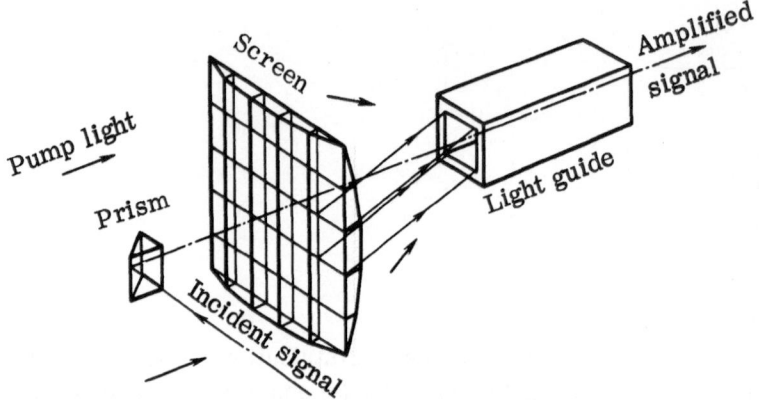

Fig. 24. Spatially uniform pumping of a Raman laser using
a screen in combination with a square-cross-section light
guide located in the active medium.

The dimensions of the focal spot can be changed by displacing the screen within a Galilean
telescope (forward to enlarge the spot and backward to reduce it). In such a system it is possible to smoothly vary the magnification (demagnification) of the light spot from unity to values
equal to the telescope's power.

Instead of the prism screen described here it is possible to use other screen focusing
systems, such as a combination of a lens screen and a lens-collective [37]. However, they are
much more complicated to make and less convenient in practice.

2. 2. The Square-Cross-Section Light Guide

As already noted in this section, spatially uniform pumping is necessary for efficient
operation of a Raman laser. A prism screen only provides a partial solution to this problem;
it creates a uniform intensity distribution in the focal plane and immediately next to it.

Homogeneous pumping over the length may be achieved by using a square-cross-section
light guide in combination with a prism screen (Fig. 24). The light guide was made of silvered
brass plates and was placed in the working chamber of the cuvette. Figure 25 shows the pump
intensity distribution across a light guide of length L = 40 cm. As can be seen from Fig. 25,
such a light guide ensures fairly high uniformity of the light field over a substantial length.

Fig. 25. The intensity distribution over the cross section
at the entrance (1), center (2), and end (3) of the light guide.

CHAPTER III

PRACTICAL DESIGNS FOR RAMAN LASERS

1. A Multiple-Pulse Liquid-Nitrogen Raman Laser

High-power, tunable monochromatic sources which produce a periodic series of pulses
are needed in a number of practical applications. One such laser that uses liquid nitrogen and
is pumped by a neodymium laser has been described in [26].

The pump source was a master oscillator and a multichannel amplifier system. The active element in the oscillator was a phosphate glass rod containing neodymium [38] (L = 240 mm,
$\emptyset$ 20 mm). The cavity consisted of a Fabry–Perot etalon (two mirrors with reflectivities
$R_1 = R_2 = 70\%$ and a gap of 1.5 mm) and a plane-parallel glass plate 0.8 mm thick. A cuvette
containing a dye (No. 3955) dissolved in nitrobenzene was placed in the cavity to further narrow
the spectrum and to control the number of spikes and their duration. A polarizer in the cavity
ensured complete polarization of the oscillator output, as is necessary for effective operation
of the Faraday decouplers between the amplifier stages. These decouplers had a nonmutual
active element consisting of a rod of TF-7 glass mounted in a solenoid. An artificial delay
line (a system of condensers and inductors) was discharged through the solenoid to provide
the required magnetic field when the neodymium laser was operating. The polarizers were
stacks of glass plates mounted at the Brewster angle. The amplifier included a preamplifier
and two parallel output stages. The beams from the output stages were brought together with
prisms and directed onto the screen focusing system. The spectrum of the pump light was
measured with a Fabry–Perot etalon and an IT-28-30 with a gap between the mirrors of 3 cm
and scanned in time with a streak camera.

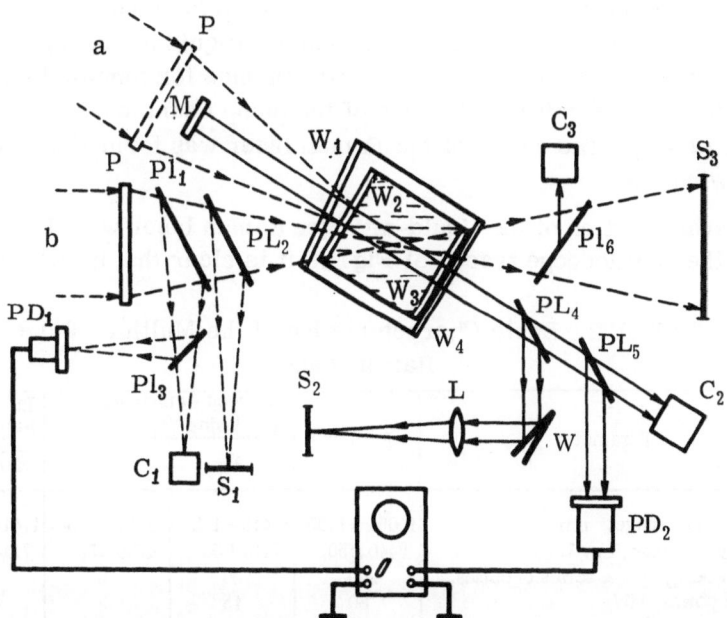

Fig. 26. Diagram of the apparatus for studying a Raman laser
with longitudinal axial (a) and nonaxial (b) pumping: P) screen
focusing system; M) opaque cavity mirror; W_1-W_4) cuvette mirrors; W_3, W_4) "output" mirror of the cavity; W) mirror wedge;
Pl_{1-6}) deflector plates; PD_1, PD_2) photodiodes; C_1-C_3) calorim-
eters; S_1-S_3) screens; O) oscilloscope. The smooth lines correspond to laser light, and the dashed lines, to pump light.

Fig. 27. The Raman laser. R, device for automatic control of the coolant level in the temperature-control bath; SV, safety valves; M, opaque cavity mirror.

Figure 26 shows a diagram of the experimental setup for studying the Raman laser shown in Fig. 27 in an overall view. The active medium (liquid nitrogen) was in a cuvette with a long working chamber (50 cm; see Fig. 15). Two schemes for longitudinal pumping were used: nonaxial (with an angle between the axis of the pump beam and the axis of the cavity of $\alpha \sim 1.5°$) and axial ($\alpha = 0$). The cavity of the Raman laser was formed by the opaque mirror M and the cuvette windows.

The main characteristics of the liquid-nitrogen Raman laser with L = 50 cm are listed in Table 2. From the oscilloscope traces of Fig. 28 it is clear that beyond 200-300 μsec after

TABLE 2. Basic Characteristics of the Multiple-Pulse
Raman Laser

Parameter	Pumping	Induced Raman scattering lasing		Induced Raman scattering per pass
		λ_S	λ_{SS}	
Wavelength range, μm	1.064÷1.053	1.416÷1.39	2.11÷2.067	1.416—1.39
Frequency range, cm^{-1}	9440±50	7114±50	4788±50	7114±50
Total energy of the series of pulses, J	240	36	6	19
Pulsed power, MW	80	15	2	8
Number of spikes	50	30	—	30
Spectrum linewidth, cm^{-1}	0.05	—	—	—
Series duration, μsec	600	500—600	—	500—600
Duration of a single spike, μsec	0.1	0.1	—	0.1
Divergence, mrad	2	0.5—1.5	—	—
Beam area, cm^2	20	1	1	—
Energy density, J/cm^2	12	36	6	—
Energy conversion efficiency, %		15	2.5	8

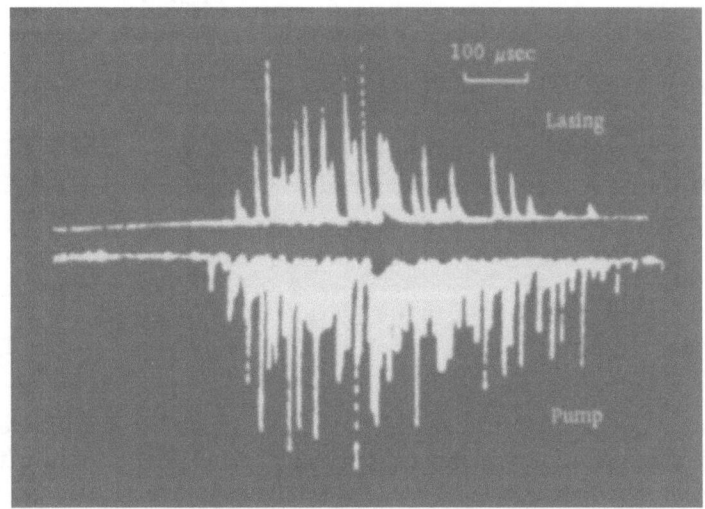

Fig. 28. Oscilloscope traces of the narrow-band pump and
of the induced Raman scattering laser.

the series of pulses begins the laser efficiency begins to decrease in time. The reason for this drop in efficiency is the nonlinear light losses described in Chapter I which lead to a reduction in the transparency of the liquid nitrogen as the intensity and duration of the pumping increase.

The angular distribution of the Raman laser energy is shown in Fig. 29. The divergence was measured with a self-calibrating technique using a mirror wedge [27] which serves as a stepwise attenuator. For a pulse series lasting 200-300 μsec the divergence (at the half energy level) was $5 \cdot 10^{-4}$ rad. When the series was extended to 700 μsec the divergence increased to $1.5 \cdot 10^{-3}$ rad. This is related to a change in the optical characteristics of the active medium with time because it is heated by Stokes radiation losses during efficient conversion of the pump light. When conversion did not occur, no changes in the optical properties of the medium were observed (this was verified using a probe beam with a divergence of $2 \cdot 10^{-4}$ rad). From Table 2 it follows that the diameter of the Raman laser beam was 1 cm. Because of this small diameter and the high directionality of the radiation at a substantial energy conversion efficiency ($\eta = 15\%$), the brightness of the Raman laser was almost two orders of magnitude greater than that of the pump source.

2. An Efficient Raman Laser with Nonaxial Pumping

In [13] it was reported that for saturation of the output signal a Raman amplifier transforms the pump into an amplified signal (at the frequency of the first Stokes component of in-

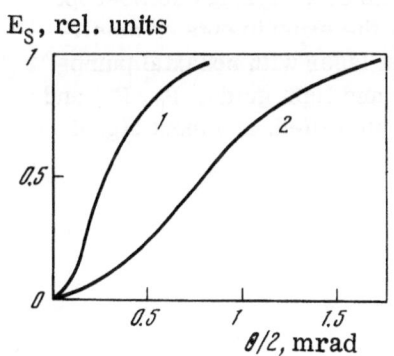

Fig. 29. The fraction of the laser output contained within the divergence angle $\vartheta/2$ as a function of ϑ: 1) t = 0.2 msec; 2) t = 0.6 msec.

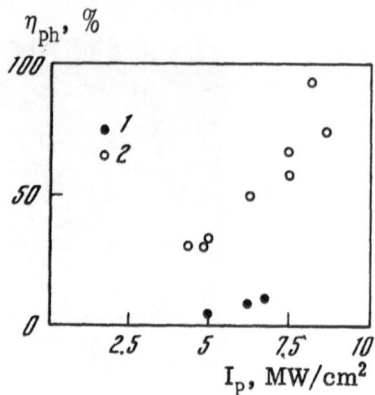

Fig. 30. Dependence of the quantum yield on the pump intensity: I_S is the intensity of the input signal; I_p is the pump intensity; I_S/I_p = 0.03-0.05; 1) nonaxial pumping (α = 6 mrad); 2) axial pumping.

duced Raman scattering) with a quantum efficiency close to 100%. It is, however, important that such efficient transformation is achieved only in an amplifier variant with axial pumping, that is, when the propagation direction of the Stokes signal is practically the same as the pump direction. As can be seen in Fig. 30, the situation changes when the pumping is nonaxial (i.e., if the symmetry axis of the beam is at an angle to the axis of the output beam). The reduction in efficiency is due to a shortening of the length of the active medium within which the Stokes signal and the exciting radiation interact. In practical pumping systems this is usually the situation. Thus, it is of decisive importance to the use of Raman amplifiers for conversion of laser light that high conversion efficiencies be obtained with nonaxial pumping.

The combination of a prism screen and a square-cross-section light guide (Fig. 24) yields spatially homogeneous pumping over a considerable length (L = 40 cm). Then the length over which the pump interacts with an output signal propagating parallel to the axis of the light guide is equal to the length of this light guide for all beams independently of the angle between them and the signal beam. This allows us to come close to ultimate conversion efficiencies at saturated gain, that is, an efficiency nearly equal to that obtained with axial pumping [13]. The results of experiments with an amplifier in the configuration of Fig. 24 are shown in Fig. 31. The pump source was a Q-switched neodymium laser. The pulse lasted 80 nsec and the spectral width of the line was 0.03-0.05 cm^{-1}.

As can be seen from the figure, the peak quantum efficiency reached 90%. The energy efficiency was about 35% in this case, which is also close to the limiting value for this pulse shape.

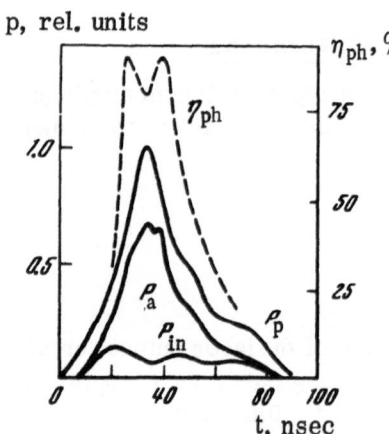

Fig. 31. The results of analyzing oscilloscope traces from one of the experiments on amplification in a Raman laser with nonaxial pumping using a screen and light guide. P_p, P_a, and P_{in} are the pump, amplified, and input signal powers.

3. Tunable Compressed Hydrogen Raman Laser and Its Application

3.1. The Pump Source

The pump source was a neodymium laser (Fig. 32) with a mean output frequency of $\nu = 9460$ cm^{-1} ($\lambda = 1.057 \mu$m) and a tuning range of ± 50 cm^{-1}. The master oscillator (with a Kerr cell Q-switch) worked by locking onto the frequency of the narrow-band output of a control oscillator [39]. The output of the master oscillator was first amplified by a system of pre-amplifiers and then by a two-channel final stage. To prevent self-excitation of the system, Faraday-effect decouplers (described in Section 1 of this chapter) and bleachable dye filters were used. The output beam had a pulse energy of up to 100 J and lasted 50 nsec. Part of the output of the master oscillator was frequency doubled with a KDP crystal and used to monitor the spectral width of the line with an STÉ-1 spectrograph and a Fabry—Perot interferometer. In all the experiments the line width was less than 0.05 cm^{-1}.

3.2. Design and Parameters of the Raman Laser

The design of the Raman laser is shown in Fig. 33. The active medium was hydrogen compressed at a pressure of p = 50-60 atm. The frequency of the vibrational transition in hydrogen is $\Omega = 4155$ cm^{-1}. Thus, lasing could occur in this case only at the two Stokes vibrational components with frequencies $\nu_S = 5305 \pm 50$ cm^{-1} ($\lambda_S = 1.867$-1.903 μm) and $\nu_{SS} = 1150 \pm 50$ cm^{-1} ($\lambda_{SS} = 8.333$-9.091 μm). It should be noted that the tuning range for the second Stokes component is about 10% of the fundamental frequency.

The pump radiation was directed onto the cuvette containing the compressed hydrogen through a glass window (clear aperture 70 mm) behind which was placed a focusing system (prism screen). This screen collected the beam incident on it in the focal plane (25 cm away) in a square spot (1 · 1 cm^2) with a uniform intensity distribution. This spot was created at the

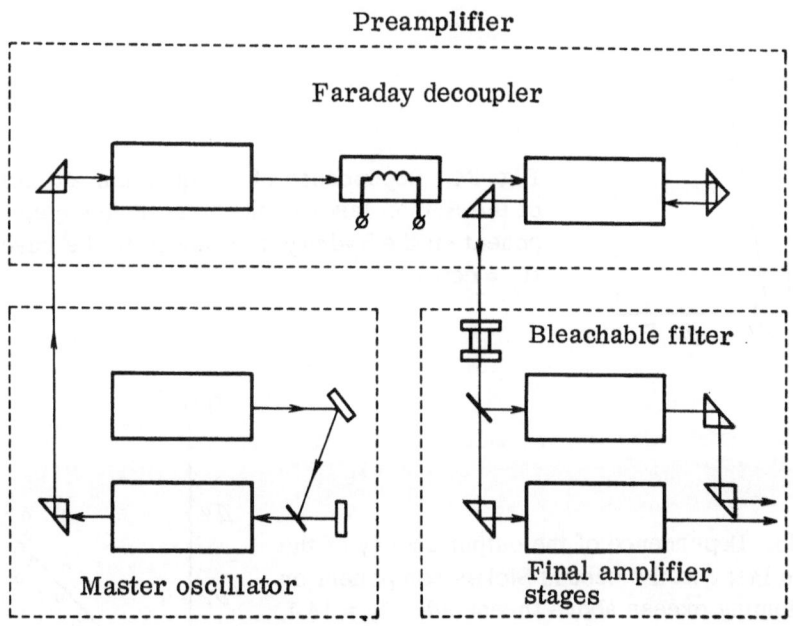

Fig. 32. A block diagram of the pump source. The parameters of the output light beam are: $E_p \approx 100$ J, $\tau_p \approx 50$ nsec, $\Delta \nu_p \leq 0.05$ cm^{-1}.

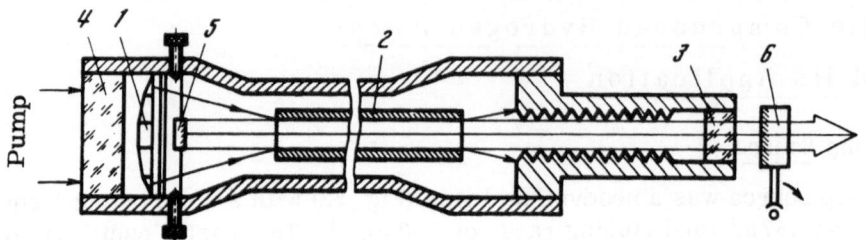

Fig. 33. A schematic diagram of the compressed hydrogen Raman laser: 1) prism screen; 2) light guide; 3) output window of BaF_2; 4) input window of K-8; 5) adjustable copper mirror; 6) selective window of BaF_2.

input of a metal light guide of the same cross section and 70 cm long. Use of the light guide made it possible to substantially increase the length of the region with a high pump intensity over its perpendicular cross section together with a uniform intensity distribution. Behind the screen there was an adjustable plane copper mirror in the shape of a 1-cm square. The 20-mm-diameter BaF_2 output window was located 40 cm from the output end of the light guide (so that the pump light would not fall on the BaF_2, which has a damage threshold of about 100 MW/cm^2). This window served as the second mirror in the cavity when efficient lasing at the first Stokes component was required.

To obtain efficient conversion to the second Stokes component a selective mirror on a BaF_2 substrate with reflectivities for the first and second Stokes components of $r_S \simeq 100\%$ and $r_{SS} = 10\%$, respectively, was mounted next to the output window. The use of a cavity which "traps" the first Stokes component reduced the threshold pumping intensity for the second component by more than an order of magnitude.

The dependence of the quantum efficiency for lasing in the first Stokes component on the hydrogen pressure in the cuvette is shown in Fig. 34. Subsequent experiments were done at a pressure of p = 50-60 atm. The energy characteristics of lasing at the second Stokes component are shown in Fig. 35. The high pump intensity led to significant conversion of pump light

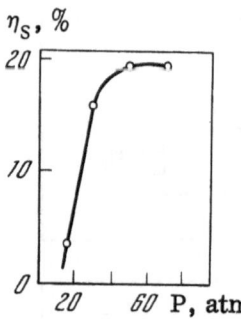

Fig. 34. Dependence of the quantum efficiency of the Raman laser at the first Stokes component on the hydrogen pressure in the cuvette (I_p = const).

Fig. 35. Dependence of the output energy of the Raman laser at the second Stokes component on the pumping excess above threshold: $E_p = 14$ J, $I_p = 0.35$ GW/cm^2.

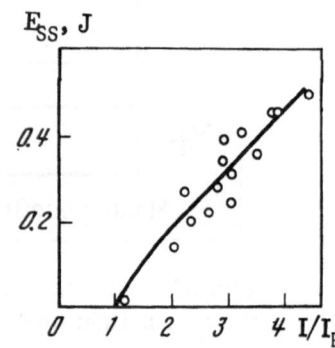

into the first Stokes component propagating along with the pump light (but not along the axis of the cavity) and to the appearance of antistokes components. These effects reduced the efficiency of the Raman laser.

The duration of the Stokes components was not measured. Previous experiments done under analogous conditions with Raman lasers using other materials and the following calculations lead us to suppose that there is a substantial shortening of the laser pulses compared to the pump.

Measurements of the divergence of light at the second Stokes component showed that 90% or more of the entire energy is concentrated in an angle $\theta_{SS} = 1.5 \cdot 10^{-3}$ rad.

3.3. Theoretical Analysis

The equations for a Raman laser operating at two Stokes components with an active medium filling the entire cavity have been given in [1, 2, 13]. In a real laser the active region is usually shorter than the cavity length. Including this fact, the system of equations for the number of photons has the form

$$\frac{dY}{d\xi} + Y = [1 - \exp(-b_S Y)] X(\xi) - \frac{L}{l} \frac{\omega_{SS}}{\omega_S} \frac{b_{SS}}{\ln(1/r_{SS})} YZ,$$

$$\frac{dZ}{d\xi} + \frac{\ln(1/r_{SS})}{\ln(1/r_S)} Z = \frac{L}{l} \frac{\omega_{SS}}{\omega_P} \frac{b_{SS}}{\ln(1/r_S)} YZ,$$

where $X(\xi)$ is the dimensionless intensity of the pump wave at the entrance to the cuvette containing the active medium, and Y and Z are the dimensionless wave intensities of the first and second Stokes components at the cavity outlet. These quantities are given by $X = I_P(\xi)/I_P^{max}$, $Y = \frac{\omega_P}{\omega_S} I_S(\xi)/I_P^{max}$; $Z = \frac{\omega_S}{\omega_{SS}} I_{SS}(\xi)/I_P^{max}$, and I_P^{max} is the maximum intensity of the exciting light pulse; $\xi = t/\tau_S$ is the dimensionless time, where $\tau_S = L/v \ln(1/r_S)$ is the lifetime of a Stokes photon with a propagation velocity v in a cavity of length L with an output mirror of reflectivity r_S; $b_{S,SS} = g_{S,SS} I_P^{max} l/\ln(1/r_S)$, where $g_{S,SS}$ is the gain coefficient (cm/MW) for the first and second Stokes components, respectively, and l is the length of the active medium ($l \neq L$).

These equations were solved on a computer with constants characteristic of the experimental conditions: $g_S = 10^{-3}$ cm/MW; $g_{SS} = 8 \cdot 10^{-5}$ cm/MW; L = 130 cm; l = 70 cm; r_S = 0.92 (as mentioned above we used an "opaque" cavity for the first Stokes component so its emission did not leave the cavity); r_{SS} = 0.2; and the pump pulse was taken to have the form $X(\xi) = \exp(-3\xi^2)$ which corresponds to a dimensional pulse length of 50 nsec which is equal to the lifetime of a Stokes photon in the cavity. From this it is clear that we are dealing with nonstationary lasing.

The results of the calculations are given in Fig. 36a-d. The variable parameter in this case is the maximum intensity of the pump pulse, which enters in the growth rate $G = g_S I_P^{max} l$. It should be noted that because of the nonstationary lasing there is a substantial narrowing of the second Stokes component pulse compared with the pump pulse.

Figure 37 shows the dependence of the quantum efficiency for the second Stokes component on the amount of pumping above threshold. The same figure includes the corresponding experimental curve obtained by analyzing the data of Fig. 35. As can be seen from Fig. 37, the calculations agree fairly well with experiment. This indicates that the theoretical model is rather close to the actual physical system.

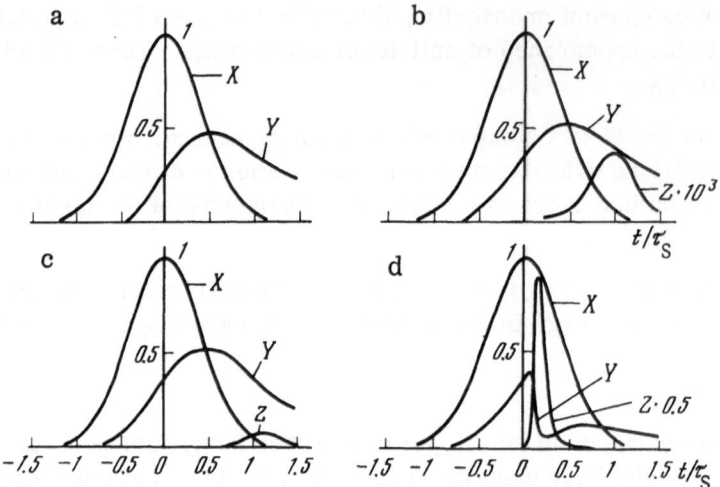

Fig. 36. Dynamics of the onset of lasing in a Raman laser at two Stokes components: τ_S is the lifetime of a first Stokes component photon in the cavity; G is the growth rate (amplification) for the first Stokes component; a) G = 6; b) 20; c) 25; d) 60.

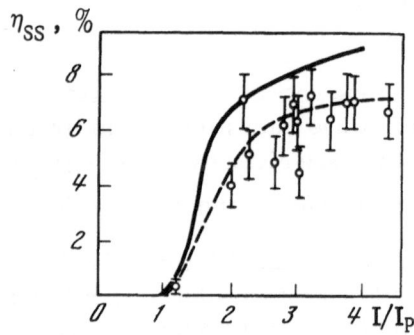

Fig. 37. Quantum efficiency of a Raman laser at the second Stokes component as a function of the pumping excess above threshold. The smooth curve gives the theoretical values, and the points, the experimental values.

3.4. Experiments with a Tunable Raman Laser

A substantial power, high monochromaticity, and continuously tunable output frequency over a relatively wide spectral interval make the Raman laser described above a convenient instrument for various kinds of physical research. As an example, we now present some measurements of one of the absorption spectrum lines of water vapor in the atmosphere near a frequency of $\nu = 5300$ cm^{-1}. The measurements were made over a path length of 4 m. During the experiments the humidity at the experiment site was monitored with a hygrometer.

A transmission spectrum of atmospheric water vapor taken on an SF-8 spectrophotometer is shown in Fig. 38. Also shown there is one of the absorption lines of water vapor obtained using light from a Raman laser operating at the wavelength of the first Stokes component. The width of the laser line in this case was less than $5 \cdot 10^{-2}$ cm^{-1}, which is almost an order of magnitude less than the observed absorption line width $\Delta\nu_{H_2O} = 0.3$ cm^{-1}. Thus, we could obtain the absolute value of the absorption coefficient in reciprocal centimeters. At the peak of the line it is $K_{H_2O} = 8 \cdot 10^{-3}$ cm^{-1} for a water vapor density of $\rho_{H_2O} = (15.5 \pm 1)$ g/m^3.

It should be noted that for the laser to operate over the entire tuning range it is necessary that the active medium (in this case hydrogen) be carefully filtered by the method described in Chapter I. The point is that the first Stokes component (as can be seen from Paragraph 4) lies in an absorption band of atmospheric water vapor (5300 cm^{-1}). In some portions

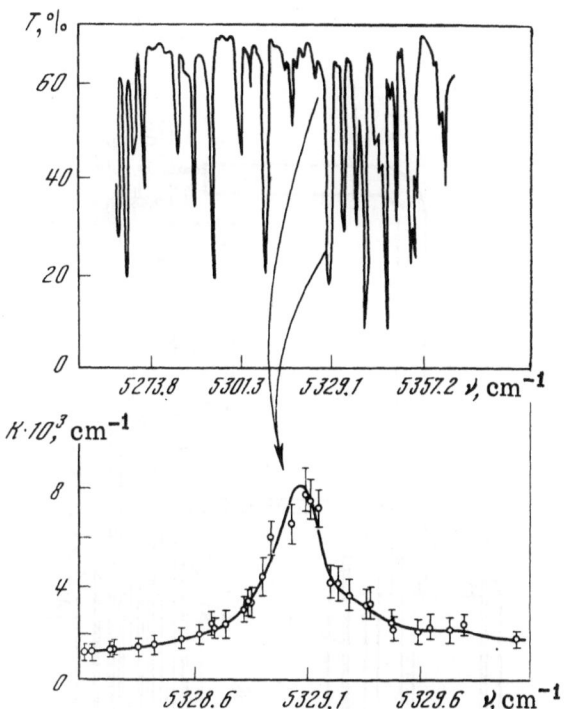

Fig. 38. Transmission spectrum of atmospheric water vapor and the shape of one of the absorption lines of atmospheric water vapor ($\rho_{H_2O} = 15.5 \pm 1 \ g/m^3$).

of this band (Fig. 38), the absorption coefficient reaches about $10^{-2} \ cm^{-1}$ at a vapor density of $\rho = 15.5 \ g/m^3$, which corresponds to normal atmospheric pressure. However, the pressure of the hydrogen in the Raman laser cuvette is 50-60 atm. The density of water vapor in compressed hydrogen that has not been specially filtered at T = 77°K may be so large that absorption in it may lead to disruption of lasing at the corresponding line.

APPENDIX

AN APPARATUS FOR AUTOMATICALLY CONTROLLING THE COOLANT LEVEL IN CRYOSTATS

The need to construct an automatic device which would monitor the coolant level in the temperature-control chamber of the cuvette and periodically fill the chamber was dictated by a number of reasons:

(1) The process of preparing the laser medium, including assembly, washing, and pumping out the cuvette and then filling it with liquid nitrogen, was a fairly long and tedious process, usually taking two or three working days.

(2) Any cryostat undergoes a finite number of filling cycles.

(3) Only a completely filled temperature-control chamber would ensure high optical uniformity of the working material.

(4) Because the coolant settled out for many days, its transparency was increased.

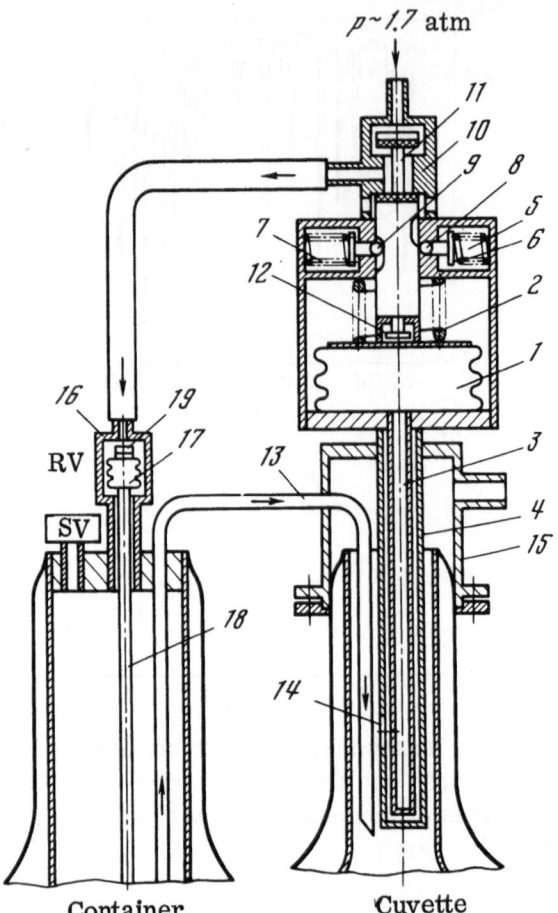

Fig. 39. Design of the apparatus for automatically controlling the liquid nitrogen level in cryostats: SV, safety valve; RV, relief
valve (made up of items 16-19); 1) working
chamber (bellows); 2) spring; 3) level sensor; 4) sensor shield; 5) catch; 6, 7) springs;
8, 9) metal balls; 10) distributor chamber;
11) piston; 12) plunger; 13) pipe for filling
with coolant; 14) opening in the sensor shield;
15) cover; 16) case; 17) bellows; 18) level sensor; 19) plug.

Automatic devices for periodic refilling of cryostats are described in [40, 41]. However,
their range of applicability is limited because they work only with cryostats having wide entrances ($\phi \sim 10$ cm). If the level sensor of one of these devices is placed in a narrow cryostat
(ϕ 1-2 cm), then the flow of vapor as the cryostat is being filled, which is greater the narrower
the port, causes this kind of automatic apparatus to respond falsely. In addition, the design in
[40] is rather complicated. It includes three working chambers and three level sensors.

We have suggested and developed an apparatus which automatically regulates the coolant
level in almost any laboratory cryostat, either with wide or narrow entrances [42] (Fig. 39).

The working chamber was filled with gaseous nitrogen. An overpressure $p \sim 2$ atm is
produced in it. Then the piston is moved to its upper stable position and is held by the ball

pressed by the spring. The upper valve to the distributor chamber is open, and the lower, closed.

The automatic apparatus works as follows. A level sensor mounted in a shield is lowered into the cryostat port along with the tube 13 for filling with the coolant. The other end of the tube goes to the bottom of a hermetically sealed liquid nitrogen container. Gaseous nitrogen at a pressure of p ~ 0.7 atm is driven into the container from the vessel through a reducer, the upper valve of the distributer chamber 10, and the relief valve RV. When the coolant level reaches the opening 14 in the shield 4, the level sensor is cooled. Gas from the working chamber condenses in it, and pressed by spring 2 the driver 12 is engaged with the piston 11 and shifts it to the lower stable position which is set by ball 8. The upper valve of the distributor chamber 10 is closed and the lower, opened. The pressure in the container falls to atmospheric and filling ceases. The next filling cycle begins after the time required for the sensor to warm up after the coolant level falls below the screen 4.

Figure 40 shows the automatic controller (dimensions ⌀ 60 × 170). The level sensor is detachable. Its length is determined by the design of the cryostat and the operating regime chosen. The features of this device include a screen around the sensor and a position fixer which establishes two stable positions for the piston in the distributor chamber corresponding to two coolant levels in the cryostat. In addition, in this design the plunger 12 in the distributor chamber is coupled to the piston 11 only when the temperature of the sensor is close to either the temperature of liquid nitrogen or room temperature. All this prevents the device from responding falsely when the sensor is cooled by vapors or by random drops of coolant which may fall on it when the cuvette is being filled.

Sometimes the design of the cryostat does not permit lowering the sensor directly into the chamber. Then it is placed in the entry port of the cryostat (see Fig. 39) and the automatic

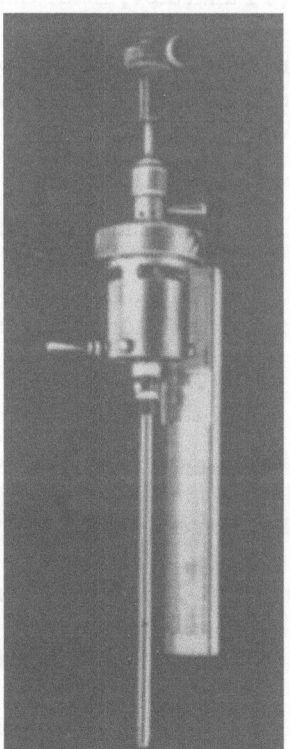

Fig. 40. Apparatus for automatically controlling the liquid-nitrogen level in cryostats.

apparatus operates as a timing relay. In this case the frequency of the filling cycles is determined by the material and dimensions of the sensor tube. When a copper sensor is used, filling takes place every 10-15 min. A tin-walled brass sensor makes it possible to fill the chamber 2-3 times in a 24-hour day.

The method of placing the sensor 3 and the filling tube 13 in the cryostat port is very significant, as is the presence of a safety (SV) and relief (RV) valve at the entrance to the nitrogen container. If the level sensor and the filling tube are lowered into a cryostat with a narrow entrance, moisture will condense on them and result in the formation of an ice plug. This can be avoided with a cover 15 with valve for escape of nitrogen vapor which condenses on the entrance to the cryostat, as shown in Fig. 39.

The excess pressure from the container is bled to the atmosphere through the safety valve SV if the pressure suddenly or accidentally goes above $p = 1$ atm.

The relief valve RV is intended to close the inlet aperture through which the excess pressure flows into the container when the coolant in the container is consumed. This prevents gaseous nitrogen from being driven through tube 13 into the temperature-controlled chamber of the cuvette.

The relief valve RV (Fig. 39) is of simple design. The bellows and level sensor are filled with gaseous nitrogen at a pressure $p \sim 2$ atm.

The automatic apparatus described here kept the cuvette in a filled (working) state for several months and, in addition, was used for periodic filling of the nitrogen traps, adsorption pumps, and Dewars while we were preparing the working material.

LITERATURE CITED

1. N. G. Basov, E. P. Markin, A. N. Oraevskii, and A. V. Pankratov, Dokl. Akad. Nauk SSSR, 198:1043 (1971).
2. N. G. Basov, A. N. Oraevsky, and A. V. Pankratov, in: Chemical and Biochemical Applications of Lasers, C. B. Moore, ed., Academic Press, New York (1974), Chapter 7, p. 207.
3. N. G. Basov, E. P. Markin, A. N. Oraevskii, and A. V. Pankratov, Pis'ma Zh. Éksp. Teor. Fiz., 14:251 (1971).
4. R. V. Ambartsumian and V. S. Letokhov, Appl. Opt., 11:354 (1972).
5. R. V. Ambartsumyan, V. S. Letokhov, E. A. Ryabov, and N. V. Chekalin, Pis'ma Zh. Éksp. Teor. Fiz., 20:597 (1974).
6. R. V. Ambartsumyan, V. S. Letokhov, G. N. Makarov, and Yu. A. Gorokhov, 69:1956 (1975).
7. V. S. Letokhov and V. P. Chebotaev, Principles of Nonlinear Laser Spectroscopy [in Russian], Nauka, Moscow (1975).
8. A. Z. Grasyuk, Kvant. Élektron., 1:485 (1974).
9. A. Z. Grasyuk, Tr. FIAN, 76:75 (1974).
10. A. Z. Grasyuk, V. F. Efimkov, I. G. Zubarev, V. I. Mishin, and V. G. Smirnov, Pis'ma Zh. Éksp. Teor. Fiz., 8:471 (1968).
11. V. V. Bocharov, M. G. Gangardt, A. Z. Grasyuk, I. G. Zubarev, and E. A. Yukov, Zh. Éksp. Teor. Fiz., 57:1585 (1969).
12. A. Z. Grasyuk, V. I. Popovichev, V. V. Ragul'skii, and F. S. Faizullov, Kvant. Élektron., No. 1, p. 70 (1971).
13. A. Z. Grasyuk, I. G. Zubarev, V. I. Mishin, and V. G. Smirnov, Kvant. Élektron., No. 5(17), p. 25 (1973).
14. A. Z. Grasyuk, I. G. Zubarev, and N. V. Suyazov, Pis'ma Zh. Éksp. Teor. Fiz., 16:237 (1972).

15. M. G. Gangardt, A. Z. Grasyuk, and I. G. Zubarev, Kvant. Élektron., No. 6, p. 120 (1971).

16. V. V. Bocharov, A. Z. Grasyuk, I. G. Zubarev, and V. F. Mulikov, Zh. Éksp. Teor. Fiz., 56:430 (1969).

17. N. G. Basov, A. Z. Grasyuk, and V. A. Katulin, Dokl. Akad. Nauk SSSR, 161:1306 (1965).

18. N. G. Basov, A. Z. Grasyuk, V. F. Efimkov, and V. A. Katulin, Fiz. Tverd. Tela, 9:88 (1967).

19. J. B. Grun, A. K. McQuillan, and B. P. Stoicheff, Phys. Rev., 180:61 (1969).

20. W. R. L. Clements and B. P. Stoicheff, Appl. Phys. Lett., 12:8 (1968).

21. A. Z. Grasyuk, O. A. Logunov, and V. G. Smirnov, Fiz. Tekh. Poluprovodn., 1:1502 (1967).

22. V. A. Zubov, A. V. Kraiskii, M. M. Sushchinskii, M. I. Fedyanin, and I. K. Shuvalov, Zh. Éksp. Teor. Fiz., 57:1585 (1969).

23. P. D. McWane, and D. A. Sealer, Appl. Phys. Lett., 8:278 (1966).

24. N. B. Vargaftik, Handbook on the Thermophysical Properties of Gases and Liquids [in Russian], Fizmatgiz, Moscow (1963).

25. V. I. Dianov-Klokov, Opt. Spektrosk., 4:448 (1958).

26. V. V. Bocharov, A. Z. Grasyuk, I. G. Zubarev, A. V. Kotov, and V. G. Smirnov, Kvant. Élektron., 1:2185 (1974).

27. V. V. Ragul'skii and F. S. Faizullov, Opt. Spektrosk., 27:707 (1969).

28. G. A. Askaryan, A. M. Prokhorov, and G. P. Shipulo, Zh. Éksp. Teor. Fiz., 44:2180 (1963).

29. A. A. Chastov and O. L. Lebedev, Zh. Éksp. Teor. Fiz., 58:848 (1970).

30. J. G. Meadors and M. A. Poirier, IEEE Quant. Electron., 8:427 (1972).

31. I. P. Fastovskii, ed., Cryogenic Techniques [in Russian], Énergiya, Moscow (1974).

32. Handbook on the Physical and Technical Foundations of Cryogenics [in Russian], Énergiya, Moscow (1973).

33. N. B. Delone, ed., Bubble Chambers [in Russian], Atomizdat, Moscow (1963).

34. A. Z. Grasyuk, I. G. Zubarev, and S. I. Mikhailov, Kratk. Soobshch. Fiz., No. 2, p. 3 (1976).

35. A. Z. Grasyuk, I. G. Zubarev, A. V. Kotov, S. I. Mikhailov, and V. G. Smirnov, Kvant. Élektron., 3:1062 (1976).

36. A. Z. Grasyuk, V. F. Efimkov, and V. G. Smirnov, Prib. Tekh. Éksp., No. 1, p. 174 (1976).

37. I. A. Valyus, Raster Optics [in Russian], Mashinostroenie, Moscow (1969).

38. A. Z. Grasyuk, I. G. Zubarev, and V. F. Mulikov, Zh. Prikl. Spektrosk., 15:806 (1971).

39. I. G. Zubarev and S. I. Mikhailov, Kvant. Élektron., 1:625 (1974).

40. B. D. Holt, Rev. Sci. Instrum., 33:121 (1962).

41. V. P. Bykov and V. N. Kostryukov, Prib. Tekh. Éksp., No. 3, p. 154 (1969).

42. A. Z. Grasyuk and V. G. Smirnov, Prib. Tekh. Éksp., No. 3, p. 245 (1975).

THE LIMITING CHARACTERISTICS OF POWER
RESONANCES IN RING GAS LASERS

V. A. Alekseev and L. P. Yatsenko

The ultimately attainable widths and contrasts of power resonances in ring gas lasers are considered. It is shown that the width of the resonances depends on both the concentration of the absorbing gaseous medium and on the macroscopic system parameters, such as the coupling of the traveling waves through back reflection, the laser intensity, and the cavity Q. The resonance width is found to have a nonlinear dependence on the gas density in the absorbing cell.

Introduction

At present there are two ways of using ring gas lasers in precision physical instruments. The first approach, the use of a ring laser as a gyroscope, has been realized, and a large number of theoretical and experimental papers deal with the limiting characteristics of these devices (see, for example, [1, 2]). The other approach, the use of a ring gas laser as an optical standard and for high-resolution spectroscopy, is presently being developed intensively. Already in the first work [3] on this topic it was observed that the power resonance of a ring gas laser makes it possible to determine the center of a transition line of an absorbing medium with great accuracy. There the width of the resonance was a tenth of the width of the absorption line of the medium. (The experiment was done on a He−Ne laser with a methane absorption cell.) Use of this resonance to stabilize the frequency of a He−Ne laser has made it possible to obtain stability of the order of 10^{-13} [4, 5]. Later [6-8] it was observed that there are generally two resonances in the power at frequencies ω near the frequency ω_{0+} of the center of the laser (amplified) transition − one at the center of the amplified transition, ω_{0+}, and the other at the center of the absorbing transition, ω_{0-}. The intensities of these resonances are related to the coupling coefficients of the traveling waves due to volume scattering [6, 8] or to the reflection coefficient of one of the waves at an auxiliary mirror outside the ring cavity [7]. In work on the frequency stabilization of ring lasers the most accurate superposition of the resonances at the centers of the amplifying and absorbing media is usually achieved by varying the pressure of the active medium, and then the maximum contrast and minimum width of the resonance are sought by varying the pressure of the absorbing medium [6].

This paper is devoted to a study of the dependence of the location of the centers of the resonances and of their limiting widths and contrasts on the pressure of the absorbing and amplifying media, the coupling coefficients, the laser intensity, and other macroscopic parameters.

The Equations for the Oscillations of a Ring Laser
Near the Center of the Absorption Line

In describing single mode lasing in a ring laser we shall proceed from the customary equations for the amplitudes E_1 and E_2 and the phases φ_1, φ_2 of the opposite waves (see, for example, [1]):

$$\dot{E}_{1,2} + \frac{\omega}{2Q_{1,2}} E_{1,2} = -2\pi\omega\varkappa_{1,2}'' E_{1,2} \mp m_{1,2}\sin(\Phi + \xi_{1,2}) E_{2,1}, \tag{1}$$

$$\dot{\varphi}_{1,2} = -2\pi\omega\varkappa_{1,2}' + \Omega_{1,2} - \omega - \frac{E_{2,1}}{E_{1,2}}\frac{m_{1,2}}{2}\cos(\Phi + \xi_{1,2}), \quad \Phi = \varphi_1 - \varphi_2. \tag{2}$$

Here $\omega + \varphi_{1,2}$ is the frequency of the traveling waves, $Q_{1,2}$ is the cavity Q, $\varkappa_{1,2}' + \varkappa_{1,2}''$ is the nonlinear polarization of the medium, $m_{1,2}$ and $\xi_{1,2}$ are the amplitude and phase of the coupling coefficients of the traveling waves, $\hat{m}_{1,2} = m_{1,2}e^{\pm i\xi_{1,2}}$, and $\Omega_{1,2}$ is the eigenfrequency of the cavity. In the following we shall examine the lasing regimes in a relatively narrow range of frequencies near the center of the amplified line, $|\omega - \omega_{0+}| \ll \gamma_+$, where γ_+ is the homogeneous width of the amplification line. In this frequency region we may assume that the total intensity of the traveling waves is constant and introduce a new variable $\psi = \psi(t)$ related to the intensities of the traveling waves by

$$E_1^2 = J\cos^2\frac{\psi}{2}, \qquad E_2^2 = J\sin^2\frac{\psi}{2}, \qquad J = E_1^2 + E_2^2 = \text{const.} \tag{3}$$

Using the substitution (3) and the explicit form of the polarization of the medium to terms of fifth order in the field (inclusively) [1], we obtain the following equations for ψ and Φ:

$$\frac{d\psi}{d\tau} = p\sin\psi\cos\psi - \delta\sin\psi + \frac{1}{2}[\tilde{m}_1\sin(\Phi + \xi_1) + \tilde{m}_2\sin(\Phi + \xi_2)] - $$
$$- \frac{1}{2}\cos\psi[\tilde{m}_1\sin(\Phi + \xi_1) - \tilde{m}_2\sin(\Phi + \xi_2)], \tag{4}$$

$$\frac{d\Phi}{d\tau} = \Omega - \Delta\cos\psi - \frac{1}{2}\left[\tilde{m}_1\tan\frac{\psi}{2}\cos(\Phi + \xi_1) - \tilde{m}_2\cot\frac{\psi}{2}\cos(\Phi + \xi_2)\right]. \tag{5}$$

Here $\tau = (\omega I/2Q)t$ is the dimensionless time, $I = (E_1^2 + E_2^2)\frac{d_+^2}{2\hbar\gamma_+\Gamma_+}$ is the dimensionless laser intensity, d_+ is the dipole moment of the transition in the amplifying medium, Γ_+ is the longitudinal relaxation constant, $\hbar$ is Planck's constant, $\Omega = (\Omega_1 - \Omega_2)\frac{2Q}{\omega}\frac{1}{I}$ is the frequency nonreciprocity of the cavity, $\delta = 2\frac{Q_1 - Q_2}{Q_1 + Q_2} > 0$, $\tilde{m}_{1,2} = m_{1,2}\frac{2Q}{\omega}\frac{1}{I}$,

$$p = \frac{\omega_+^2}{\gamma_+^2} - \frac{\gamma_+\Gamma_+}{(\Delta\omega_D)^2} + \theta_+ I - \mu\xi\left(\frac{\omega_-^2}{\omega_-^2 + \gamma_-^2} + \theta_-\xi I\right), \tag{6}$$

$$\Delta = \frac{\omega_+}{\gamma_+} - \mu\xi\frac{\omega_-\gamma_-}{\omega_-^2 + \gamma_-^2}. \tag{7}$$

The plus and minus signs denote parameters characterizing the amplifying and absorbing media, $\Delta\omega_D$ is the Doppler linewidth, $\frac{\gamma_\pm}{\Delta\omega_D} \ll 1$, $\mu = \frac{N_-L_-}{N_+L_+}\frac{d_-^2}{d_+^2}$, $N_\pm$ are the differences in the levels populations, $L_\pm$ are the lengths of the cuvettes with amplifying and absorbing media, $\xi = \frac{d_-^2}{d_+^2}\frac{\gamma_+\Gamma_+}{\gamma_-\Gamma_-}$, and $\omega_\pm = \omega - \omega_{0\pm}$.

From the requirement that the pumping exceed the threshold value, we find $\mu\xi < 1$. In the following we shall everywhere assume that the inequalities γ_-, $\Gamma_- \ll \gamma_+$, Γ_+ are satisfied.

The parameters $\theta_\pm$ are nonzero near the center of the line $|\omega_\pm| \lesssim \gamma_\pm$. Their specific values in this region depend on the ratio between the longitudinal and transverse relaxation times and lie within $\theta \simeq 0.2\text{-}0.3$.

The ratio between the coupling coefficients of the waves $m_{1,2}e^{\pm i\xi_{1,2}}$ depends on the specific experimental conditions [1]. When volume scattering on fluctuations in the real part of the dielectric constant ε is the main cause of coupling the traveling waves, these coefficients are complex conjugates: $m_1 = m_2$, $\xi_1 = \xi_2$. In the case of volume scattering on fluctuations in the imaginary part of ε and also in those cases in which the waves are coupled due to diffraction at the cavity mirrors, the coupling coefficients are anticomplex conjugate: $m_1 = -m_2$, $\xi_1 = \xi_2$. When the frequency of the gas laser is being stabilized for peak power of one of the waves, an auxiliary mirror outside the cavity is often used to ensure reflection of one of the waves into the other [7]. In this case it is natural to assume that one of the coupling coefficients is much greater than the other.

Power Resonances at Low Intensity

When there is no coupling between the traveling waves $m_{1,2} = 0$ and the condition $\frac{\gamma_+ \Gamma_+}{(\Delta\omega_D)^2} - \theta_+ I + \mu\xi > 0$ is fulfilled, there is a range of frequencies ω_+ near the center of the amplification transition line for which the stability requirement for lasing in a single traveling wave, $p < 0$, is satisfied. The frequency region for which this regime is realized is easily found from Eq. (6) to be

$$|\omega_+| \leqslant \sqrt{\chi\gamma_+}, \quad \chi = \frac{\gamma_+ \Gamma_+}{(\Delta\omega_D)^2} - \theta + I + \mu\xi. \tag{8}$$

When the cavity Q for both waves is the same ($Q_1 = Q_2$) the damping of one of the waves in region (8) is due to random causes. Thus, in an experiment an artificial difference in the cavity Q's is usually introduced so that the frequency region (8) is realized for the traveling wave with the greater Q. In the following we shall assume that $Q_1 > Q_2$ so that the wave E_2 is damped in region (8). Thus, for $m_{1,2} = 0$ the lasing is described by the solution

$$\psi = 0, \quad \Phi = (\Omega - \Delta)\tau. \tag{9}$$

Therefore, for sufficiently small coupling coefficients we may assume that the condition $\psi \ll 1$ is fulfilled and in this approximation find a stationary solution to Eqs. (4) and (5). Setting $\dot\psi = \dot\Phi = 0$, $\sin\psi \simeq \psi$, and $\cos\psi \simeq 1$, we find from Eqs. (4) and (5) that

$$\psi^2 = \frac{m_2^2}{(\Delta - \Omega)^2 + (p - \delta)^2}, \quad \cos(\Phi - \xi_2) = -\frac{\Delta - \Omega}{\sqrt{(\Delta - \Omega)^2 + (p - \delta)^2}}. \tag{10}$$

To study the shape of the resonance of the weak wave described by Eq. (10) we shall first ignore the narrow-frequency band $|\omega_-| \lesssim \gamma_-$. For $|\omega_-| > \gamma_-$ the parameters p and Δ may be assumed to be given by

$$p = \frac{\omega_+^2}{\gamma_+^2} - \chi, \quad \Delta = \frac{\omega_+}{\gamma_+}. \tag{11}$$

From Eq. (10) for ψ^2 it is clear that in this case the weak wave has a narrow power resonance near the point $\Delta = \Omega$, so the dependence of p on ω_+ in the resonance region may be neglected. Finally we obtain

$$E_2^2 \simeq J\frac{\psi^2}{4} = Jg\frac{\tilde\Gamma^2}{(\omega_+ - \Omega\gamma_+)^2 + \tilde\Gamma^2}, \tag{12}$$

$$g = \frac{1}{4}\frac{\tilde m_2^2}{(\chi - \Omega^2 + \delta)^2}, \quad \tilde\Gamma = (\chi - \Omega^2 + \delta)\gamma_+. \tag{13}$$

We note first that, as can be seen from Eqs. (10) and (13), the intensity of the resonance in the weak wave is determined only by the reflection coefficient m_2 of the strong wave into the weak wave and is independent of the ratio of the coefficients m_2 and m_1. A similar lasing regime was observed experimentally in [7].

The resonance (12) has a Lorentz shape with a width $\Gamma \simeq \chi\gamma_+$ much less than the homogeneous linewidth, and when there is no frequency nonreciprocity of the cavity (i.e., $\Omega = 0$), it exactly matches the center of the amplified transition line. The difference in the Q for the opposed waves causes a small variation in the resonance width which, obviously, is fairly difficult to observe experimentally. The frequency displacement of the cavity leads to a more important effect, a shift in the maximum of the resonance. Thus, for example, when $2Q/\omega = 10^{-7}$ sec, $\Omega_1 - \Omega_2 = 10^4$ sec^{-1}, which is 10^{-3} of the width of a cavity mode, and $I = 10^{-2}$. The shift in the line peak relative to the center of the amplifying transition is equal to $\Delta\omega_+ = 0.1\gamma_+$. We recall that the entire region in which lasing occurs on waves with unequal intensities is of order $\delta\omega_+ = 2\sqrt{\chi}\gamma_+$ and, for example, when $\chi = 0.1$, $\delta\omega_+ \simeq 0.6\,\gamma_+$. Thus, for these parameters the shift $\Delta\omega_+ = \frac{1}{6}\delta\omega_+$ will increase as the intensity is reduced even for small saturated power levels. A shift in the resonance relative to the center of the amplified line is easily detected experimentally since because of this effect the resonance (12) is shifted to one of the edges of the region where lasing with unequal intensities occurs [7, 9]. Yet another possible reason for this shift is discussed in [7].

We now consider the shape of the resonance at frequencies $\omega_- \lesssim \gamma_-$. The presence of an absorbing cell leads to a rapid change in this region in the values of p and Δ which causes a narrow structure to develop against the background of the relatively wide profile (12). If the parameter $\mu\xi$ and the ratio γ_-/γ_+ are of the same order of magnitude (i.e., $\mu\xi \simeq \gamma/\gamma_+$), this structure is rather complicated. More realistic, however, is the case $\mu\xi \gg \gamma_-/\gamma_+$. In this case the changes in p and Δ in the region $|\omega_-| \lesssim \gamma_-$ due to the amplifying medium may be neglected and the intensity of the weak wave may be written in the form

$$E_2^2 = \frac{1}{4} J \frac{\bar{m}_2^2}{\left(\dfrac{\omega_+ - \omega_-}{\gamma_+} - \Omega - \mu\xi\,\dfrac{\omega_-\gamma_-}{\omega_-^2 + \gamma_-^2}\right)^2 + \left(\chi' + \mu\xi\,\dfrac{\omega_-^2}{\omega_-^2 + \gamma_-^2}\right)^2}, \tag{14}$$

$$\chi' = \frac{\gamma_+\Gamma_+}{(\Delta\omega_D)^2} - \frac{(\omega_+ - \omega_- - \Omega\gamma_+)^2}{\gamma_+^2} - (\theta_+ - \xi\theta_-)\,I. \tag{15}$$

A simpler form of Eq. (14) results when the inequalities

$$\frac{\omega_+ - \omega_-}{\gamma_+} - \Omega \ll \mu\xi, \qquad \chi' \ll \mu\xi \tag{16}$$

hold. In this case we may set $\dfrac{\omega_-\gamma_-}{\omega_-^2 + \gamma_-^2} \simeq \dfrac{\omega_-}{\gamma_-}$ and $\dfrac{\omega_-^2}{\omega_-^2 + \gamma_-^2} \simeq \dfrac{\omega_-^2}{\gamma_-^2}$ in the denominator of Eq. (14). We finally obtain

$$E_2^2 = \frac{1}{4} J \frac{\bar{m}_2^2\,\dfrac{\gamma_-^2}{(\mu\xi)^2}}{(\omega_- - \alpha\gamma_-)^2 + \left(\dfrac{\chi'}{\mu\xi} + \alpha^2\right)^2 \gamma_-^2}, \tag{17}$$

$$\alpha = \frac{\omega_+ - \omega_- - \Omega\gamma_+}{\gamma_+\mu\xi}.$$

Thus, when conditions (16) are satisfied a narrow resonance of width $\left(\dfrac{\chi'}{\mu\xi} + \alpha^2\right)\gamma_- \ll \gamma_-$ appears against a background of a relatively wide profile (12). When $\alpha = 0$ this resonance locates the center of the absorption line exactly.

In using the power resonance (17) for stabilizing the laser frequency, an important characteristic is the contrast of the resonance $a = [(E_2^2)_{max} - (E_2^2)_{bgnd}]/(E_2^2)_{bgnd}$. To estimate this quantity, we note that condition (16) implies $\chi \simeq \mu\xi$, so that $\chi \gg \chi'$. Thus the contrast of the resonance (17) against the background of the wider profile (12) is $a = \left(\frac{\chi'}{\mu\xi} + a^2\right)^{-2} \gg 1$. We note that the width of the resonance which locates the center of the absorption line of the medium is equal to γ_- and its contrast is $a \simeq \mu\xi/2 \ll 1$. Thus, the width of the resonance in this operating regime of a ring laser is much less and its contrast is much greater than the corresponding features of the resonance in the standing wave regime. This allows us to hope that the frequency of a ring laser may be stabilized more simply and more accurately than in a laser with a Fabry–Perot cavity.

As can be seen from Eq. (17), unequal transition frequencies of the amplifying and absorbing media or the existence of a frequency nonreciprocity in the cavity $\Omega \neq 0$ result in a shift in the resonance with respect to the center of the line in the absorbing medium by $\Delta\omega = \alpha\gamma_- \ll \gamma_-$. From the standpoint of using ring lasers as frequency standards this type of shift in the resonant frequency from the center of the absorption line is extremely undesirable since it causes the lasing frequency to depend on the macroscopic parameters of the system. Thus, in stabilizing the laser frequency the shift $\Delta\omega = \alpha\gamma_-$ must be reduced to a minimum. It is clear from Eqs. (17) and (12) that the shift does not occur ($\alpha = 0$) when the centers of the absorption line and resonance (12) coincide exactly. It is interesting that when this is so the resonance (12) itself depends on the frequency nonreciprocity of the cavity and may not coincide with the center of the amplifying line. Therefore, the centers of the resonances (17) and (12) may be made to coincide in two ways: by changing the pressure of the active medium (the method usually used now) and by changing the frequency nonreciprocity of the cavity. Obviously it can be expected experimentally that this overlapping of the line centers can be realized to an accuracy of the order of 10^{-4} of the width of resonance (12), which in this case is equal to $\chi\gamma_+ \simeq \mu\xi\gamma_+$. Then $\alpha = 10^{-4}$ and $\Delta\omega = 10^{-4}\gamma_-$. When a laser with a Fabry–Perot cavity is used to stabilize the frequency, the analogous shift in the power resonance from the center of the absorption line caused by imprecise overlapping of the frequencies ω_+ and ω_- is $\Delta\omega = \frac{\gamma_-}{\mu\xi\Delta\omega_D} \frac{\omega_+ - \omega_-}{\Delta\omega_D}\gamma_-$. When $\frac{\omega_+ - \omega_-}{\Delta\omega_D} \simeq 10^{-4}$, the shift is $\Delta\omega = 10^{-4}\frac{\gamma_-^2}{\mu\xi\Delta\omega_D} \ll 10^{-1}\gamma_-$. Thus, the resonance (17) is very much more sensitive to imprecise overlapping of ω_- and ω_+ than the resonance of a laser with a Fabry–Perot cavity.

We recall that the resonance has the simple form (17) only when condition (16) is satisfied. As the parameters $\chi' \gtrsim \mu\xi$ and $\alpha \gtrsim 1$ increase, Eq. (17) ceases to be applicable. In this case we can see from the more accurate expression (14) that the width of the resonance is of order γ_-, the contrast is $a \simeq (\mu\xi/\chi')^2 \lesssim 1$, and the shift is $\Delta\omega \simeq \gamma_-$. Resonances with these parameters would be awkward for frequency stabilization.

The Shape of the Power Resonances for Strong Nonreciprocal Coupling of the Opposing Waves ($m_2 \gg m_1$)

It can be seen from Eqs. (12) and (13) that the intensity of the power resonance of the weak wave increases as the reflection coefficient of the strong wave into the weak one increases. In practice this is often used to increase the intensity of the resonance and is done by mounting the external mirror which partially reflects the strong wave on the weak wave [5]. It may seem strange that the wave E_2 remains weak even when $m_2 \gg m_1$. To explain this effect let us examine the frequency region $\omega_+ \simeq \pm\sqrt{\chi\gamma_+}$, in which the transition takes place from standing-wave lasing to lasing with waves of greatly different intensities.

At frequencies such that $\omega_+ \simeq \pm\sqrt{\chi\gamma_+}$ there may exist three different stationary solutions corresponding to three different values of ψ: $|\psi^{(1)} - \pi| \ll 1$, $\psi^{(2)} \ll 1$, and $\left|\psi^{(3)} - \frac{\pi}{2}\right| \ll 1$.

Then the solutions $\psi^{(1)}$ and $\psi^{(2)}$ describe the lasing regimes in which the waves E_1 and E_2 are weak, respectively, and $\psi^{(3)}$ describes lasing with traveling waves of almost equal intensity. For large coupling coefficients the solution (10) obtained in the previous section is generally inapplicable (the condition $\psi \ll 1$ is violated). However, as can be seen from Eqs. (12) and (13), the condition $\psi \ll 1$ or the similar condition $|\pi - \psi| \ll 1$ is satisfied, as before, in the frequency range $\omega_+ \simeq \sqrt{\chi \gamma_+}$, where the intensity of the weak wave is small for practically any coupling coefficients. Thus we may assume that in the region $\omega_+ \simeq \sqrt{\chi \gamma_+}$ the solution $\psi^{(2)}$ is described by Eqs. (12) and (13) and that $\psi^{(1)}$ is obtained from Eqs. (12) and (13) on making the substitutions $E_2 \to E_1$, $\tilde{m}_2 \to m_1$, $\Omega \to -\Omega$, and $\delta \to -\delta$. A study of the stability of $\psi^{(1)}$ and $\psi^{(2)}$ shows that $\psi^{(1)}$ is stable for $p < -\delta$ and $\psi^{(2)}$ is stable for $p < \delta$. It is also easy to find the stability region of the laser regime $\psi^{(3)}$. For $\tilde{m}_{1,2} \ll \delta$, the solution $\psi^{(3)}$ is stable for $p > \delta$ and for $\tilde{m}_2 \gg \delta$, $\tilde{m}_1$ is stable when $p > -\delta \left(1 - \frac{\tilde{m}_2 - \tilde{m}_1}{|\Delta|}\right)$.

Therefore, the solution $\psi^{(3)}$ becomes unstable in the region $p > -\delta$ where only the solution $\psi^{(2)}$ is stable. Thus, in the region $-\delta\left(1 - \frac{\tilde{m}_2 - \tilde{m}_1}{|\Delta|}\right) < p < \delta$ there is a transition from the lasing regime $\psi^{(3)}$ to $\psi^{(2)}$. As the parameter p is reduced further when the inequality $p < -\delta$ begins to be satisfied, both $\psi^{(2)}$ and $\psi^{(1)}$ become stable; however, there is no reason for $\psi^{(2)}$ to go into $\psi^{(1)}$. For $p < \delta$, the wave E_2 is still weak although $m_2 \gg m_1$. The actual reason for this situation is that the transition from regime $\psi^{(3)}$ into regime $\psi^{(2)}$ occurs in the frequency region $\omega_+ \simeq \pm \sqrt{\chi}\gamma_+$, where the intensities of the weak waves are practically zero in both regime $\psi^{(2)}$ and regime $\psi^{(1)}$; that is, backscatter generally does not play a role, and the transition from $\psi^{(3)}$ to $\psi^{(2)}$ occurs only because of the difference in the Q-factors, δ.[†]

We now consider the shape of the resonance in the E_2 wave when the coupling coefficient $\tilde{m}_2$ becomes sufficiently large that the condition $\psi \ll 1$ is violated. We shall assume that $\tilde{m}_1 = 0$, $\tilde{m}_2 = \tilde{m}$, and for simplicity we set $\Omega = 0$ and $\delta = 0$.[‡] Equations (4) and (5) take the form

$$\frac{d\psi}{d\tau} = p \sin \psi \cos \psi + \tilde{m} \sin \varphi \cos^2 \frac{\psi}{2}, \qquad (18)$$

$$\frac{d\varphi}{d\tau} = -\Delta \cos \psi + \tilde{m} \frac{\cos \varphi}{\sin \psi} \cos^2 \frac{\psi}{2}, \qquad (19)$$

$$\varphi = \Phi + \xi_2.$$

For the stationary solution of this system we obtain

$$\tan \varphi = -\frac{p}{\Delta},$$

$$\frac{1}{4} \frac{\tilde{m}^2}{p^2 + \Delta^2} = \frac{x(1 - 2x)^2}{1 - x} = f(x), \qquad (20)$$

$$x = \sin^2 \frac{\psi}{2}.$$

For $x \ll 1$, $f(x) \simeq x$ and solution (20) transforms to Eq. (10). The form of the function $f(x)$ is shown in Fig. 1. From this figure it is clear that for sufficiently small values of the parameter $q = \frac{1}{4} \frac{\tilde{m}^2}{p^2 + \Delta^2} < 0.09$ there are three solutions of Eq. (20). A stability analysis shows, however, that for $P < 0$ the only stable solution is x_1 which for $q \ll 1$ goes into the solution (10) $E_2^2 = Jq$. For $q > 0.09$ there are no stable stationary solutions for Eqs. (18) and (19) for $p < 0$. From the discussion of the preceding section it is clear that as the coupling coefficient $\tilde{m}$ of

---

[†] We recall that we assume $Q_1 > Q_2$ and, therefore, $\delta > 0$. When $\delta < 0$ the transition is from $\psi^{(3)}$ into $\psi^{(1)}$.

[‡] In this case we are interested in the frequency region $|p| \gg \delta$ where the difference in the Q-factors is not important.

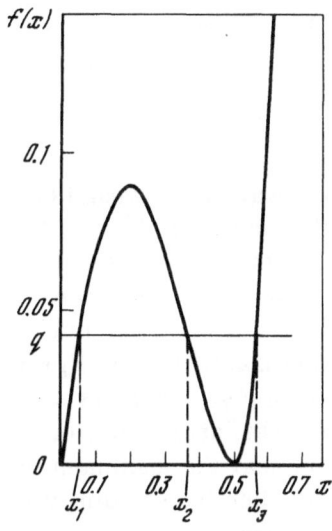

Fig. 1. The form of the function $f(x)$.

the waves increases the condition $q < 0.09$ begins to fail first in the frequency region corresponding to the maxima of the resonances of the weak wave E_2.

Let us examine the form of the power resonance with a relatively large coupling coefficient in the absence of an absorbing cell. In this case the resonance $\mu\xi = 0$ is still stable over the entire region E_2^2 when the condition $p < 0$ or $\tilde{m}^2/4\chi^2 < 0.09$ is fulfilled. For $\tilde{m} = \tilde{m}_{max}$ the limiting intensity of the resonance is attained at the maximum $E_2^2 = 0.19J$ for which the lasing regime remains stable. This limiting value for stability was obtained in another way and experimentally confirmed in [7]. It is interesting to note that when $\tilde{m} \leq \tilde{m}_{max}$ the form of the power resonance differs substantially from Lorentzian. Thus, for $\tilde{m} = \tilde{m}_{max}$ we can use the expansion $f(x) \simeq 0.09 - 2.1(x - 0.19)^2$ to study the shape of the resonance near the maximum. Setting $p^2 = \chi^2$ and $\Delta^2 = \omega_+^2/\gamma_+^2$, we obtain

$$E_2^2 \simeq J\left(0.19 - 0.21\,\frac{|\omega_+|}{\gamma_+\,\chi}\right) \simeq J\left(0.19 - 0.12\,\frac{|\omega_+|}{\gamma_+\tilde{m}_{max}}\right). \tag{21}$$

As can be seen from Eq. (21) with the substitution $\tilde{m} \to \tilde{m}_{max}$ the power resonance is peaked and the derivative $dE_2^2/d\omega_+$ has a discontinuity when $\omega_+ = 0$. Far from the maximum the condition $\psi \ll 1$ holds so that Eq. (12) $E_2^2 \simeq J\gamma_+^2\tilde{m}_{max}^2/\omega_+^2$ is applicable. The complete form of the profile is shown in Fig. 2. The width of the resonance, Γ, is roughly $\Gamma \simeq 0.8\tilde{m}_{max}\gamma_+$.

For $\tilde{m} > m_{max}$ near the line center there appears a frequency region in which the solution x_1 does not exist. In this frequency region a transition occurs to another lasing regime in

Fig. 2. The shape of the power resonance in a ring laser without an absorbing cell for $\tilde{m} = \tilde{m}_{max} = 0.6\chi$.

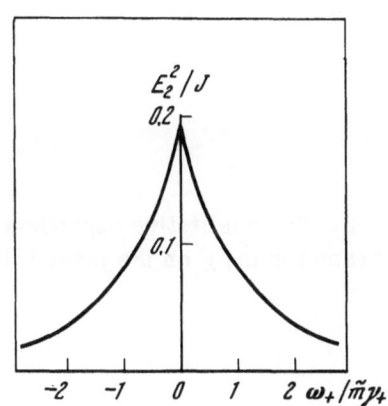

which the wave E_1 is weak. The solution in this regime has a form similar to Eq. (12):

$$E_1^2 = \frac{1}{4} J \frac{\tilde{m}_1^2}{\chi^2 + \frac{\omega_+^2}{\gamma_+^2}}.$$
(22)

Since in this case $\tilde{m}_1 \ll \tilde{m}_2$ the lasing regime in which the wave E_2 is weak is stable for the entire frequency region $|\omega_+| \lesssim \sqrt{\chi\gamma_+}$. When $\tilde{m}_2 \to \tilde{m}_{max}$ the resonance E_2^2 is peaked and its form is quite different from Lorentzian. For $\tilde{m}_2 \gtrsim \tilde{m}_{max} \gg \tilde{m}_1$ the lasing pattern appears as follows. At frequencies $\omega_+ \simeq -\sqrt{\chi\gamma_+}$ a transition occurs from the standing wave regime to lasing with waves of unequal intensities and weak E_2. Further on, near the maximum of the resonance, there is a jump from lasing with a weak wave E_2 to regime (22) with weak E_1 which remains stable up to frequencies $\omega_+ \simeq +\sqrt{\chi\gamma_+}$. At frequencies $\omega_+ \simeq \sqrt{\chi\gamma_+}$ there is a transition from regime (22) to the standing wave regime. Scanning in reverse, at frequencies $\omega_+ \simeq +\sqrt{\chi\gamma_+}$ there is a transition into a regime with weak E_2 and later, near the maximum in the resonance, the lasing enters regime (22) which for $\omega_+ \simeq -\sqrt{\chi}\gamma_+$ goes into the standing wave regime. Thus, for $\tilde{m}_2 > m_{max} \gg \tilde{m}_1$ a hysteresis effect will be observed.

In the previous section we have found the width of the resonance in E_2^2 to be $\Gamma = \chi\gamma_+$. When the parameter χ is reduced, as can be done, for example, by increasing the output power, the width of the resonance, Γ, falls. However, as can be seen from this discussion this narrowing is limited by the condition $0.6\chi > \tilde{m}_2$. Then the minimum attainable resonance width for a given value of $\tilde{m}_2$ is $\Gamma_{min} \simeq 0.8\tilde{m}_2\gamma_+$ and increases with the reflectivity $\tilde{m}_2$.

A similar situation also occurs with a methane absorption cell. The minimum width of resonance (17) is $\Gamma_{min} \simeq 0.8 \frac{\tilde{m}_2}{\mu\xi} \gamma_-$, $\mu\xi \gg \tilde{m}_2$. Then the summit of peak (17) is sharper and the contrast also reaches the limiting value $a \simeq 0.36 \left(\frac{\mu\xi}{\tilde{m}_2}\right)^2$.

The dependence of the limiting width and contrast of the resonance on the pressure in a methane absorbing cell is of some interest. At sufficiently low pressures $p \lesssim 10^{-2}$ Torr the width γ_- is determined, as a rule, by the time of flight and is independent of the pressure. In this case the parameter $\mu\xi$ is proportional to the methane pressure, so the width of the resonance Γ_{min} decreases as the pressure rises, while the contrast increases. At high pressures $p \gtrsim 10^{-2}$ Torr the width γ_- becomes proportional to the methane density N_- and the parameter $\mu\xi \propto 1/N_-$. As a result the width of the resonance $\Gamma \propto N_-^2$ and the contrast $a \propto N_-^2$. Then the optimum methane pressure for operation of the laser is that at which the collisional width of the line equals its transit width.

We now consider the effect of the laser power on the shape of the resonances for a fixed coupling coefficient m. We recall that including the fifth-order terms in the field in the perturbation theory calculation of the polarization of the medium results in the parameters χ and χ' depending on the field intensity. Figure 3 shows the qualitative dependence of the parameter

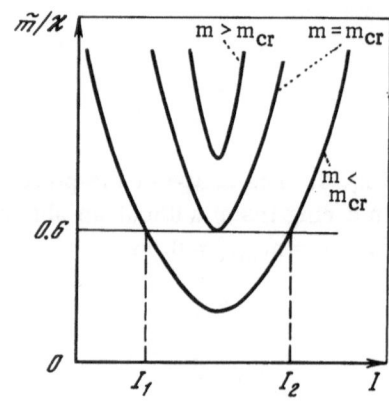

Fig. 3. The qualitative dependence of the parameter $\tilde{m}/\chi$ on the laser intensity.

$$\frac{\tilde{m}}{\chi} = \frac{2Q}{\omega}\ \frac{m}{I\left(\frac{\gamma_+\Gamma_+}{(\Delta\omega_D)^2} - \theta_+ I\right)}$$ on the laser power in the absence of an absorbing cell for various val-

ues of the coupling coefficient between the opposing waves. For $m > m_{cr} = \frac{\omega}{2Q}\ \frac{0.15}{\theta_+}\left[\frac{\gamma_+\Gamma_+}{(\Delta\omega_D)^2}\right]^2$

there will be a jump to regime (22) for all values of the intensity near the center of the resonance. As the coupling coefficient is reduced, $m < m_{cr}$, there appears a range of intensities $I_1 < I < I_2$ in which a regime with weak E_2 exists and is stable over the entire region $|\omega_+| < \sqrt{\chi\gamma_+}$. As already noted, the intensity E_2 at the maximum of the resonance grows with increasing $\tilde{m}/\chi$. Thus in the region $I_1 < I < (I_1 + I_2)/2$ the intensity E_2^2 falls, while in the region $(I_1 + I_2)/2 < I < I_2$ it increases as the output power is increased. The width of the resonance falls monotonically as the power is increased from a value of $\Gamma_1 \simeq 0.8\ \frac{2Q}{\omega}\ \frac{m}{I_1}\ \gamma_+$ at $I = I_1$ to $\Gamma_2 \simeq 0.8\ \frac{2Q}{\omega}\ \frac{m}{I_2}\ \gamma_+$. We note that $\Gamma_1/\Gamma_2 = I_2/I_1$.

As the laser power is changed, the behavior of the resonance (17), which fixes the center of the absorption line, may be quite different with an absorbing cell than in the absence of an absorbing cell. In this case the stability criterion is $\tilde{m}/\chi' < 0.6$. If $\theta_+ < \theta_- \xi$ [see Eq. (15)] then as the output power is increased to $I > I_1$ the intensity of the resonance (17) at its maximum decreases monotonically from its limiting value $E_2^2 = 0.19J$ to zero. Then the width of the resonance and the background intensity increase but the contrast decreases. In the opposite case, $\theta_+ > \theta_- \xi$, the behavior of the resonance (17) is qualitatively the same as its behavior in the absence of an absorbing cell. If $\theta_+ \simeq \theta_- \xi$, then the laser power depends weakly on the shape and width of the resonance.

The Shape of the Power Resonances when the Coupling

Coefficients of the Opposite Waves Are Complex Conjugates

In the first work on the observation and use of power resonances in ring lasers for frequency stabilization [6, 8] no external mirror was used to increase the contrast of the resonance. In that case, apparently, the main contribution to the coupling coefficients of the traveling waves is from volume scattering, that is, $m_1 = m_2 = m$ and $\xi_1 = \xi_2 = \xi_0$. Equations (4) and (5) take the form

$$\frac{d\psi}{d\tau} = p\sin\psi\cos\psi + \tilde{m}\sin\varphi, \tag{23}$$

$$\frac{d\varphi}{d\tau} = \left[-\Delta + \tilde{m}\ \frac{\cos\varphi}{\sin\psi}\right]\cos\psi, \qquad \varphi = \Phi + \xi_0. \tag{24}$$

As in the previous paragraph, we take $\Omega = \delta = 0$. The system of equations (23) and (24) has two stationary solutions. The first solution,

$$\cos\psi = 0, \quad \sin\varphi = 0, \tag{25}$$

corresponds to a standing wave regime and is stable for $p > 0$. The second solution,

$$\sin^2\frac{\psi}{2} = \frac{1}{2} - \frac{1}{2}\left\{1 - \frac{1}{2}\left[1 + \frac{\Delta^2}{p^2} - \sqrt{\left(1 + \frac{\Delta^2}{p^2}\right)^2 - \left(\frac{2\tilde{m}}{p}\right)^2}\right]\right\}^{1/2}, \tag{26}$$

describes a lasing regime with waves of unequal intensity. For it to be stable, the following conditions must hold:

$$p < 0, \frac{2\tilde{m}}{|p|} < 1 + \left(\frac{\Delta}{p}\right)^2. \tag{27}$$

We first consider the shape of the resonance in the absence of an absorbing cell, $\mu\xi = 0$. Near the center of the absorbing line we can use Eqs. (11) for p and Δ. In this case the maximum of

Eq. (26) occurs at $\omega_+ = 0$ and equals

$$\sin^2 \frac{\psi}{2} = \frac{1}{2} - \frac{1}{2\sqrt{2}}\left(1 + \sqrt{1 - \left(\frac{2\tilde{m}}{\chi}\right)^2}\right)^{1/2}. \qquad (28)$$

As the frequency shift ω_+ increases, $\sin^2(\psi/2)$ monotonically decreases. Thus, Eq. (26) describes a power resonance with a center at the point $\omega_+ = 0$. The intensity of the resonance increases with the parameter $2\tilde{m}/\chi$ and reaches a maximum of

$$\left(\sin^2 \frac{\psi}{2}\right)_{max} = \frac{\sqrt{2} - 1}{2\sqrt{2}} \simeq 0.146 \qquad (29)$$

when $2\tilde{m}/\chi = 1$.

The resonance profile in this case differs from Lorentzian as in the case of Eq. (21): The derivative $dE_2^2/d\omega_+$ has a discontinuity at $\omega_+ = 0$ and the overall shape of the profile is similar to that shown in Fig. 2. The width of the resonance is roughly $\Gamma \simeq \tilde{m}\gamma_+$.

As the parameter $2\tilde{m}/\chi > 1$ is further increased, condition (27) ceases to be satisfied near the peak of the resonance, and at frequency differences $|\omega_+| < \chi[(2\tilde{m}/\chi) - 1]^{1/2}\gamma_+$ a fluctuating intensity regime sets in [10]. On leaving this frequency regime the laser returns to regime (26).

Therefore, for $2\tilde{m}/\chi < 1$ the width of the resonances and their limiting intensity and shape differ only quantitatively from the same parameters of the resonances examined in the preceding section. Thus, the power peaks which locate the center of the absorption line behave roughly as when $\tilde{m}_2 \gg \tilde{m}_1$.

Conclusion

These considerations show that in ring gas lasers with an absorbing cell ($\gamma_- \ll \gamma_+$) it is possible to obtain power resonances which fix fairly closely on the center of the absorbing transition of the medium. Then the contrast of the resonance, $a = (I_{max} - I_{bgnd})/I_{bgnd}$, may be very large ($a \gg 1$) while in lasers with Fabry–Perot cavities $a \lesssim 0.1$. At the same time the width of the power peak may be much less than the homogeneous width of the absorption line. We now estimate these quantities for $m_2 \gg m_1$, $m_2 = 10^{-5}(c/L) \simeq 3 \cdot 10^3$ sec^{-1}, $I \simeq 10^{-2}$, $\omega/2Q = 10^{+7}$ sec^{-1}, $\tilde{m}_2 = m_2(2Q/\omega)(1/I) \simeq 3 \cdot 10^{-2}$, and $\mu\xi \simeq 0.3$. The limiting width of the resonance in this case is $\Gamma_{min} \simeq 0.8 \frac{\tilde{m}_2}{\mu\xi}\gamma_- \simeq 0.1\gamma_-$, the contrast is $a \simeq 0.36\left(\frac{\mu\xi}{\tilde{m}_2}\right)^2 \simeq 4 \cdot 10^3$ (that is, the background is practically undetectable in general), and the intensity of the resonance at its peak is about 19% of the total intensity.

Recently the effect of the hyperfine structure of the methane absorption line on the frequency reproducibility of a gas laser has been extensively discussed (see, for example, [11, 12]). In fact, the different components of the hyperfine structure of the methane absorption line saturate in different ways, so the location of the center of the power peak depends on the saturation parameter of the methane line, $\xi I = \left(\frac{d_- E}{\hbar}\right)^2 \frac{1}{\gamma_-\Gamma_-}$. In lasers with a Fabry–Perot cavity, to achieve good contrast in the resonance, $a \simeq \mu\xi/2 \simeq 0.1$, requires working under conditions such that $\xi I \simeq 1$. Meanwhile it is very difficult to control ξI experimentally since it depends on the output power, the cavity geometry, the field distribution in the perpendicular cross section of the beam, and other macroscopic parameters. Thus the oscillations in the value of ξI for different devices may be of the order of $\sim 0.5\xi I$. It has been theoretically estimated [11] that a change in the saturation parameter by 0.5 when $\xi I \simeq 1$ causes a shift in the power resonance center by a value of the order of 10^3 Hz. Reducing the saturation parameter by increasing γ_- is inconvenient since then the resonance is broadened and the accuracy of the coupling to its center is reduced. A ring laser may greatly simplify solving this complicated problem. Thus, for example, if the width of the line γ_- is determined by collisions and the width of the resonance of the ring laser is $\Gamma_{min} \simeq 0.1\gamma$ (as in the above example), then it is possible to go ahead and increase the width γ_- somewhat. As the density of methane is increased from $N_-^{(1)}$ to $N_-^{(2)} = 3.2 N_-^{(1)}$ the quantity $\gamma_- \propto N_-$ also increases by 3.2 times and the quantity $\mu\xi \propto 1/N_-$ decreases by 3.2 times. Finally, the width of the resonance in a ring laser, Γ_{min}, is equal to $\Gamma_{min} = 0.32\gamma_-(N_-^{(2)}) = \gamma_-(N_-^{(1)})$; that is, it reaches a width which equals that of the power resonance in a gas laser with a Fabry−Perot cavity at a density of $N^{(1)}$. Thus; the accuracy of the coupling to the center of the resonance in a ring laser at a density of $N_-^{(2)}$ becomes the same as that in a laser with a Fabry−Perot cavity at the lower density $N_-^{(1)}$. However, the saturation parameter at a density of $N_-^{(2)}$ is 10 times less than at a density of $N_-^{(1)}$ since $\xi I \propto 1/N^2$. Thus, in a ring laser it is possible to obtain a resonance of the same width with greater contrast than in a laser with a Fabry−Perot cavity, but with a much smaller saturation parameter. Then the effect of the hyperfine structure on the frequency reproducibility may be greatly reduced.

LITERATURE CITED

1. Yu. L. Klimontovich, Wave and Fluctuation Processes in Lasers [in Russian], Fizmatgiz, Moscow (1974).
2. F. Aronowitz, Laser Gyroscopes, in: Applications of Lasers [Russian translation], Mir, Moscow (1974).
3. N. G. Basov, É. M. Belenov, M. V. Danileiko, V. V. Nikitin, and A. N. Oraevskii, Pis'ma Zh. Éksp. Teor. Fiz., 12:145 (1970).
4. N. G. Basov, É. M. Belenov, M. I. Vol'nov, M. V. Danileiko, and V. V. Nikitin, Pis'ma Zh. Éksp. Teor. Fiz., 15:659 (1972).
5. N. G. Basov, É. M. Belenov, M. V. Danileiko, and V. V. Nikitin, Kvant. Élektron., No. 1, p. 42 (1971).
6. V. A. Alekseev, N. G. Basov, E. M. Belenov, M. I. Vol'nov, N. A. Gubin, V. V. Nikitin, and A. N. Nikolaenko, Kvant. Élektron., 1:1089 (1974).
7. M. A. Andronova, I. L. Bershtein, and N. A. Markelov, Kvant. Élektron., 1:645 (1974).
8. E. M. Belenov, M. V. Danileiko, V. R. Kozubovskii, A. P. Nedavnii, and M. T. Shpak, Pis'ma Zh. Éksp. Teor. Fiz., 20:696 (1974).
9. A. V. Gnatovskii, M. V. Danileiko, V. R. Kozubovskii, T. V. Rozhdestvenskaya, V. P. Fedin, and M. T. Shpak, Ukr. Fiz. Zh., 19:1808 (1974).
10. É. M. Belenov, M. V. Danileiko, V. R. Kozubovskii, A. P. Nedavnii, A. N. Nikolaenko, and M. T. Shpak, Kvant. Élektron., 1:2647 (1974).
11. J. L. Hall and C. Bordé, Phys. Rev. Lett., 30:1101 (1973).
12. S. N. Bagaev, E. V. Baklanov, E. L. Timov, and V. P. Chebotaev, Pis'ma Zh. Éksp. Teor. Fiz., 20:292 (1974).

LASER RANGING OF THE MOON

Yu. L. Kokurin

A new method for studying the earth—moon system based on direct laser ranging measurements of distances to the moon is developed. The theory for determining the parameters of the earth—moon system from ranging measurements is worked out. On this basis a number of problems in geodesy, geophysics, astronomy, and selenodesy are posed which can be solved by laser ranging of the moon. An automatic laser ranging system was built and is used to make regular measurements of the distances to corner reflectors on the moon with an accuracy of ±90 cm. About 900 measurements of the distances to lunar reflectors have been made. Some parameters of the earth—moon system are determined using these measurements.

1. INTRODUCTION

The development of space studies and space flight, especially over the last two decades, has posed a whole series of new problems in celestial mechanics, astrometry, geophysics, and geodesy. A need has arisen for substantial improvements in the accuracy of the laws of motion of the heavenly bodies, including the interaction of the earth and moon. The accuracy with which the geometry and dynamics of the earth—moon system were known was insufficient in the light of these new requirements. The traditional goniometric methods of astronomical measurements which were used for a long time to measure the relative motion of the earth and moon could not provide substantial further progress in this area. Thus there was great interest in finding new methods of studying the earth—moon system. The development of quantum electronics and the invention of the laser [1] opened up the possibility of highly accurate measurements of the distance to the moon by a new method, laser ranging.

It was of interest to see what a direct measurement of the third component, the distance to the moon, would yield as opposed to and in combination with the two angular coordinates measured by the goniometric methods.

The first approximate calculations showed [2] that laser ranging of the moon makes it possible to improve the accuracy of the basic parameters of the earth—moon system by several orders of magnitude and thereby to obtain some significant results in a number of problems in selenodesy, astrometry, geophysics, and geodesy.

The structure of the moon, its shape, dynamic properties, and orbital and rotational motions contain traces of the factors which made the moon a satellite of the earth. Studies of the moon in these aspects makes it possible to obtain information on its nature, evolution, and internal structure. In particular, the study of the physical libration makes it possible to determine the dynamic compressions of the moon, which are function of its principal moments of inertia. A study of the shape of the moon and its relation to the center of mass and a measurement of the degree of elongation of the moon in the direction of the earth will make it possible to draw conclusions about the structure of the moon and its cosmological evolution.

The traditional method of measuring the physical libration of the moon consisted of studying the motion of some point on the surface of the moon relative to its edge. The accuracy with which the libration angles are determined in this way is small (about 10 selenocentric seconds for an overall physical libration of about 2 minutes) [3]. Thus, the dynamic compressions of the moon were not determined accurately enough. The shape of the moon is usually examined on the basis of astrophotographs taken at different libration angles. The small magnitude of the optical libration (about 8°) limits the accuracy of goniometric measurements of the moon's shape and the accuracy of selenodesic constructions. The elongation of the moon in the direction of the earth, for example, is measured in this way with an error of not less than 1 km [4].

Laser ranging makes it possible to measure the parameters of the physical libration, study the moon's shape, and determine the relationship of the points on its surface to one another and to its center of mass more accurately than the traditional goniometric methods by at least two orders of magnitude. This is a significant contribution to constructing an overall picture of the moon's evolution and to study its structure.

The computations involved in space flight require information on the moon's orbital motion along with data on its rotation and shape. The ephemeris is calculated on the basis of the Hill−Brown−Eckert theory of the lunar orbital motion. There the errors in the geocentric ecliptic coordinates of the moon are determined by the magnitudes of the terms left out in the expansions of these coordinates and by the errors in the expansion coefficients. For example, the distance to the moon is computed with this theory with an error of the order of 1 km. More complete theories of the lunar orbital motion are now being developed. New, more accurate measurements of the parameters of the orbital motion, such as the mean geocentric distance of the moon, the eccentricity constants and the obliquity of the moon's orbit, and many others, are needed to determine the constants in these theories. Estimates show that with the laser ranging technique we can measure the constants of the orbital motion of the moon with an accuracy two or three orders of magnitude better than that of the other methods. This will aid in the creation of new theories of the moon's motion and in the construction of more accurate ephemerides.

Another group of scientific problems which can be studied using laser ranging of the moon lies in the areas of geodesy and geophysics. The solution of these problems is based on the fact that with laser ranging it is possible to establish the relationship between distant points on the earth's surface and between them and the earth's center of mass. The accuracy attained in these measurements (about 20 cm) is much greater than that of ordinary astronomical and geodesic methods. These measurements can serve as a basis for a more precise study of the earth's shape and its relation to the center of mass, for establishing the relationship among various geodesic systems, for calibrating their scaling, and for determining their orientation.

An exact determination of the coordinates of various points on the earth's surface opens the possibility of studying the irregularity of the earth's rotation and the motion of the pole [5]. Of particular interest is the study of temporal variations in the relative coordinates of points on the earth's surface. This provides a basis for the study of such geodesic phenomena as tidal deformations [5], seasonal displacement of distinct blocks of the earth's core [6], and the continental drift [7-10], whose speed and mechanism are still unknown. Thus, laser ranging of the moon will aid research on these problems [11].

To solve these problems will require high accuracy in measurements of distances to the moon, of the order of a few meters or better. Such accuracy is possible only if the reflection point on the moon is localized, that is, if we can place special targets on the moon which efficiently reflect light back toward the observer (corner reflectors). In addition, it is necessary to build laser transmitters that emit powerful short light pulses (lasting 10^{-8} sec or less) and electronic equipment for measuring the time of flight of these pulses.

Research on laser ranging of the moon has involved the following basic areas: (1) theoretical development of the scientific and applied problems which may be solved using laser ranging of the moon; (2) the design, manufacture, and placing of corner reflectors on the moon; (3) the building of laser transmitters and receiver-measurement apparatus; and (4) organizing and making the measurements.

II. DETERMINING THE PARAMETERS OF THE EARTH−MOON SYSTEM FROM
THE RESULTS OF LASER−RANGING MEASUREMENTS

1. Introduction

Laser ranging makes it possible to measure directly the distance between a point on the earth's surface and a point on the moon's surface. By comparing the measured quantity with the distance between these points calculated for the same time, it is possible to gain an idea of the accuracy of the theories used for the calculation. These results are of particular value when these points are fixed and the distance between them is measured over a long time which encompasses the periods of all the basic components of the mutual motion of the earth and moon. It then seems possible to relate the measured distances to the fundamental constants of the earth−moon system and to determine these constants.

We shall derive the relationship between the parameters of the earth−moon system and the distance between points on the surfaces of the earth and moon and consider the prospects for improving the accuracy of these parameters by means of differential adjustments. With this we shall discover a group of scientific and applied problems which can be solved using laser ranging of the moon.

Before going on to derive the basic equations, we note that the result of a ranging measurement is a distance expressed in light seconds. On going to units of length an error is introduced due to the inaccuracy of the presently accepted value of the speed of light [12]. Neglecting this error, we shall measure all linear characteristics of the earth-moon system in units proportional to c.

2. The Relation between the Measured
Locator − Reflector Distance and the Parameters
of the Earth − Moon System

Let us consider the mathematical scheme for improving the accuracy of the basic parameters of the earth−moon system. This scheme was first published by us in simplified form in [2, 13, 14]. Figure 1 shows the geometrical arrangement and illustrates the relationship of the measured distance D between the locator O and the reflector R to the distance between the centers of mass of the earth T and the moon L and to the other linear and angular characteristics of the earth−moon system. The equatorial geocentric rectangular system of coordinates XYZ, the geocentric celestial sphere with a radius equal to the geocentric radius-vector ρ_0 of the locator, and the selenocentric celestial sphere with a radius $R_{\mathbb{C}}$ equal to the selenocentric radius-vector of the reflector are also shown here. The following notation is used in this figure: z is the geocentric zenith distance of the moon; λ, the longitude of the locator (angle between the planes of the local meridian and the Greenwich meridian which pass through the rotation axis of the earth); s, the local sidereal time with respect to the plane of the local meridian shown here; φ', the geocentric latitude of the locator; $\alpha_{\mathbb{C}}$, $\delta_{\mathbb{C}}$, t, the right ascension, declination,

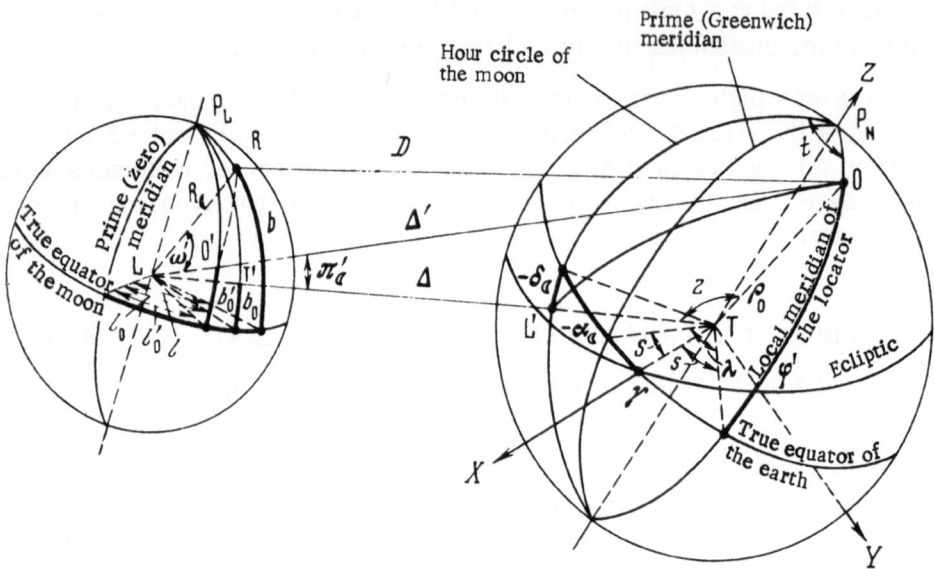

Fig. 1. The geometry of the earth–moon system. The heavy lines indicate the arcs corresponding to the angles denoted by the letters b, b_0, b_0'', $-\delta_\mathbb{C}$, $-\alpha_\mathbb{C}$, φ'.

and hour angle of the moon; l, b, the selenographic longitude and latitude of the reflector; l_0, b_0, the librations of the moon in longitude and latitude (sum of the optical and physical librations); l_0', b_0', the librations of the moon in longitude and latitude including the topocentric corrections; ω', the selenocentric zenith distance of the locator; $\pi_\mathbb{C}'$ the topocentric parallax of the moon; Δ and Δ', the instantaneous distances between the center of mass of the moon and earth, respectively, and the laser range finder (locator); P_N, the north pole of the actual equator of the earth; P_L, the north pole of the actual equator of the moon; and T' and O', the location on the selenocentric sphere of the center of mass of the earth and the locator, respectively, as "seen" from the center of the moon.

The formulas for evaluating the distance D follow directly from the diagram of Fig. 1. From the triangles OLR and OLT we have

$$D = (\Delta'^2 + R_\mathbb{C}^2 - 2\Delta' R_\mathbb{C} \cos \omega')^{1/2}; \qquad \Delta' = (\Delta^2 + \rho_0^2 - 2\Delta\rho_0 \cos z)^{1/2}. \qquad \textbf{(1)}$$

From the spherical triangles $OP_N L'$ and $P_L RO'$, we find

$$\cos z = \sin \varphi' \sin \delta_\mathbb{C} + \cos \varphi' \cos \delta_\mathbb{C} \cos t;$$
$$\cos \omega' = \sin b \sin b_0' + \cos b \cos b_0' \cos (l - l_0'). \qquad \textbf{(2)}$$

The hour angle t can be expressed as

$$t = s - \alpha_\mathbb{C} = S(T) + \lambda - \alpha_\mathbb{C}, \qquad \textbf{(3)}$$

where s is the local sidereal time at the measurement time, S(T) is the sidereal time at the actual Greenwich meridian at the same time, and T is the universal time at the same moment.

In a number of cases it is convenient to replace the spherical geocentric coordinates of the locator, ρ_0, λ, φ', with cylindrical coordinates, where λ is the longitude defined above, w = $\rho_0 \cos \varphi'$ is the radius of the parallel, and h = $\rho_0 \sin \varphi'$ is the distance of the locator from the instantaneous equatorial plane. Then

$$\Delta' = (\Delta^2 + w^2 + h^2 - 2\Delta w \cos \delta_{\mathbb{C}} \cos t - 2\Delta h \sin \delta_{\mathbb{C}})^{1/2}. \tag{4}$$

Therefore, the distance D depends on (a) the geocentric equatorial coordinates of the moon, Δ, $\alpha_{\mathbb{C}}$, $\delta_{\mathbb{C}}$, (b) the geocentric coordinates of the locator, Δ, $\alpha_{\mathbb{C}}$, $\delta_{\mathbb{C}}$; (c) the selenocentric coordinates of the reflector, $R_{\mathbb{C}}$, l, b, (d) the librations of the moon in longitude and latitude, l_0', b_0', including the topocentric corrections, and (e) the sidereal time at the Greenwich meridian, S(T).

In order to express D in terms of the constants and time alone, it is necessary to determine the dependence of the following quantities on the constants and time: Δ, $\alpha_{\mathbb{C}}$, $\delta_{\mathbb{C}}$, ρ_0, λ, φ', $R_{\mathbb{C}}$, l, b, l_0', b_0', S. Let us consider these dependences in turn.

(a) To determine the geocentric equatorial coordinates of the moon's center of mass, Δ, $\alpha_{\mathbb{C}}$, and $\delta_{\mathbb{C}}$, we can use the expansions of the sine of the lunar parallax, $\sin \pi_{\mathbb{C}}$, and of the ecliptical longitude $\lambda_{\mathbb{C}}$ and latitude $\beta_{\mathbb{C}}$ from Brown's theory of lunar motion [15-22]:

$$\sin \pi_{\mathbb{C}} = \sum a_n \cos (il + jl' + kF + rD) + \sum k_i \cos (a_i + b_i t),$$
$$\lambda_{\mathbb{C}} = \mathbb{C} + \sum a_n^{\lambda} \sin (il + jl' + kF + rD) + \sum k_i^{\lambda} \sin (a_i + b_i t), \tag{5}$$
$$\beta_{\mathbb{C}} = \sum a_n^{\beta} \sin (il + jl' + kF + rD) + \sum k_i^{\beta} \sin (a_i + b_i t),$$

where $l = \mathbb{C} - \pi$ is the mean anomaly of the moon, $l' = L' - \pi'$ is the mean anomaly of the sun, $F = \mathbb{C} - \Omega$ is the argument of the moon's latitude, $D = \mathbb{C} - L'$ is the mean elongation of the moon from the sun, $\mathbb{C}$, L' are the mean longitudes of the moon and sun, π, π' are the mean longitudes of the lunar and solar perigees, Ω is the mean longitude of the node of the moon's orbit, i, j, k, r = 0, ±1 ... up to $|i| \leq 6$, $|j| \leq 4$, $|k| \leq 5$, and $|r| \leq 10$.

The sums with coefficients a_n are perturbations of the moon's motion induced by the sun. The sums with coefficients k_i characterize the perturbations from the planets. The arguments $\mathbb{C}$, π, Ω, L', π', l, l', F, D, and the mean longitudes of the planets in the solar system (T, Earth; V, Venus; M, Mars; Sn, Saturn; J, Jupiter; Q, Mercury) are known as the fundamental arguments.

The fundamental arguments have the form

$$(\mathbb{C}, \pi, \Omega, L', \pi', l, l', F, D, T, V, M, Sn, J, Q) = A_0 + A_1 T + A_2 T^2 + A_3 T^3 + \sum k_i \sin (a_i + b_i t + c_i t^2), \tag{6}$$

where A_0, A_1, A_2, A_3 are coefficients and T and t are the time in Julian centuries and days, respectively, reckoned from the fundamental epoch of 0 January, 1900, and the mean Greenwich noon to the moment of the measurement. The coefficients a_i, b_i, and c_i of Eqs. (5) and (6) are linear combinations of the A_0, A_1, A_2, A_3 in the expansions of the fundamental arguments.

If we write the largest terms in the expansions leaving out terms with coefficients in $\lambda_{\mathbb{C}}$ and $\beta_{\mathbb{C}}$ less than 5″ and in $\sin \pi_{\mathbb{C}}$ less than 1″, then the equations for the lunar coordinates will have the form

$$\sin \pi_{\mathbb{C}} = 3422''.700 + 28''.2333 \cos 2D + 186''.5398 \cos l +$$
$$+ 34''.3117 \cos (l - 2D) + 3''.0861 \cos (l + 2D) +$$
$$+ 1''.9178 \cos (l' - 2D) + 10''.1657 \cos 2l + 1''.4437 \cos (l + l' - 2D) + 1''.1528 (l - l') + \ldots$$

$$\lambda_{\mathbb{C}} = \mathbb{C} + 22639''.500 \sin l + 191''.953 \sin (l + 2D) - 4586''.465 \sin (l - 2D) -$$
$$- 38''.428 \sin (l - 4D) + 13''.902 \sin 4D + 2369''.912 \sin 2D -$$
$$- 125''.154 \sin D - 668''.146 \sin l' - 24''.420 \sin (l' + 2D) +$$
$$+ 769''.016 \sin 2l + 14''.387 \sin (2l + 2D) - 211''.656 \sin (2l - 2D) -$$
$$- 30''.773 \sin (2l - 4D) - 109''.673 \sin (l + l') + 28''.475 \sin (l - l' - 2D) -$$

$$- 5\overset{''}{.}741 \sin{(2F + 2D)} - 411\overset{''}{.}608 \sin{2F} - 55\overset{''}{.}173 \sin{(2F - 2D)} -$$
$$- 8\overset{''}{.}466 \sin{(l + D)} - 18\overset{''}{.}609 \sin{(l - D)} + 18\overset{''}{.}023 \sin{(l' + D)} +$$
$$+ 36\overset{''}{.}124 \sin{3l} - 13\overset{''}{.}193 \sin{(3l - 2D)} - 7\overset{''}{.}649 \sin{(2l + l')} -$$
$$- 8\overset{''}{.}627 \sin{(2l + l' - 2D)} + 9\overset{''}{.}703 \sin{(2l - l')} -$$
$$- 7\overset{''}{.}412 \sin{(l + 2l' - 2D)} - 45\overset{''}{.}099 \sin{(l + 2F)} +$$
$$+ 39\overset{''}{.}528 \sin{(l - 2F)} + 9\overset{''}{.}366 \sin{(l - 2F - 2D)} + \ldots \tag{7}$$

$$\beta_{\mathbb{C}} = 18461\overset{''}{.}480 \sin{F} + 117\overset{''}{.}262 \sin{(F + 2D)} - 623\overset{''}{.}658 \sin{(F - 2D)} +$$
$$+ 15\overset{''}{.}122 \sin{(l + F + 2D)} + 1010\overset{''}{.}180 \sin{(l + F)} -$$
$$- 166\overset{''}{.}577 \sin{(l + l' - 2D)} - 6\overset{''}{.}580 \sin{(l + l' - 4D)} -$$
$$- 199\overset{''}{.}485 \sin{(-l + l' + 2D)} - 999\overset{''}{.}695 \sin{(-l + l')} -$$
$$- 33\overset{''}{.}359 \sin{(-l + l' - 2D)} - 6\overset{''}{.}492 \sin{(l + l')} -$$
$$- 29\overset{''}{.}689 \sin{(l' + F - 2D)} + 8\overset{''}{.}001 \sin{(-l' + F + 2D)} +$$
$$+ 12\overset{''}{.}140 \sin{(l - l' - 2D)} - 5\overset{''}{.}357 \sin{(F + D)} - 6\overset{''}{.}299 \sin{3F} +$$
$$+ 61\overset{''}{.}913 \sin{(2l + F)} - 15\overset{''}{.}565 \sin{(2l + F - 2D)} -$$
$$- 31\overset{''}{.}763 \sin{(-2l + F)} - 5\overset{''}{.}331 \sin{(l + l' + F)} -$$
$$- 7\overset{''}{.}463 \sin{(l + l' + F - 2D)} + 8\overset{''}{.}902 \sin{(-l + l' + F + 2D)} +$$
$$+ 5\overset{''}{.}096 \sin{(-l - l' + F)} + 6\overset{''}{.}756 \sin{(l - l' + F)} -$$
$$- 5\overset{''}{.}655 \sin{(-l + l' + F)} + \ldots$$

The quantity Δ which we require is related to $\sin{\pi_{\mathbb{C}}}$ by

$$\Delta = a_e / \sin{\pi_{\mathbb{C}}}, \tag{8}$$

where a_e is the equatorial radius of the earth.

To find the geocentric equatorial coordinates of the moon, $\alpha_{\mathbb{C}}$, $\delta_{\mathbb{C}}$, in terms of the ecliptic $\lambda_{\mathbb{C}}$ and $\beta_{\mathbb{C}}$ we have to use the known formulas for transforming from the ecliptic coordinate system to the equatorial [23]:

$$\cos{\delta_{\mathbb{C}}} \cos{\alpha_{\mathbb{C}}} = \cos{\beta_{\mathbb{C}}} \cos{\lambda_{\mathbb{C}}},$$
$$\cos{\delta_{\mathbb{C}}} \sin{\alpha_{\mathbb{C}}} = - \sin{\beta_{\mathbb{C}}} \sin{\varepsilon} + \cos{\beta_{\mathbb{C}}} \sin{\lambda_{\mathbb{C}}} \cos{\varepsilon}, \tag{9}$$
$$\sin{\delta_{\mathbb{C}}} = \sin{\beta_{\mathbb{C}}} \cos{\varepsilon} + \cos{\beta_{\mathbb{C}}} \sin{\lambda_{\mathbb{C}}} \sin{\varepsilon},$$

where ε is the true obliquity of the ecliptic to the earth's equator.

We now determine which constants from the equations for the moon's orbital motion enter into Eq. (1) for D.

The general form of the term in the trigonometric series for the lunar coordinates $\sin{\pi_{\mathbb{C}}}$, $\lambda_{\mathbb{C}}$, $\beta_{\mathbb{C}}$ is

$$a A_{i, j, k, r}^{i', j', k'} e^{i + 2i'} (e')^{j + 2j'} \gamma^{k + 2k'} \alpha^s \frac{\sin}{\cos} (il + jl' + kF + rD), \tag{10}$$

where a is a scale factor equal to the parallax constant for the expansion of $\sin{\pi_{\mathbb{C}}}$ and to unity for the expansions of $\lambda_{\mathbb{C}}$ and $\beta_{\mathbb{C}}$; $A_{i, j, k, r}^{i', j', k'}$ is a constant coefficient obtained from a solution of the differential equations for the moon's motion, e' is the eccentricity of the sun's orbit, e and γ are eccentricity constant and the declination constant of the moon's orbit, and α is a constant which depends on the ratio of the solar and lunar parallaxes.

The coefficients A_1, A_2, A_3 in Eq. (6) are expressed in terms of the parameter m = $n'/(n - n')$, where n and n' are the average motions (average angular velocities) of the moon and sun, respectively. The coefficients A_0 for each fundamental argument in Eq. (6) are constants of integration.

Thus, the coordinates of the moon depend on ten constants if we include only the solar perturbations. These are e, e', γ, α, l_0, l_0', F_0, D_0, m, and Δ_0, where l_0, l_0', F_0, D_0 are the values of the basic fundamental arguments at the initial epoch (i.e., at the initial moment from which time is reckoned when expanding the coordinates in a series).

(b) The true longitude λ and latitude φ' of the locator are determined by the equations for the changes in the longitudes and latitudes due to motion of a rotation axis in the body of the earth [24]:

$$\lambda = \lambda_0 + \Delta\lambda = \lambda_0 + \frac{1}{15}(x \sin\lambda_0 - y \cos\lambda_0)\tan\varphi',$$

$$\varphi' = \varphi_0' + \Delta\varphi' = \varphi_0' + \frac{1}{15}(x \cos\lambda_0 + y \sin\lambda_0),$$

(11)

where $\Delta\lambda$ and $\Delta\varphi'$ are corrections to the average values of the longitude λ_0 and latitude φ_0' from motion of the pole, λ, φ' are the true values of the longitude and latitude, and x, y are the coordinates of the instantaneous pole relative to the average. (The X axis is oriented along the tangent to the mean meridian of Greenwich and the Y axis is along the tangent to the meridian with an average longitude of 90°.) The quantities x and y are slowly varying functions of time. (The Euler period, in particular, is about 305 days and the period of the Chandler oscillations is about 14 months [24].) Thus, over short time intervals, for example, of the order of days, all three coordinates of the locator, ρ_0, λ, and φ', may be assumed constant.

(c) For this discussion we shall assume that the selenocentric coordinates $R_{\mathbb{C}}$, l, b of the reflector are constant.

(d) The quantities l_0' and b_0' are computed from the total librations of the moon in longitude l_0 and latitude b_0 and the topocentric corrections which take into account the angular displacement of the locator from the direction LT that can be "seen" from the moon's center.

In the theory of rotation the moon [25-27] is regarded as an absolutely solid body. This makes it possible to use the differential equations for rotation of a solid to describe its motion. The solution of these equations yields the components of the rotation as functions of time and the integration constants.

The rotation of the moon can be represented as a sum of several average rotations and the physical libration.

The average rotation is described by the three laws of Cassini: (1) the ascending node of the lunar equator on the ecliptic coincides with the descending node of the lunar orbit on the ecliptic; (2) the obliquity of the lunar equator to the ecliptic is constant; and (3) the angular speed of rotation equals the mean sidereal motion of the moon in its orbit.

The actual rotation of the moon differs from a rotation according to the Cassini laws. The deviation from these laws, known as the physical libration, is specified by three components: τ, the physical libration in the longitude; ρ, the physical libration in the obliquity of the lunar equator to the ecliptic; and σ, the physical libration in the longitude of the ascending node of the lunar equator at the ecliptic.

The components of the physical libration are found as solutions of the differential equations for the moon's rotation. They are trigonometric series with coefficients dependent on the dynamic compressions of the moon, $\alpha = (C - B)/A$, $\beta = (C - A)/B$, and $\gamma = (B - A)/C$, where A, B, and C are the principal moments of inertia of the moon, and the obliquity of the lunar equator to the ecliptic is $\mathcal{I}$.

From the rotation equations we also obtain an additional coupling of α and β with $\mathcal{I}$. Since $\alpha - \beta + \gamma - \alpha\beta\gamma = 0$, the expansion coefficients can be expressed fully in terms of two parameters, for example, α and β.

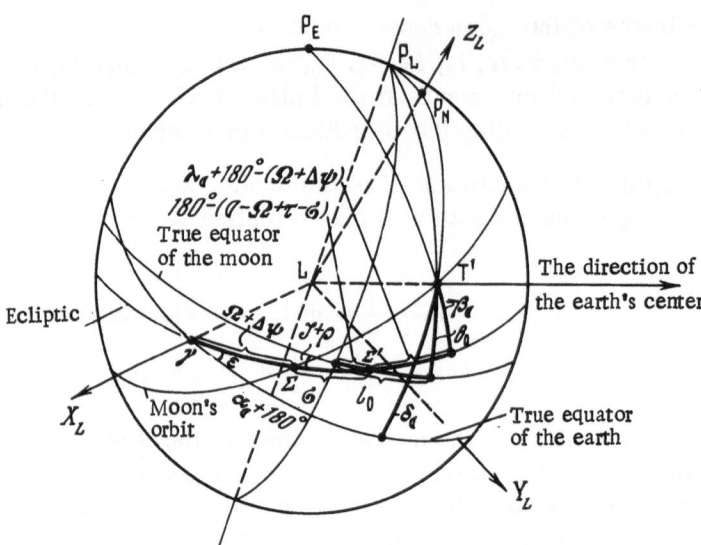

Fig. 2. The lunocentric celestial sphere. L is the center of
mass of the moon, γ is the point of the vernal equinox, and
P_E, P_N, P_L are the poles of the ecliptic, earth's equator, and
moon's equator, respectively.

Often the parameters in the solution are taken to be $\mathcal{I}$ and the physical libration constant
$f = \alpha/\beta$. We shall use them below.

In order to find the quantities we require, l_0, b_0, we shall use the method described in [28].
Let us examine Fig. 2, which shows the lunocentric celestial sphere. In addition to the notation
of Fig. 1 here we have used the following: P_E, the north pole of the ecliptic; Σ the ascending
node of the moon's orbit on the ecliptic; Σ' the descending node of the true equator of the moon
on the ecliptic; Ω the longitude of the node Σ; $\Omega + \sigma$, the longitude of node Σ'; $\Delta\psi$, the nutation
along the longitude; $\mathcal{I}$, the constant obliquity of the mean lunar equator to the ecliptic; $\mathcal{I} + \rho$,
the obliquity of the true lunar equator to the ecliptic; ε, the true obliquity of the ecliptic to the
earth's equator; τ, the physical libration of the moon along the longitude; σ, the libration of
the ascending node of the lunar equator on the ecliptic; ρ, the libration of the obliquity of the
true lunar equator to the ecliptic; and X_L, Y_L, Z_L, the geoequatorial selenocentric system of
rectangular coordinates with an axis LX_L directed toward the point of the vernal equinox γ.

The librations in the selenographic longitude l_0 and latitude b_0 are found from the spherical
triangle $P_E T' P_L$ (see Fig. 2) by the formulas

$$\cos b_0 \sin(l_0 + \mathbb{C} - \Omega + \tau - \sigma) = -\sin\beta_{\mathbb{C}} \sin(\mathcal{I} + \rho) + \cos\beta_{\mathbb{C}} \cos(\mathcal{I} + \rho) \sin(\lambda_{\mathbb{C}} - \Omega - \Delta\psi),$$

$$\cos b_0 \cos(l_0 + \mathbb{C} - \Omega + \tau - \sigma) = \cos\beta_{\mathbb{C}} \cos(\lambda_{\mathbb{C}} - \Omega - \Delta\psi),$$

$$\sin b_0 = -\sin\beta_{\mathbb{C}} \cos(\mathcal{I} + \rho) - \cos\beta_{\mathbb{C}} \sin(\lambda_{\mathbb{C}} - \Omega - \Delta\psi). \tag{12}$$

Since the locator is not situated on the line joining the centers of mass of the earth and
moon, the topocentric libration l_0', b_0' required to calculate the angle ω' in Eq. (2) differs from
the geocentric libration l_0, b_0. This difference, which reduces to a parallel displacement of
point T' to location O' (see Fig. 1), can be taken into account by replacing the geocentric ecliptic
coordinates of the moon, $\lambda_{\mathbb{C}}$, $\beta_{\mathbb{C}}$, in Eq. (12) by the topocentric $\lambda_{\mathbb{C}}'$, $\beta_{\mathbb{C}}'$. The relation between the

topocentric and geocentric ecliptic coordinates of the moon is given by [29]

$$\Delta' \cos\beta'_{\mathbb{C}} \sin\lambda'_{\mathbb{C}} = \Delta \cos\beta_{\mathbb{C}} \sin\lambda_{\mathbb{C}} - \rho_0 \cos\Phi \sin\theta,$$

$$\Delta' \cos\beta'_{\mathbb{C}} \cos\lambda'_{\mathbb{C}} = \Delta \cos\beta_{\mathbb{C}} \cos\lambda_{\mathbb{C}} - \rho_0 \cos\Phi \cos\theta,$$

$$\Delta' \sin\beta'_{\mathbb{C}} = \Delta \sin\beta_{\mathbb{C}} - \rho_0 \sin\Phi,$$

$$(13)$$

where Φ and θ are the geocentric ecliptic coordinates of the locator. They are given by

$$\cos\Phi \cos\theta = \cos\varphi' \cos s,$$
$$\cos\Phi \sin\theta = \cos\varphi' \sin s \cos\varepsilon + \sin\varphi' \sin\varepsilon,$$
$$\sin\Phi = -\cos\varphi' \sin s \sin\varepsilon + \sin\varphi' \cos\varepsilon.$$

$$(14)$$

When cylindrical coordinates of the locator, λ, w, h, are used, Eqs. (14) will have the form

$$\rho_0 \cos\Phi \cos\theta = w \cos s,$$
$$\rho_0 \cos\Phi \sin\theta = w \sin s \cos\varepsilon + h \sin\varepsilon,$$
$$\rho_0 \sin\Phi = -w \sin s \sin\varepsilon + h \cos\varepsilon.$$

$$(15)$$

Here s is the local sidereal time and ε is the true obliquity of the earth's equator to the ecliptic.

The components of the physical libration τ, ρ, σ can be taken in the form of expansions in the time series with coefficients depending on the parameters f and $\mathcal{J}$ [25]. Thus, for $f = 0.70$ and $\mathcal{J} = 1°32'20''$ these expansions have the form

$$\tau = -14\overset{..}{.}2\sin g - 0\overset{..}{.}4\sin 2g + 73\overset{..}{.}2\sin g' + 1\overset{..}{.}5\sin(2g'+2\omega') - $$
$$- 2\overset{..}{.}8\sin(g'-\omega-\omega') + 0\overset{..}{.}7\sin(g'-2\omega+2\omega') + $$
$$+ 8\overset{..}{.}1\sin(2g'-2\omega-2\omega') - 3\overset{..}{.}4\sin(g-2g'-2\omega-2\omega') + $$
$$+ 0\overset{..}{.}5\sin(2g-2g'+2\omega-2\omega') + 16\overset{..}{.}4\sin 2\omega + 1\overset{..}{.}2\sin(\omega-\omega') + \cdots$$

$$\rho = -104\overset{..}{.}0\cos g - 0\overset{..}{.}6\cos 2\omega + 31\overset{..}{.}4\cos(g+2\omega) - 10\overset{..}{.}4\cos(2g+2\omega) - $$
$$- 0\overset{..}{.}4\cos(3g+2\omega) + 0\overset{..}{.}6\cos(g+2g'+2\omega') - 2\overset{..}{.}1\cos(g-2g'+2\omega-2\omega') - $$
$$- 3\overset{..}{.}3\cos(2g'+2\omega') + 0\overset{..}{.}5\cos g' - 1\overset{..}{.}0\cos 2g + \cdots$$

$$(16)$$

$$\mathcal{J}\sigma = -104\overset{..}{.}0\sin g - 0\overset{..}{.}6\sin 2\omega + 31\overset{..}{.}4\sin(g+2\omega) - 10\overset{..}{.}4\sin(2g+2\omega) - $$
$$- 2\overset{..}{.}7\sin(g-2g'+2\omega-2\omega') - 3\overset{..}{.}3\sin(2g+2\omega') + $$
$$+ 0\overset{..}{.}6\sin(g+2g'-2\omega') - 1\overset{..}{.}0\sin 2g + 2\overset{..}{.}1\sin g' + 0\overset{..}{.}4\sin 2\omega + \cdots$$

where (in Hain's notation) g = l is the mean lunar anomaly, g' = l' is the mean solar anomaly, and w = $\pi - \Omega$ and w' = $\pi' - \Omega$ are the distances of the lunar and solar perigees from the node of the lunar orbit.

(e) The sidereal time at the Greenwich meridian is determined by an equation describing the earth's rotation about its axis [30]:

$$S(T) = 12^h + T + \alpha(T) + N_\alpha,$$

$$(17)$$

where N_α is the nutation in the right ascension and $\alpha(T)$ is the right ascension of the mean sun at universal time T.

The nutation in the right ascension N_α is computed from time series [31] for the nutation in the longitude $\Delta\psi$ including the obliquity ε of the earth's equator to the ecliptic; i.e., $N_\alpha = \Delta\psi \cos\varepsilon$. An analysis of Eq. (17) shows that S depends on the following parameters: the mean longitude of the sun L' and the mean tropic motion of the sun μ at the initial epoch, as well as the mean obliquity of the ecliptic to the equator ε and the nutation constant N. Thus, we have a number of equations (3), (5)-(9), (11)-(14), (16), and (17) which, when substituted in a certain order in Eq. (1), yield an equation for the distance D as a function of a single variable (the universal time T) and a number of constants (the parameters of the earth—moon system). We shall not make these substitutions here since our final aim is to obtain an equation coupling D to the system constants in differential form, so the method of differential correction of the constants may be used. Thus, using the equations given above, we proceed at once to derive the equation which relates the corrections in the parameters of the earth—moon system to the corrections in D determined by the laser ranging measurements.

3. Equations for the Differential Correction of the
Parameters of the Earth — Moon System

The difference $dD = D_0 - D_c$ measured at some time in the distance D_0 between the locator and reflector and the distance D_c calculated for the same time characterizes the error in the calculations due to the inaccuracy in the theory of the motions of the bodies making up the system.

Neglecting the question of the imperfections in one or another analytic theory of the motion and rotation of the earth and moon (i.e., motions which have not been taken into account or have been included incorrectly, simplifications in solving the differential equations of the motion, etc.) and assuming the mathematical model of the system consisting of Eqs. (3), (5)-(9), (11)-(14), (16), and (17) to be correct, we shall examine the magnitude of the deviation $D_0 - D_c$ as a result of, first, errors in the parameters used in this model and, second, random measurement errors.

We write the dependence of D on the system parameters in general form:

$$D = F\ (T, \Delta_0, e,\ e',\ \ldots,\ R_{\mathbb{C}},\ l,\ b). \tag{18}$$

We shall determine the parameters in Eq. (18) in a way similar to the method of differential correction of planet orbits [29, 32, 33]. Let us expand D in a Taylor series in increments of the parameters. Using the fact that these parameters arc known or can be computed with small errors [24, 28, 13] which in relative units are of order 10^{-5}-10^{-7} (for example, $\Delta(\Delta_0)/\Delta_0 \sim 10^{-5}$-$10^{-6}$; $\Delta e/e \sim 10^{-7}$; $\Delta\rho_0/\rho_0 \sim 10^{-5}$, etc.), we shall limit ourselves to expanding terms of the first order of smallness. Formally this expansion thus reduces to finding the total differential of D,

$$dD = \frac{\partial F}{\partial \Delta_0}\, d\Delta_0 + \frac{\partial F}{\partial e}\, de + \ldots + \frac{\partial F}{\partial b}\, db. \tag{19}$$

Within the limits of our assumption that the increments are small, we can assign this equation a different meaning. Thus, if we understand dD to be the correction to the computed distance obtained as a difference $dD = D_0 - D_c$, then the increments in the parameters will have the significance of corrections to the parameter values used in computing D_c.

If, however, dD is taken to mean the error in determining D_0, for example, due to random measurement errors, then the increments in the parameters should be taken to mean the errors in their determination caused by the error in D_0.

Thus, Eq. (19) may be used in two ways: to find the corrections to the parameters from the differences $D_0 - D_c$ and to estimate the errors in determining the individual parameters from the measurement errors and the errors in the remaining parameters.

In the following we shall call linear equations of the type (19) conditional equations (in the terminology of the method of least squares). Constructing the conditional equations for each measurement, we obtain a system (overdetermined when the number of measurements exceeds the number of unknowns) whose least-squares solutions yield the most probable values of the corrections to the parameters.

The conditional equation is obtained from Eq. (1) as follows. We differentiate D and find

$$dD = k_1 d\Delta' + k_2 \cos \omega' dR_{\mathbb{C}} + k_3 R_{\mathbb{C}} d(\cos \omega'),$$

$$d\Delta' = \left(\frac{\Delta}{\Delta'} - \frac{\rho_0}{\Delta'} \cos z\right) d\Delta + \left(\frac{\rho_0}{\Delta'} - \frac{\Delta}{\Delta'} \cos z\right) d\rho_0 + \frac{\Delta}{\Delta'} \rho_0 d(\cos z). \tag{20}$$

Differentiating $\cos \omega'$ and $\cos z$ we find from Eq. (2) that

$$d(\cos \omega') = k_4 dl_0' + k_5 db_0' + k_6 dl + k_7 db, \tag{21}$$

$$d(\cos z) = (\cos \varphi' \sin \delta_{\mathbb{C}} - \sin \varphi' \cos \delta_{\mathbb{C}} \cos t) d\varphi' + (\sin \varphi' \cos \delta_{\mathbb{C}} - \cos \varphi' \sin \delta_{\mathbb{C}} \cos t) d\delta_{\mathbb{C}} - \cos \varphi' \cos \delta_{\mathbb{C}} \sin t d(s - \alpha_{\mathbb{C}}).$$

Here

$$k_1 = \frac{\Delta'}{D} \left(1 - \frac{R_{\mathbb{C}}}{\Delta'} \cos \omega'\right); \qquad k_2 = -\frac{\dot{\Delta}'}{D} \left(1 - \frac{R_{\mathbb{C}}}{\Delta' \cos \omega'}\right);$$

$$k_3 = -\frac{\Delta'}{D} = \left[1 + \left(\frac{R_{\mathbb{C}}}{\Delta'}\right)^2 - \frac{2R_{\mathbb{C}}}{\Delta'} \cos \omega'\right]^{-1/2}; \qquad k_4 = \cos b \cos b_0' \sin(l - l_0');$$

$$k_5 = \sin b \cos b_0' - \cos b \sin b_0' \cos(l - l_0'); \qquad k_6 = -k_4;$$

$$k_7 = \cos b \sin b_0' - \sin b \cos b_0' \cos(l - l_0').$$

In cylindrical coordinates λ, w, h, we have

$$d\Delta' = \left(\frac{\Delta}{\Delta'} - \frac{w}{\Delta'} \cos \delta_{\mathbb{C}} \cos t - \frac{h}{\Delta'} \sin \delta_{\mathbb{C}}\right) d\Delta + \left(\frac{w}{\Delta'} - \frac{\Delta}{\Delta'} \cos \delta_{\mathbb{C}} \cos t\right) dw +$$

$$+ \left(\frac{h}{\Delta'} - \frac{\Delta}{\Delta'} \sin \delta_{\mathbb{C}}\right) dh + \frac{\Delta}{\Delta'} (w \sin \delta_{\mathbb{C}} - h \cos \delta_{\mathbb{C}}) d\delta_{\mathbb{C}} + \frac{\Delta}{\Delta'} w \cos \delta_{\mathbb{C}} \sin t d(s - \alpha_{\mathbb{C}}). \tag{22}$$

Since in the following we shall use the geocentric ecliptic coordinates of the moon, $\lambda_{\mathbb{C}}$, $\beta_{\mathbb{C}}$, it is necessary to go from $d\alpha_{\mathbb{C}}$, $d\delta_{\mathbb{C}}$ to $d\lambda_{\mathbb{C}}$, $d\beta_{\mathbb{C}}$ in Eqs. (21) and (22). The relation between them is found from Eqs. (9) to be

$$d\alpha_{\mathbb{C}} = \frac{1}{\cos \delta_{\mathbb{C}}} (A' d\lambda_{\mathbb{C}} + B' d\beta_{\mathbb{C}} + C' d\varepsilon),$$

$$d\delta_{\mathbb{C}} = \frac{1}{\cos \delta_{\mathbb{C}}} (A'' d\lambda_{\mathbb{C}} + B'' d\beta_{\mathbb{C}} + C'' d\varepsilon), \tag{23}$$

where

$$A' = \cos \beta_{\mathbb{C}} (\sin \lambda_{\mathbb{C}} \sin \alpha_{\mathbb{C}} + \cos \lambda_{\mathbb{C}} \cos \alpha_{\mathbb{C}} \cos \varepsilon),$$

$$A'' = \cos \beta_{\mathbb{C}} \cos \lambda_{\mathbb{C}} \sin \varepsilon, \qquad B' = \sin \beta_{\mathbb{C}} (\cos \lambda_{\mathbb{C}} \sin \alpha_{\mathbb{C}} - \sin \lambda_{\mathbb{C}} \cos \alpha_{\mathbb{C}} \cos \varepsilon),$$

$$B'' = \cos \beta_{\mathbb{C}} \cos \varepsilon - \sin \beta_{\mathbb{C}} \sin \lambda_{\mathbb{C}} \sin \varepsilon, \qquad C' = -\sin \delta_{\mathbb{C}},$$

$$C'' = -\sin \beta_{\mathbb{C}} \sin \varepsilon + \cos \beta_{\mathbb{C}} \sin \lambda_{\mathbb{C}} \cos \varepsilon.$$

The differentials dl_0' and db_0' in Eq. (21) may be obtained from Eqs. (12). If we replace the geocentric coordinates of the moon, $\lambda_{\mathbb{C}}$, $\beta_{\mathbb{C}}$, in Eq. (12) by the topocentric coordinates, $\lambda_{\mathbb{C}}'$, $\beta_{\mathbb{C}}'$, then we find equations for determining the component topocentric librations l_0', b_0'. Writing $\eta = l_0' + \mathbb{C} - \Omega + \tau - \sigma$ and $\xi = \lambda_\mathbb{C}' - \Omega - \Delta\psi$ in Eq. (12) and differentiating these formulas, we find dl_0' and db_0' to be

$$dl_0' = \frac{1}{\cos b_0'}\left[(\mathscr{E}'\cos\eta - \mathscr{E}''\sin\eta)\,d\lambda_\mathbb{C}' + (\mathscr{G}'\cos\eta - \mathscr{G}''\sin\eta)\,d\beta_\mathbb{C}' + \mathscr{H}'\cos\eta\,d\rho\right] - d(\tau - \sigma),$$

$$(24)$$

$$db_0' = \frac{1}{\cos b_0'}(\mathscr{E}'''d\lambda_\mathbb{C}' + \mathscr{G}'''d\beta_\mathbb{C}' + \mathscr{H}'''d\rho),$$

where

$$\mathscr{E}' = \cos\beta_\mathbb{C}'\,\cos(\mathscr{I} + \rho)\cos\xi; \qquad \mathscr{E}'' = -\cos\beta_\mathbb{C}'\,\sin\xi;$$

$$\mathscr{E}''' = -\cos\beta_\mathbb{C}'\,\sin(\mathscr{I} + \rho)\cos\xi;$$

$$\mathscr{G}' = -\cos\beta_\mathbb{C}'\,\sin(\mathscr{I} + \rho) - \sin\beta_\mathbb{C}'\,\cos(\mathscr{I} + \rho)\sin\xi;$$

$$\mathscr{G}'' = -\sin\beta_\mathbb{C}'\,\cos\xi; \qquad \mathscr{G}''' = -\cos\beta_\mathbb{C}'\,\cos(\mathscr{I} + \rho) + \sin\beta_\mathbb{C}'\,\sin(\mathscr{I} + \rho)\sin\xi;$$

$$\mathscr{H}' = -\sin\beta_\mathbb{C}'\,\cos(\mathscr{I} + \rho) - \cos\beta_\mathbb{C}'\,\sin(\mathscr{I} + \rho)\sin\xi;$$

$$\mathscr{H}''' = \sin\beta_\mathbb{C}'\,\sin(\mathscr{I} + \rho) - \cos\beta_\mathbb{C}'\,\cos(\mathscr{I} + \rho)\sin\xi.$$

The differentials of the components of the physical libration $d\tau$, $d\rho$, and $d\sigma$ in Eqs. (24) may be expressed in terms of f and $\mathscr{I}$ as follows:

$$d\tau = \frac{\partial\tau}{\partial f}\,df + \frac{\partial\tau}{\partial\mathscr{I}}\,d\mathscr{I};$$

$$d\rho = \frac{\partial\rho}{\partial f}\,df + \frac{\partial\rho}{\partial\mathscr{I}}\,d\mathscr{I}; \qquad\qquad (25)$$

$$d\sigma = \frac{\partial\sigma}{\partial f}\,df + \frac{\partial\sigma}{\partial\mathscr{I}}\,d\mathscr{I}.$$

The coefficients in front of df and $d\mathscr{I}$ are found by differentiating the series for τ, ρ, σ [25]. Since in these series only the coefficients of the trigonometric functions depend on f and $\mathscr{I}$, calculating the coefficients in Eq. (25) reduces to differentiating the coefficients of the corresponding series. As the dependence of the latter on f and $\mathscr{I}$ is given in the form of tables, their derivatives can be found only by numerical differentiation.

To find the differentials of the topocentric ecliptic coordinates of the moon, $d\lambda_\mathbb{C}'$ and $d\beta_\mathbb{C}'$, in Eq. (24), we differentiate Eq. (13):

$$\cos\beta_\mathbb{C}'\cos\lambda_\mathbb{C}'\,d\Delta' - \Delta'\sin\beta_\mathbb{C}'\cos\lambda_\mathbb{C}'\,d\beta_\mathbb{C}' - \Delta'\cos\beta_\mathbb{C}'\sin\lambda_\mathbb{C}'\,d\lambda_\mathbb{C}' =$$

$$= \cos\beta_\mathbb{C}\cos\lambda_\mathbb{C}\,d\Delta - \Delta\sin\beta_\mathbb{C}\cos\lambda_\mathbb{C}\,d\beta_\mathbb{C} - \Delta\cos\beta_\mathbb{C}\sin\lambda_\mathbb{C}\,d\lambda_\mathbb{C} - d(\rho_0\cos\Phi\cos\theta),$$

$$\cos\beta_\mathbb{C}'\sin\lambda_\mathbb{C}'\,d\Delta' - \Delta'\sin\beta_\mathbb{C}'\sin\lambda_\mathbb{C}'\,d\beta_\mathbb{C}' + \Delta'\cos\beta_\mathbb{C}'\cos\lambda_\mathbb{C}'\,d\lambda_\mathbb{C}' =$$

$$(26)$$

$$= \cos\beta_\mathbb{C}\sin\lambda_\mathbb{C}\,d\Delta - \Delta\sin\beta_\mathbb{C}\sin\lambda_\mathbb{C}\,d\beta_\mathbb{C} + \Delta\cos\beta_\mathbb{C}\cos\lambda_\mathbb{C}\,d\lambda_\mathbb{C} - d(\rho_0\cos\Phi\sin\theta),$$

$$\sin\beta_\mathbb{C}'\,d\Delta' + \Delta'\cos\beta_\mathbb{C}'\,d\beta_\mathbb{C}' = \sin\beta_\mathbb{C}\,d\Delta + \Delta\cos\beta_\mathbb{C}\,d\beta_\mathbb{C} - d(\rho_0\sin\Phi).$$

Then, differentiating Eqs. (15), substituting the resulting expressions for $d(\rho_0\cos\Phi\cos\theta)$, $d(\rho_0\cos\Phi\sin\theta)$, and $d(\rho_0\sin\Phi)$ in Eq. (26), and solving Eq. (26) for $d\lambda_\mathbb{C}'$ and $d\beta_\mathbb{C}'$, we obtain

$$d\lambda_\mathbb{C}' = \frac{\cos\beta_\mathbb{C}}{\Delta'\cos\beta_\mathbb{C}'}\sin(\lambda_\mathbb{C} - \lambda_\mathbb{C}')\,d\Delta - \frac{\Delta\sin\beta_\mathbb{C}}{\Delta'\sin\beta_\mathbb{C}'}\sin(\lambda_\mathbb{C} - \lambda_\mathbb{C}')\,d\beta_\mathbb{C} +$$

$$+ \frac{\Delta\cos\beta_\mathbb{C}}{\Delta'\cos\beta_\mathbb{C}'}\cos(\lambda_\mathbb{C} - \lambda_\mathbb{C}')\,d\lambda_\mathbb{C} - \frac{1}{\Delta'\cos\beta_\mathbb{C}'}(\sin\lambda_\mathbb{C}'\cos s -$$

$$- \cos \lambda'_{\mathbb{C}} \sin s \cos \varepsilon) \, dw - \frac{\cos \lambda'_{\mathbb{C}} \sin \varepsilon}{\Delta' \cos \beta'_{\mathbb{C}}} \, dh - \frac{w}{\Delta' \cos \beta'_{\mathbb{C}}} (\sin \lambda'_{\mathbb{C}} \sin s +$$

$$+ \cos \lambda'_{\mathbb{C}} \cos s \cos \varepsilon) \, ds + \frac{\cos \lambda'_{\mathbb{C}}}{\Delta' \cos \beta'_{\mathbb{C}}} (w \sin s \sin \varepsilon - h \cos \varepsilon) \, d\varepsilon,$$

(27)

$$d\beta'_{\mathbb{C}} = - \operatorname{tg} \beta'_{\mathbb{C}} \frac{d\Delta'}{\Delta'} + \frac{\sin \beta_{\mathbb{C}}}{\Delta' \cos \beta'_{\mathbb{C}}} \, d\Delta + \frac{\Delta \cos \beta_{\mathbb{C}}}{\Delta' \cos \beta'_{\mathbb{C}}} \, d\beta_{\mathbb{C}} + \frac{\sin s \sin \varepsilon}{\Delta' \cos \beta'_{\mathbb{C}}} \, dw -$$

$$- \frac{\cos \varepsilon}{\Delta' \cos \beta'_{\mathbb{C}}} \, dh + \frac{w \cos s}{\Delta' \cos \beta'_{\mathbb{C}}} \, ds + \frac{1}{\Delta' \cos \beta'_{\mathbb{C}}} (w \sin s \cos \varepsilon + h \sin \varepsilon) \, d\varepsilon.$$

Finally, in Eqs. (20) and (27) we must put $d\Delta$, $d\lambda_{\mathbb{C}}$, and $d\beta_{\mathbb{C}}$. To do this we use the expansions of the parallax and the ecliptic longitude and latitude of the moon from Brown's theory [20] and take the coefficients of these expansions in the form (10).

We write the expansion for the parallax in the form

$$\sin \pi_{\mathbb{C}} = \frac{a}{\Delta} = \frac{a}{\Delta_0} \, Q \, (T, e, e', \gamma, a, l_0, l'_0, F_0, D_0, \mathbb{C}_0),$$

(28)

where Q is a series obtained from the expansion of the parallax after dividing by the linear factor a/Δ_0.

Writing the angular coordinates of the moon in general form as well, $\lambda_{\mathbb{C}} = \lambda_{\mathbb{C}} (T, e, e' \dots)$ and $\beta_{\mathbb{C}} = \beta_{\mathbb{C}} (T, e, e' \dots)$, we obtain

$$d\Delta = \frac{d\Delta_0}{Q} - \frac{\Delta_0}{Q^2} \sum_1^9 \frac{\partial Q}{\partial p_i} \, dp_i,$$

$$d\lambda_{\mathbb{C}} = d \, \mathbb{C}_0 + \sum_1^9 \frac{\partial \lambda_{\mathbb{C}}}{\partial p_i} \, dp_i,$$

(29)

$$d\beta_{\mathbb{C}} = \sum_1^9 \frac{\partial \beta_{\mathbb{C}}}{\partial p_i} \, dp_i,$$

where $p_i = e, e', \gamma, a, l_0, l'_1, F_0, D_0, \mathbb{C}_0$.

To obtain the differentials of the moon's coordinates (29) we have to take the expansion for them in the general form. For example, the initial terms of the expansion for Q are

$$Q = 1 + 0.9927 e \cos l + 0.1826 \, e \cos (l - 2D) + 0.033 e' \cos (l' - 2D) -$$

$$- 0.1137 a \cos D - 0.0152 \gamma^2 \cos (2F - 2D) + \dots$$

(30)

From Eqs. (29), (27), (24), and (20)-(23) we obtain a conditional equation in the final form

$$dD = D_{\Delta_0} d\Delta_0 + \sum_1^9 D_{p_i} dp_i + D_w dw + D_h dh +$$

$$+ D_s ds + D_\varepsilon d\varepsilon + D_l dl + D_b db + D_{R_{\mathbb{C}}} dR_{\mathbb{C}} + D_f df + D_{\mathcal{J}} d\mathcal{J},$$

(31)

where $p_i = e, e'\gamma, a, l_0, l'_0, F_0, D_0, \mathbb{C}_0$. The coefficients of this equation are

$$D_{\Delta_0} = \frac{1}{Q} \left\{ \left\{ k_1 + \frac{R_{\mathbb{C}}}{\Delta'} \tan \beta'_{\mathbb{C}} \frac{k_3}{\cos b'_0} [k_4 (\mathcal{G}' \cos \eta - \mathcal{G}'' \sin \eta) + k_5 \mathcal{G}'''] \right\} \times \right.$$

$$\times \left(\frac{\Delta}{\Delta'} - \frac{w}{\Delta'} \cos \delta_{\mathbb{C}} \cos t - \frac{h}{\Delta'} \sin \delta_{\mathbb{C}} \right) + \frac{R_{\mathbb{C}}}{\Delta'} \cdot \frac{k_3}{\cos \beta'_{\mathbb{C}} \cos b'_0} \times$$

$$\times \left. \{ \cos \beta_{\mathbb{C}} \sin (\lambda_{\mathbb{C}} - \lambda'_{\mathbb{C}}) [k_4 (\mathcal{G}' \cos \eta - \mathcal{G}'' \sin \eta) + k_5 \mathcal{G}'''] + \sin \beta_{\mathbb{C}} [k_4 (\mathcal{G}' \cos \eta - \mathcal{G}'' \sin \eta) + k_5 \mathcal{G}'''] \} \right\},$$

$$D_{p_i} = -\frac{D_{\Delta_0 \Delta_0}}{Q}\frac{\partial Q}{\partial p_i} + \left\{ -\frac{\Delta}{\Delta'} w A' \sin t + \left(\frac{\Delta}{\Delta'} w \tan \delta_{\mathbb{C}} - h\right) A'' + \right.$$

$$+ \frac{R_{\mathbb{C}}}{\Delta'}\frac{k_3 \cos\beta_{\mathbb{C}}\cos(\lambda_{\mathbb{C}} - \lambda'_{\mathbb{C}})}{\cos\beta'_{\mathbb{C}}\cos b'_0}\Delta\left[k_4(\mathscr{E}'\cos\eta - \mathscr{E}''\sin\eta) + k_5\mathscr{E}'''\right] \bigg\} \times$$

$$\times \frac{\partial\lambda_{\mathbb{C}}}{\partial p_i} + \left\{ -\frac{\Delta}{\Delta'} w B' \sin t + \left(\frac{\Delta}{\Delta'} w \tan \delta_{\mathbb{C}} - h\right) B'' + \right.$$

$$+ \frac{R_{\mathbb{C}}}{\Delta'}\frac{k_3\Delta}{\cos\beta'_{\mathbb{C}}\cos b'_0}\{-\sin\beta_{\mathbb{C}}\sin(\lambda_{\mathbb{C}} - \lambda'_{\mathbb{C}})[k_4(\mathscr{E}'\cos\eta - \mathscr{E}''\sin\eta) +$$

$$+ k_5\mathscr{E}'''] + \cos\beta_{\mathbb{C}}[k_4(\mathscr{G}'\cos\eta - \mathscr{G}''\sin\eta) + k_5\mathscr{G}''']\}\bigg\} \frac{\partial\beta_{\mathbb{C}}}{\partial p_i},$$

$$D_w = k_1\left(\frac{w}{\Delta'} - \frac{\Delta}{\Delta'}\cos\delta_{\mathbb{C}}\cos t\right) + \frac{R_{\mathbb{C}}}{\Delta'}\frac{k_3}{\cos\beta'_{\mathbb{C}}\cos b'_0}(\sin\lambda'_{\mathbb{C}}\cos s -$$

$$- \cos\lambda'_{\mathbb{C}}\sin s\cos\varepsilon)[k_4(\mathscr{E}'\cos\eta - \mathscr{E}''\sin\eta) + k_5\mathscr{E}'''] + \frac{R_{\mathbb{C}}}{\Delta'}\frac{k_3\cos s\sin\varepsilon}{\cos\beta'_{\mathbb{C}}\cos b'_0} \times$$

$$\times [k_4(\mathscr{G}'\cos\eta - \mathscr{G}''\sin\eta) + k_5\mathscr{G}'''],$$

$$D_h = k_1\left(\frac{h}{\Delta'} - \frac{\Delta}{\Delta'}\sin\delta_{\mathbb{C}}\right) - \frac{R_{\mathbb{C}}}{\Delta'}\frac{k_3}{\cos\beta'_{\mathbb{C}}\cos b'_0}\{\cos\lambda'_{\mathbb{C}}\sin\varepsilon \times$$

$$\times [k_4(\mathscr{E}'\cos\eta - \mathscr{E}''\sin\eta) + k_5\mathscr{E}'''] + \cos\varepsilon[k_4(\mathscr{G}'\cos\eta - \mathscr{G}''\sin\eta) + k_5\mathscr{G}''']\},$$

$$D_s = k_1\frac{\Delta}{\Delta'} w\cos\delta_{\mathbb{C}}\sin t + \frac{R_{\mathbb{C}}}{\Delta'}\frac{k_3 w}{\cos\beta'_{\mathbb{C}}\cos b'_0}\{-\sin\lambda'_{\mathbb{C}}\sin s +$$

$$+ \cos\lambda'_{\mathbb{C}}\cos s\cos\varepsilon)[k_4(\mathscr{E}'\cos\eta - \mathscr{E}''\sin\eta) + k_5\mathscr{E}'''] + \cos s[k_4(\mathscr{G}'\cos\eta - \mathscr{G}''\sin\eta) + k_5\mathscr{G}''']\},$$

$$D_\varepsilon = -k_1\frac{\Delta}{\Delta'}[w\sin t C' - (w\tan\delta_{\mathbb{C}} - h) C''] + \frac{R_{\mathbb{C}}}{\Delta'}\frac{k_3}{\cos\beta'_{\mathbb{C}}\cos b'_0} \times$$

$$\times \{(w\sin\varepsilon\sin s - h\cos\varepsilon)[k_4(\mathscr{E}'\cos\eta - \mathscr{E}''\sin\eta) + k_5\mathscr{E}'''] +$$

$$+ (w\sin s\cos\varepsilon + h\sin\varepsilon)[k_4(\mathscr{G}'\cos\eta - \mathscr{G}''\sin\eta) + k_5\mathscr{G}''']\},$$

$$D_l = -k_3 k_4 R_{\mathbb{C}}, \qquad D_b = k_3 k_5 R_{\mathbb{C}}, \qquad D_{R_{\mathbb{C}}} = k_2\cos\omega',$$

$$D_f = k_3 k_4\left(\frac{\partial\tau}{\partial f} - \frac{\partial\sigma}{\partial f}\right) + \frac{R_{\mathbb{C}}}{\cos b'_0}(k_4\mathscr{H}'\cos\eta + k_5\mathscr{H}''')\frac{\partial\rho}{\partial f},$$

$$D_{\mathscr{G}} = k_3 k_4\left(\frac{\partial\tau}{\partial\mathscr{G}} - \frac{\partial\sigma}{\partial\mathscr{G}}\right) + \frac{R_{\mathbb{C}}}{\cos b'_0}(k_3 k_4\mathscr{H}'\cos\eta + k_3 k_5\mathscr{H}''')\frac{\partial\rho}{\partial\mathscr{G}}.$$

    The conditional equations of the form (31) have the advantage that they make it possible to present clearly the basic causes and nature of the change in the coefficients as the parameters are readjusted in time. This in turn makes it possible to justify several simplified variants of the general pattern of the motion. In the following we shall use this fact in numerically evaluating some of the parameters of the earth—moon system from simplified systems of conditional equations.

    Equations of type (1) may be used to precalculate the distances D exactly, as is required to make laser ranging measurements, and Eq. (31) may be used to determine the corrections to the earth—moon system parameters which arise as a result of the actual laser ranging measurements and to make a priori estimates of the accuracy with which these parameters have been determined.

    To study the systems of conditional equations with the aid of a computer it is convenient to write them in matrix form, as has been done in [34].

In order to make an a priori estimate of the accuracy in determining the corrections to the parameters of this problem (or, equivalently, the accuracy of determining the parameters themselves), we analyzed a system of these equations.[†]

The equations were constructed for different times and intervals which encompassed all the periods in the mutual motion of the earth and moon. Differentiation of the components of the physical libration τ, ρ, σ was reduced to differentiating with respect to the parameters of the physical libration (the physical libration constant f and the obliquity of the equator to the ecliptic $\mathscr{I}$), and differentiation of the ecliptic coordinates of the moon, Δ, $\lambda_{\mathbb{C}}$, $\beta_{\mathbb{C}}$, was with respect to the following parameters: the mean longitudes of the moon $\mathbb{C}$ and the ascending node of the lunar orbit Ω, the mean longitudes of the perigees of the lunar and solar orbits π and π', the mean longitude of the sun L', the eccentricities of the lunar and solar orbits e and e', the constant γ of the obliquity of the lunar orbit to the ecliptic, and the mean geocentric distance of the moon Δ_0.

The corresponding systems of normal equations were constructed, and their solutions were studied for measurement series of various lengths: 1 day, 1 month, 9.3 years, and 18.6 years. For each of these periods the density of measurements (the interval between successive measurements) as well as the number and mutual locations of the observation points on the earth and of the reflectors on the moon were varied. The analysis yielded the following results.

1. For the daily series of measurements the majority of coefficients for the correction of the parameters in the conditional equations remain practically constant. Thus it becomes possible to solve only the very simplified problem of determining the coordinates of the observation points. In that case it appears that only the longitude λ and the radius w of the parallel are determined with good accuracy while the error in the distance to the equatorial plane, h, is large.

Increasing the duration of the observations within the limits of a single passage across the sky and increasing the density of the measurements greatly increases the accuracy of the determination of λ and w. Increasing the number of observation points to two or three does not increase the accuracy of λ and w, but synchronous measurements from two points makes it possible to determine the difference $h_2 - h_1$ of their distances from the equatorial plane with good accuracy.

2. Measurements over a month offer many more possibilities than those over a daily interval. In addition to the coordinates of the observation points (which are determined with roughly the same accuracy as in the daily measurements), month-long observations offer the possibility of determining a number of constants of the rotational and orbital motion of the moon with an accuracy comparable to the best optical determinations. The errors in such constants as e, e', γ, Δ_0, l, b, $R_{\mathbb{C}}$, f, and $\mathscr{I}$ when measured from a single point are considerably reduced when the number of reflectors on the moon is increased from one to three. Further increasing the number of reflectors yields no gain in accuracy.

The optimum appears to be achieved when three reflectors are placed at the points with selenographic coordinates given by $l_1 = +40°$, $b_1 = +45°$; $l_2 = 0°$, $b_2 = -45°$; $l_3 = -40°$, $b_3 = +45°$. The accuracy in determining the constants may be further increased by observing the same three reflectors from two points on the earth and also by increasing the rate of the measurements.

3. By regular measurements over a year such that the measurement times are uniformly distributed over this period it seems possible to increase the accuracy of such constants of the

[†] The calculations were done at the Institute of Theoretical Astronomy of the Academy of Sciences of the USSR in the "Astronomical Yearbook of the USSR" department, and the results are published in Astronomicheskii Zhurnal [35].

moon's orbital motion as $\mathbb{C}, \Omega, \pi$. The accuracy with which the physical libration parameters f and $\mathcal{J}$ are determined is increased by roughly an order of magnitude compared to the accuracy obtained with monthly measurements.

As in the previous case the optimum appears to be two to three reflectors on the moon and two observation points. The parameters of the heliocentric motion of the earth L' and π' are determined with relatively low accuracy over a yearly interval.

4. The accuracy in determining the corrections to the parameters increases when the measurement interval is increased to 9.3 years, that is, to half the regression period of the node lines of the lunar orbit. Further increasing the measurement interval to the full period (18.6 years) yields roughly the same results as doubling the rate of the measurements.

Based on this analysis we may make the following basic conclusions. To determine the maximum possible number of parameters of the earth−moon system with maximum accuracy it is necessary to (a) make regular distance measurements over a period of at least 9.3 years, (b) uniformly distribute the times of the measurements both over the entire period and over the yearly, monthly, and daily subintervals within it in order to span the bulk of the periods of variation of the coefficients of the conditional equations with measurements, (c) measure the distances to three reflectors suitably positioned on the moon, and (d) make the measurements

TABLE 1

Parameter	Measurement interval	No. of observation points	No. of reflectors on moon	Error in laser ranging determination	Error in optical determination
Coordinates of the observation point:					
longitude λ	12 h	1	1	$0".01$	$0".1$
distance to the rotation axis w	12 h	1	1	0.2 m	7 m
distance to the equatorial plane h	9.3 yr	1	1	0.2 m	7 m
Selenographic coordinates of the reflector					
longitude l	1 yr	1	1	$0".6$	$300"$
latitude b	1 yr	1	1	$0".6$	$180"$
radius-vector $R_{\mathbb{C}}$	1 yr	1	1	5.0 m	1000 m
Parameters of the physical libration					
the constant f	9.3 yr	1	2	0.0001	0.02
obliquity of the linear equator to the ecliptic $\mathcal{J}$	1 mo	1	3	$0".2$	$3"-8"$
Parameters of the moon's orbital motion:					
average distance to the moon Δ_0	1 yr	1	1	4.5 m	1000 m
eccentricity e	1 yr	1	1	$1 \cdot 10^{-9}$	$1 \cdot 10^{-7}$
mean longitude of the moon $\mathbb{C}$	1 yr	1	1	$0".004$	$0".02$
mean longitude of the perigee of the moon's orbit π	9.3 yr	1	1	$0".002$	$0".2$
mean longitude of the ascending node of the lunar orbit along the ecliptic Ω	9.3 yr	1	1	$0".06$	$0".3$
constant of the obliquity of the moon's orbit to the ecliptic γ	1 yr	3	1	$2 \cdot 10^{-8}$	$1 \cdot 10^{-7}$
Parameters of the orbital motion of the earth:					
orbital eccentricity e'	1 yr	1	1	$5 \cdot 10^{-9}$	$7 \cdot 10^{-7}$
mean longitude of the sun L'	9.3 yr	1	1	$0".01$	—
mean longitude of the perigee of the sun's orbit π'	9.3 yr	1	1	$0".06$	—
Obliquity of the earth's equator to the ecliptic ε	1 yr	1	2	$0".02$	—

from two observatories. It is best if they are located at one latitude (about 45°) with a spacing of about 60-90° in longitude.

Table 1 shows the expected root-mean-square errors in 18 parameters of the earth—moon system. It includes data corresponding to various measurement conditions (duration, number of observing points and reflectors) which ensure the greatest accuracy in these parameters. Also shown for comparison are the root-mean-square errors of the best measurements of some parameters by other methods [36]. We note that because of the linearity of the conditional equations the errors in the parameters are proportional to the measurement errors. The data in Table 1 were obtained assuming that the root-mean-square error in the measurement of the locator—reflector distance is 50 cm, which is approximately the accuracy of current equipment. However, work has already begun on the real technical possibility of increasing the accuracy of the measurement to 3-5 cm. (Further reduction in the measurement error is limited by the difficulty of accounting for fluctuations in the optical path of the laser signal in the atmosphere.) Thus, it is possible to plan on increasing the accuracy of measuring the parameters by at least an order of magnitude above the levels indicated in the table.

4. Scientific Problems Which May Be Solved
Using Laser Ranging of the Moon

These estimates of the accuracy of measurements of various characteristics of the earth—moon system show that laser ranging measurements of distances from the earth to fixed points on the moon's surface will enable us to substantially improve the accuracy of our description of the geometry and, with sufficiently long measurements, the dynamics of the earth—moon system. Based on these estimates we can formulate the basic research areas in which laser ranging of the moon may offer significant progress. Here we briefly describe the main problems.

1. **Establishing a Network of Reference Geodesic Points and Studying the Coupling of Various Geodesic Systems.** If the geocentric coordinates of the observation points can be determined with an accuracy of about 20 cm, then they can be used as reference geodesic points of very high precision. If these points were located within various geodesic systems, then it would be possible to measure the coupling of these systems to one another with the accuracy mentioned above.

2. **Improving the Astronomical-Geodesic Network of the USSR.** If two or more geodesic points determined in this way were to be located in the boundaries of the astronomical-geodesic network (of the USSR) and sufficiently far apart from one another (3000-10,000 km), then on linking these points to the nearest points of the network it should be possible to refine the scale of the network to a relative error of 10^{-7}-10^{-8}, refine its orientation, and to relate it to the center of mass of the earth. This would result in improved accuracy in the determination of the fundamental geodesic parameters, including the dimensions of the terrestrial ellipsoid which are based on data from the astronomical-geodesic network.

3. **Studies of the Motion of the Earth's Poles.** With two observation points it would be possible to organize a service which would observe the motion of the poles and determine their instantaneous coordinates to within about 30 cm, that is, 1 or 1.5 orders of magnitude more accurately than with other methods. As opposed to other astronomical methods the location of the local verticals at the observation points will not affect the results of the observations.

Constant monitoring of the motion of the poles would make it possible to study the motion of the observation points with respect to the center of mass and would make it possible to include the effect of polar motion on the orbits of artificial earth satellites.

   4. More Refined Laws of the Earth's Rotation. Simultaneous observations from two points make it possible to measure their relative longitudes with an accuracy of about 0.010 angular seconds. These data could be used to study the nonuniform rotation of the earth due to the displacement of mass in the atmosphere and within the earth and to tidal effects.

   5. The Study of Continental Drift. If the observation points were placed on different tectonic blocks, the mutual horizontal motions of these blocks could be studied. At their probable speeds of 1-10 cm per year, the time required to measure these speeds would be 5-20 years.

   6. Refining the Orbits and Laws of Orbital Motion of the Moon. Measuring the distances to reflectors on the moon over a time interval which encompasses the main periodic components of the moon's orbital motion makes it possible to improve the accuracy of the constants of this motion which enter the Hill—Brown theory by two or three orders of magnitude. These measurements may indicate perturbations in the lunar motion that are not included in this theory, for example, damping of the moon's rotation due to a secular reduction in the gravitational constant [37].

   7. Studies of the Moon's Rotation. By observing three reflectors on the moon we can measure the physical libration of the moon with an accuracy one or two orders of magnitude better than now. This may allow us to choose among alternative models of the internal structure and hypotheses about the moon's origin.

   8. The Study of the Moon's Shape. Laser ranging measurements make it possible to measure the selenographic coordinates with an accuracy two or three orders of magnitude above that of other methods. This opens up the possibility of studying the moon's shape and dynamics if enough corner reflectors are placed on the moon. Knowledge of the dynamics will allow studies to be made of the effect of tidal deformations on the dynamics of the earth—moon system. The corner reflectors can be used as a system of reference points for exact selenodesic plots, in particular, for determining the scale of lunar maps.

   It should be noted that the greatest promise is offered by laser ranging in combination with other methods (optical goniometry, radio interferometry on extremely long bases, and so on). This would increase the accuracy and reliability with which many parameters of the earth—moon system are determined.

   In this regard it is appropriate to note that laser ranging makes it possible to simultaneously measure the angular coordinates of the points being probed and the distances to them. To do this it is necessary to make synchronous measurements to one reflector on the moon from several points on the earth. Each pair of such points forms a side (base) of a triangle, the other two sides of which are distances to the reflector. Solving this triangle can yield all three of its angles. A simple calculation shows that the accuracy of angular measurements with this method is the same as with radio interferometry with an extremely long baseline (or the same baseline length) when centimeter waves are used (for which the effect of the ionosphere may be neglected). This is natural since in both methods the accuracy is limited by fluctuations in the optical path length of light or radio waves in the atmosphere.

III. THE PARAMETERS AND DESIGN OF CORNER REFLECTORS

1. Introduction

   The measurements of distances to the moon with an error of less than a few meters can be of scientific value. The difficulty in achieving such accuracy is due to the fact that the laser beam directed toward the moon has some divergence. This is due both to the properties of the laser beam itself, which has a divergence of order 3″-5″ at the output of the transmitting telescope and, more fundamentally, to scattering of the laser light on atmospheric inhomogeneities. Under normal atmospheric conditions the magnitude of this effect is also 3″-5″. Thus, on leaving the atmosphere the laser beam has a total divergence averaging 5″-7″; that is, it illuminates an area of diameter 8-12 km on the moon. The nonuniform relief inside this area and its slope relative to the laser beam cause blurring of the reflected signal in time that increases with the roughness of the surface in this area and its distance from the center of the lunar disc. For example, for an area located 500 km from the center of the lunar disc, the blurring of the signal due to the sphericity of the moon is 5-8 km.

The only way to increase the accuracy of the measurements is to place artificial targets of small area on the moon which efficiently reflect light back to the observer. This localizes the reflection point on the moon with an accuracy determined by the size of the targets and thus ensures the possibility of spatial and temporal selection of the signal when it is received. Such targets can be corner reflectors (triple prisms, cataphotes) which, if chosen to have the appropriate parameters and made with high precision, can yield considerable gain in the magnitude of the reflected signal compared to the signal scattered by the moon's surface.

We now consider the choice of parameters for corner reflectors.

2. The Energy of the Laser Signal Reflected

from the Moon's Surface

The energy of the laser signal reflected from the moon's surface is determined by the ranging formula in which the fourth power of the distance to the target has been replaced by the square since all the energy of the laser beam falls on the target,

$$W_D^L = \frac{W_{tr} S_T}{\pi R^2} \rho k_{tr} k_D k_{atm}^2. \tag{32}$$

Here W_D^L is the energy of the reflected signal falling in the receiver telescope, W_{tr} is the energy emitted by the laser, S_T is the aperture area of the receiver telescope, R is the distance to the moon, ρ is the moon's albedo, k_{tr} and k_D are the transmission coefficients of the transmitter and detector optics, and k_{atm} is the transmission coefficient of the atmosphere.

As a numerical estimate we take the following values of the quantities in Eq. (32): S_T = 4.4 m² (the ZTSh-2.6 telescope at the Crimean observatory); R = 380,000 km; ρ = 0.1 [38]; k_D = 0.6 (see Chapter V, Section 5); k = 0.2 (see Chapter IV, Section 2); k_{atm} = 0.8 (see Chapter IV, Section 1).

Then

$$W_D^L \cong 0.75 \cdot 10^{-19} W_{tr}.$$

For a ruby laser (λ = 6943 Å) whose energy when used as a pulsed ranging transmitter is usually about 3 J, the number of photons in the received signal is

$$n_D^L = \frac{W_D^L \lambda}{hc} \simeq 0.8.$$

If we take the photodetector efficiency at this wavelength to be about 0.03 (a photomultiplier with a multialkali cathode), then the number of signal pulses at the photodetector output per laser pulse is about 0.02, or one response pulse in roughly 50 laser shots. This shows clearly how important it is to increase the energy of the reflected signal. One possible way of increasing the energy efficiency of a laser ranging system is to use light reflectors on the moon.

3. The Energy of the Reflected Signal when Reflectors on the Moon Are Used

An elementary light reflector is an equilateral four-faced pyramid OABC which is obtained if one cuts the corner of a cube with the plane ABC which then serves as the input face of the prism (Fig. 3a). A light ray incident on the reflector through plane ABC undergoes triple reflection from the side faces and leaves in the opposite direction parallel to the incident ray. The incident and reflected rays are symmetric with respect to the vertex O and are parallel regardless of the angle of incidence (Fig. 4). For normal incidence of the rays on the input face the reflecting area is a regular hexagon (see Fig. 3b) whose area is

$$s = a^2/\sqrt{3}, \tag{33}$$

where a is the cube edge.[†]

The principal limitation on the parameters of a reflector is its weight. Thus, before specifying the allowable weight of a reflector, we shall calculate the number of elements necessary to obtain a maximum reflected signal at the receiver point. The weight and area of a corner reflector are given by

$$P_0 = \frac{a^3}{6}\delta N, \qquad S_0 = sN, \tag{34}$$

where δ is the density of the prism material and N is the number of prisms. The area illuminated on earth by the reflected signal from an ideal reflector is determined by the diffractive divergence of the reflected beam. In a strict calculation of the diffractive divergence it would be necessary to take into account the fact that the diffraction aperture is hexagonal. As an adequate approximation we take the reflector aperture to be the circle inscribed in the hexagon with diameter

$$D_0 = 2a/\sqrt{6}. \tag{35}$$

If we assume that the elements are placed randomly in the incident flux, then with a large number of elements this kind of reflector may be regarded as a set of spatially incoherent reflectors. In this case the diffractive divergence of the entire reflector is equal to the diffractive divergence of a single element,

$$\alpha_0 = 2\frac{1.2\lambda}{D_0} = \frac{2.94\lambda}{a}. \tag{36}$$

The energy density at the receiver point is $w_D \sim S_0/S_L S_E$.

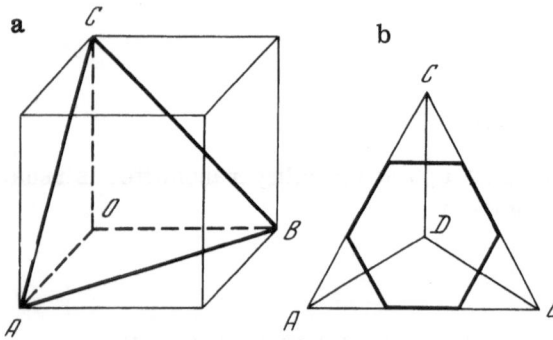

a

b

Fig. 3. A diagram of an elementary reflector: a) input face of the prism; b) the reflecting area for normally incident rays.

[†] In the following we shall call the size of the corner element the length of the edge of the cube from which it has been cut.

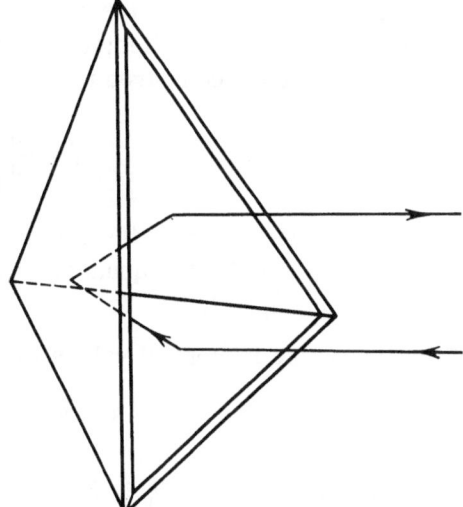

Fig. 4. The ray path in a corner reflector.

The area of the light spot on the moon, S_L, is determined by the divergence of the laser beam after it has passed through the atmosphere and is equal to

$$S_L = \pi \left(R \frac{\alpha_L}{2} \right)^2,$$ (37)

where α_L is the divergence angle of the laser beam after it has passed through the atmosphere and R is the distance to the moon.

The area of the spot on the earth is

$$S_E = \pi \left(R \frac{\alpha_0}{2} \right)^2 = 2.16\,\pi R^2 \frac{\lambda^2}{a^2}.$$ (38)

For a given reflector weight $S_E \sim N^{-2/3}$. Thus, the energy flux at the receiver is $w_D \sim N^{-1/3}$. From this it follows that to obtain the maximum reflected signal it is necessary to use a single-element reflector. We now express the amount of reflected signal as a function of the reflector weight when N = 1 and compare this with the signal obtained without a reflector. At normal incidence the energy flux on a corner reflector is

$$w_{tr} = \frac{W_{tr} D_{tr}}{4\pi R^2}.$$ (39)

Here w_{tr} is the energy flux, W_{tr} is the energy emitted by the laser, and D_{tr} is the directionality of the transmitter system, given by $D_{tr} = 4\pi R^2 S_L^{-1}$. The energy flux at the receiving telescope in the direction of the emission maximum is

$$w_D = \frac{w_{tr} S_{eff}}{4\pi R^2},$$ (40)

where S_{eff} is the effective area of the corner reflector [39], given by

$$S_{eff} = \frac{4\pi}{\lambda^2} s^2 = \frac{4\pi a^4}{3\lambda^2}.$$ (41)

Thus the ranging formula now has the form

$$W_D^0 = W_{tr} \frac{a^4 S_T}{3\lambda^2 R^2 S_L} \gamma^2 k_{tr} \, k_D \, k_{atm}^2 k_{refl},$$ (42)

where k_{refl} is the reflectivity of the prism and γ is a coefficient which takes into account the reduction in the area of the reflector for oblique incidence.[†] The gain in the reflected signal when a reflector is used compared to diffuse scattering from the surface of the moon is

$$\beta = \frac{W_D^0}{W_D^L} = \frac{\pi a^4 k_{refl} \gamma^2}{3\lambda^2 \rho S_L}.\qquad(43)$$

Including triple reflection from the prism faces and possible dirtying of the reflector on the moon, we take $k_{refl} = 0.5$. The quantity γ may vary between 1.0 and 0.8. For our estimates we take $\gamma = 0.8$ and $S_L = 25$ km^2 ($\alpha = 3''$).

Substituting the numerical values in Eq. (43) we obtain

$$\beta = 2.7 \cdot 10^{-3}\ a^4 \cong 7.7 \cdot 10^{-3}\ P_0^{4/3}.\qquad(44)$$

Here a is in centimeters and P_0 is in grams for a prism with a density of $\delta = 2.7$ g/cm^3. The diffractive divergence of the reflected beam at half energy is (in angular seconds)

$$\alpha_0^{(0.5)} = 1.02\frac{\lambda}{4.95 \cdot 10^{-6} D_0} = 18a^{-1} = 14P_0^{-3}.\qquad(45)$$

Here are the values of $\alpha_0^{(0.5)}$ and β for several values of the weight P_0:

P_0, kg	a, cm	$\alpha_0^{(0.5)''}$	β
0.5	10.4	1.7	32
1.0	13	1.4	79
2.0	16.5	1.1	205
5.0	22	0.8	640
10.0	28	0.6	1690

4. The Effect of Fabrication Errors on the Magnitude of the Reflected Signal

The values and equations given above were obtained assuming that the corner element is an ideal pyramid, with absolutely perfect right angles between the faces, absolutely plane faces, and optically homogeneous material. The diffractive divergence of this reflector is determined by its entire area which in this case is in phase. If one of the prism angles deviates from $\pi/2$ by θ, then a reflected ray is split into two rays lying in a plane perpendicular to the edge corresponding to this angle. The divergence angle of the beams is [40]

$$\alpha_\theta = \frac{4\sqrt{6}}{3}n\theta = 3.27n\theta,\qquad(46)$$

where n is the refractive index of the prism.

When two or three of the prism angles have deviations the reflected beams is broken up into four or six separate beams, respectively, whose mutual positions are determined by the magnitudes and signs of the deviations. If the angles between these greatly exceed the divergence of each of them (which is now determined by new diffractive apertures consisting of quadrangles a, a', b, b', c, and c' of Fig. 5), six separate diffraction patterns are formed each of which has a central intensity equal to 1/36 the intensity for an ideal reflector. Such a re-

----
[†] The axis of a reflector located on the moon may deviate from the direction of the incident beam due to libration by roughly $\pm 8°$.

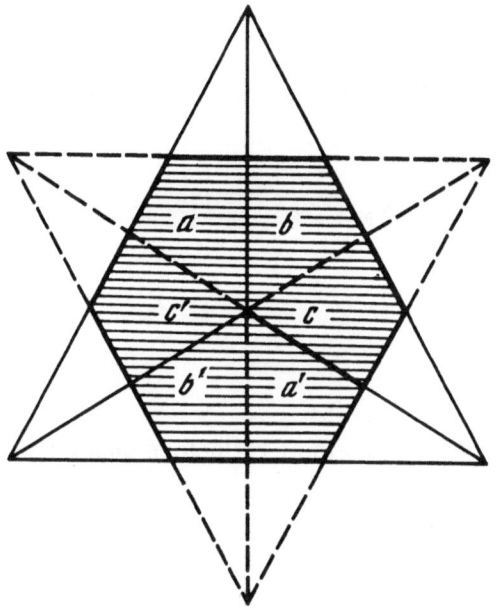

Fig. 5. The input apertures of the reflector.

Fig. 6. A diffraction diagram of a corner reflector with identical deviations from 90° in all three angles.

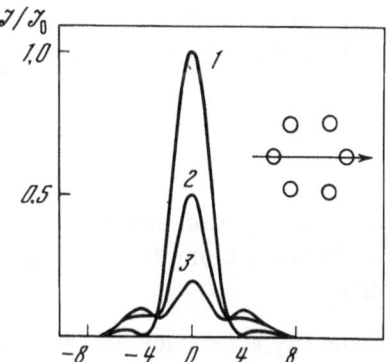

flector is not suitable for laser ranging, however, since the reflected rays do not fall on the point where the light source is located. The condition for conservation of the central maximum $(\alpha_\theta \leq \alpha_0^{(0.5)})$ places rigid requirements on the accuracy of the reflector angles. It is easy to see that the allowable errors in these angles are at most a few tenths of an angular second. In this case the overall diffraction pattern is obtained as a result of the interference of the separate beams corresponding to the areas a, a', b, b', c, and c'. As an illustration, Fig. 6 shows cross sections of the diffraction pattern of a reflector with identical deviations in all three angles. Curves 1, 2, and 3 correspond to $\theta_1 = 0$, $\theta_2 = \lambda/\pi r$, and $\theta_3 = 3\lambda/2\pi r$, where r is the radius of the circle inscribed in the hexagonal aperture [41, 42]. If we take the allowable reduction at the center to be that corresponding to curve 2, then for r = 5 cm and λ = 0.7 μm the allowable error in the angles is $\theta \sim 1''$.

Distortions in the face planes lead to more complicated distortions of the diffraction diagram of the reflector. An analysis shows that the faces must be held flat to within hundredths of a micron.

The directional diagram of a real reflector working in real lunar conditions, where additional deformations may arise due to temperature gradients, is extremely complicated and it is impossible to calculate it. Hence we shall use several simplifying assumptions to estimate the reflected signal from a real reflector. In deriving Eqs. (42) and (43) we used the concept of an effective area based on the notion that the reflecting surface was in phase. The idea of an effective area is not applicable to a real reflector. A quantitative estimate of the reflected signal may be made in this case if we characterize the reflector by some effective directional diagram Θ with a uniform energy distribution within it. Then the magnitude of the reflected signal may be determined from

$$W'_D = \frac{16 W_{tr} s S_T}{\pi^2 R^4 \alpha_L^2 \Theta^2} \gamma^2 k_{tr} k_D k_{refl} k_{atm}^2 \tag{47}$$

and the gain is

$$\beta' = \frac{4 a^2 k_{refl} \gamma^2}{\sqrt{3} S_L \rho \Theta^2}. \tag{48}$$

The quantity Θ is determined experimentally on the basis of tests of reflectors under conditions simulating those on the moon.

5. The Effect of Aberration of Light on the Reflected Signal

The (velocity) aberration of light which leads to a shift in the diffraction pattern on the earth has a significant effect on the magnitude of the reflected signal at the receiving point. This aberration is due to the mutual motion of the earth and moon and is given by the expression

$$\varphi = 2v/c,$$

where φ is the aberration angle and v is the component of the moon's velocity relative to the earthbound observer along the normal to the line of sight and lying in the orbit plane; c is the speed of light.

The quantity φ depends on the hour angle, the moon's declination, and the latitude of the observation site and may lie between 0".7 and 1".4. The displacement of the reflected beam on the earth is determined by the projection on the surface of the earth of the vector φR, where R is the distance to the moon. For middle latitudes and moderate hour angles of the moon (about ± 45° from the meridian), that is, under the conditions in which laser ranging measurements are usually made, the displacement of the reflected beam on the earth with respect to the transmission point is 1.5 to 3.0 km, depending on the hour angle, and the direction of the displacement varies during the lunar month by ± 45° from eastward [43].

The effect of aberration can be fully compensated if the receiver and transmitter are separated from one another in accordance with the aberration displacement. However, besides the fact that this requires the use of two separate telescopes for the transmitter and receiver, this demands that they be moved relative to one another in accordance with the time variation of the aberration displacement. This solution is unacceptable from a technical point of view.

The effect of aberration can be greatly reduced if we make the element of the corner reflector sufficiently small that the half-angle of its diffraction pattern exceeds the aberration angle. For an ideal reflector Eq. (44) indicates that $\beta \sim a^4 N$, where N is the number of elements, and for a given maximum weight $P_{refl} = N P_0$ the entire reflector has $\beta \sim a$. This means that we must choose an element with maximum size a_{max} and weight $P_{0\,max}$ for which the condition $\alpha_0/2 > \varphi$ still holds. The loss in gain must be compensated by taking the number of elements to be $N = P_{refl}/P_{0\,max}$.

Two situations may occur with a real reflector: (1) $\Theta/2 > \varphi$ only for elements that are so small that the directional pattern is mainly determined by the diffractive properties and the

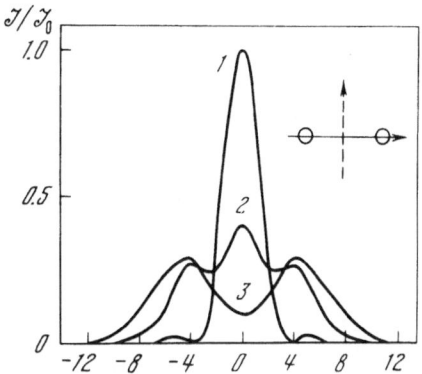

Fig. 7. The diffraction pattern of a corner reflector with a deviation from 90° in one of its angles.

role of fabrication errors is small; then the same recommendations on choosing the element sizes are valid as for an ideal reflector; and (2) $\Theta/2 > \varphi$ is also true for elements for which the role of diffraction is small and the directional pattern is mainly determined by fabrication errors; then the width of the pattern may be assumed independent of the dimensions of the reflector. In case (2) Eq. (48) implies that the gain $\beta \sim a^2 N$, or for a given weight of the entire reflector, $\beta \sim 1/a$. Thus, it is necessary to choose an element of minimum dimension $a_{\min}$ for which it is still true that $\Theta > \alpha_0$ and again compensate the loss in gain by increasing the number of elements. Clearly, the dimension $a_{\min}$ corresponds to an element for which the contributions of diffraction and fabrication errors to the formation of the directional pattern are roughly the same. For these values of a the dependence $\beta \sim a$ takes the form $\beta \sim 1/a$.

With modern technology and monitoring techniques the errors in the angles of the element can be made equal to about $0".2-0".3$, which corresponds to a broadening of the diagram to $2"-3"$. This means that in practice the second of these situations occurs and $a_{\min}$ must be chosen equal to 6-9 cm in accordance with Eq. (45).

Another means of partially compensating the aberration is to artificially elongate the reflection patterns in the plane of the aberration angle by introducing a deviation from $\pi/2$ in one of the angles of the reflecting element and its corresponding orientation on the moon. The deviation is chosen to be equal to $\theta = 2\varphi/3 \cdot 27n$, where φ is the mean aberration angle ($\sim 1"$). The possibilities of this method are illustrated in Fig. 7 in which curves 1, 2, and 3 correspond to the same values of θ as in Fig. 6. Compared with the previous method, this one makes it possible to use larger but fewer elements, which is favorable from a weight standpoint. Therefore, fabrication errors and aberration do not permit realization of the most favorable (from the standpoint of weight) single-element design. All existing reflectors have been made up of many elements.

6. Practical Construction of Corner Reflectors

The difficulties in developing and fabricating reflectors due to the high requirements of angular tolerance and optical homogeneity of the prism material are aggravated by the need to maintain the optical characteristics of the elements in the harsh temperature conditions on the moon. Sharp temperature changes between day and night (−120°C to 150°C) and large temperature drops between the surface and the surrounding space will cause temperature gradients to develop within a prism. This causes the prisms to undergo thermal deformation and destroys its optical homogeneity. The heating on a single side of the reflector by the sun during the lunar day is a particularly harsh thermal regime. Thus, special systems for thermal protection and thermal stabilization seem necessary to reduce heat exchange between the reflector and the environment and reduce the temperature gradients in the prisms.

In addition to these requirements the reflector must have sufficient mechanical durability to resist shock and vibration loads which it will experience during space flight.

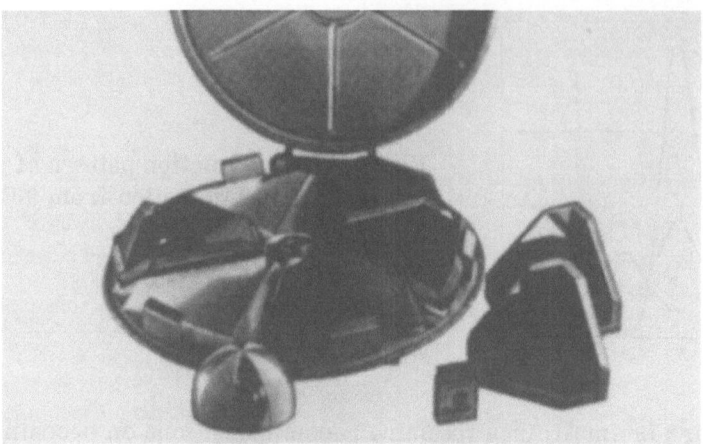

Fig. 8. The Soviet corner reflector.

Thus, the construction of reflectors for laser ranging of the moon, which must satisfy a number of contradictory requirements (high efficiency, thermal stability, low weight, mechanical durability), is a complicated engineering problem.

Up to now three types of lunar reflectors have been developed, Soviet, French, and American. They differ in design, size, precision of fabrication of the corner elements, and number of elements, as well as in the design of the thermal stabilization system. The choice of one or another design is mainly determined by weight and dimensional limits of the spacecraft used to place the reflector on the moon and by technological considerations.

The Soviet reflector is based on elements consisting of three mutually perpendicular plates whose flatness and mutual angles are maintained with high precision (0.03 μm and 0".2, respectively). The reflector is a panel of six corner elements made of glass crystal (sitall; see Fig. 8). According to its design, it should work over a wide temperature range corresponding to the night- and daytime conditions on the moon. One feature of this reflector which distinguishes it from the other designs is an elongation of the reflection pattern which makes it possible to partially compensate for the reduction in the reflected signal at the receiver point due to aberration.

Fig. 9. The French L-1 reflector.

Fig. 10. The American A-11 reflector.

The French reflector (Fig 9) consists of 14 triple prisms made of highly homogeneous "homosil" quartz. The prism angles are maintained with an accuracy of 0".2, and the face flatness, to an accuracy of 0.07 μm. The working faces are coated with a layer of silver. The prisms are mounted on a common base, or panel, which is covered on the bottom and sides with a multilayer sheath of thermally isolating material (Inconel).

The weight allowed the American reflectors (Fig. 10) was much greater than that permitted the French. Thus it was possible to make a reflector which was able to work during the lunar day although its efficiency was somewhat less than that at night.

In developing this design all measures were taken to reduce the contribution of thermal deformations and refractive index gradients in the prisms. In particular, the size of the prisms was chosen to be small (4.7 cm) to ensure sufficiently wide diffraction widths of the reflection patterns.

The thermal protection system was constructed according to a different principle from that used in the French and Soviet variants. It is a massive slab of aluminum alloy. The prisms are recessed in the pockets of this plate to ensure good heat transfer between the prisms and the plate and to reduce the temperature gradients.

TABLE 2

Reflector parameter	USSR, development	France, Lunokhod 1 and 2	USA	
			Apollo 11 and 14	Apollo 15
Type of prism	Hollow	Solid	Solid	Solid
Material	Glass crystal	Homosil	Suprasil-1	Suprasil-1
Prism size, cm	12	8.9	4.7	4.7
Angular tolerance	±0.2"	±0.2"	±0.3"	±0.3"
No. of prisms	6	14	100	300
Dimensions, cm	46×36×10	44×19×7.5	46×46×8	104×63×8
Weight, kg	7.0	3.5	20	32
Area, cm²	750	640	1100	3300
Prism pattern θ	3"×5"	5"	10"	10"
Gain $N\beta'$†	~25	~13	~5.5	~16.5
Expected signal magnitude‡ (No. of photons)	~1.4	~0.7	~0.3	~0.9

†The quantity $N\beta'$ is given for lunar night conditions; during the lunar day it drops by
  about 1.5 times for the American reflectors and about 30 times for the French.
‡This estimate was obtained using Eq. (47) with the following parameters: $W_{tr} = 2$ J,
  $S_T = 4.4$ m²; R = 380,000 km; $\alpha_L = 7"$; $\gamma = 0.8$; $k_{tr} = 0.6$; $k_D = 0.2$; $k_{refl} = 0.5$;
  $k_{atm} = 0.8$. The photodetector efficiency was taken to be $k_{PM} = 0.1$.

TABLE 3

Reflector	Date placed on moon	Selenographic coordinates	
		l	b
Lunokhod 1	17 November 1970	−34°47′	+38°24′
Lunokhod 2	15 January 1973	+30°24′	+26°07′
Apollo 11	21 July 1969	+23°26′45″	+0°41′15″
Apollo 14	5 February 1971	−17°25′	−3°36′
Apollo 15	31 July 1971	+03°39′30″	+26°04′54″

To reduce the absorption of solar energy the working faces of the prisms were not metallized.

The basic parameters of the corner reflectors are shown in Table 2.

At present there is a network of five reflectors on the moon. Three of them have been placed there by the American spacecrafts Apollo 11, Apollo 14, and Apollo 15, and two have been mounted on the Soviet Lunokhod 1 and Lunokhod 2 lunar vehicles.

Table 3 lists the times the reflectors were placed on the moon and their selenographic coordinates.

IV. LASER-RANGING DEVICES

1. The Design of a Laser-Ranging Device and Specifications for Its Parameters

A device for laser ranging of the moon is shown in Fig. 11. Such a device, with a few changes and complications, has been used in practically all experimental work in this area. A light beam produced by a laser is collimated and directed at the required point on the moon by a telescope (T). The signal reflected from the moon is received by the same telescope and

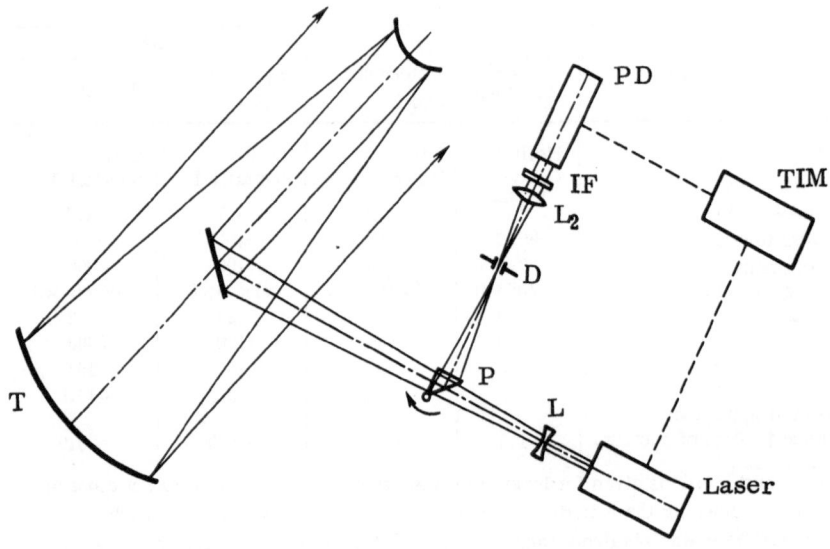

Fig. 11. Diagram of a device for laser ranging of the moon.

directed by a "send—receive switch" P through a diaphragm D and an interference filter IF onto a photodetector PD. The time for the light signal to propagate to the target and back is measured with a time interval measurement device (TIM) which is gated on by the laser pulse pulse and off by the received reflected signal.

We now formulate the basic specifications for a laser-ranging system and its individual parts and consider ways of realizing the required parameters.

1. The Laser

The basic parameters of a laser are the wavelength and spectral width of the light, the output energy, the diameter and divergence of the beam, the pulse duration, the pulse repetition rate, and the lifetime of the components.

To obtain the most reflected signal per unit time a laser with maximum energy and pulse repetition rate and minimum beam divergence must be used. To obtain the greatest precision in the measurement the pulse duration must be as short as possible.

These requirements are to a certain extent contradictory and in developing a laser for this purpose a number of compromise solutions have to be accepted.

For its overall parameters and operational characteristics the most appropriate system for laser ranging of the moon is a Q-switched ruby laser, although its wavelength is not the best from the standpoint of photodetector sensitivity. We shall discuss the basic characteristics of the laser transmitter based on a ruby laser now used in the vast majority of laser range-finders.

The wavelength of a laser lies within the luminescence band of the active medium and depends on its temperature and the spectral properties of the cavity. The spectral composition of the output may be fairly complicated although, as opposed to ordinary light sources, all the light is concentrated in a very narrow spectral interval. The wavelength of a ruby laser at $T = 300°K$ is $\lambda = 6943$ Å [44]. The substantial temperature dependence of the output wavelength on the temperature makes it possible to choose a value of λ that matches a transparency window of the atmosphere [44], while the required frequency stability can be ensured by regulating the temperature of the air or liquid which cools the ruby rod.

The width of the output spectrum of a ruby laser is usually about 0.5 Å [45]. Since the passband of the interference filter in the photodetection system of the laser rangefinder is usually much greater than this (several Angstrom units), the spectral selectivity of the range-finder is determined by the filter. If it were possible to build a filter with a passband tenths or hundredths of an Angstrom wide [46], then the spectrum of the laser would have to be narrowed correspondingly and the frequency stabilized. By mode selection it is possible to reduce the spectral width to about 0.005 Å [47].

In determining the required spectral width of the laser we must try to have the emission line lie entirely between fine-structure bands in the absorption spectrum of the atmosphere or, otherwise, to have it cover several bands. In the latter case, if the emission spectrum is continuous [48], the overall losses in the atmosphere will be small because of the narrowness of the atmospheric absorption bands [49].

The pulse energy of the laser is one of the important parameters determining the efficiency of the system as a whole. Since every laser medium has a limiting pulse energy density or optical durability beyond which the medium is rapidly destroyed, the energy of a laser pulse depends in general on the diameter of the working element in the final amplifier stage of the laser. It is very difficult to obtain ruby rods of diameter greater than 20 mm. Thus, ruby rods of diameter 10-20 mm are widely used. At an energy density of about 1.5 J/cm^2 the output energy usually does not exceed about 5 J in a pulse lasting about 10^{-8} sec. When the energy density exceeds this amount the operating lifetime of the rod drops rapidly.

The divergence of the laser beam is defined as the plane angle formed by the straight lines through a perpendicular cross section of the beam at the 0.5 intensity level. For a single-mode laser [45] the beam divergence is close to the diffractive limit $\alpha_L \sim \lambda/d_F$, where λ is the wavelength of the light and d_F is the diameter of the end face of the lasing part of the rod.

Due to such factors as insufficient quality of the rod, nonuniform pumping, and temperature gradients in the rod, the divergence of a laser beam deteriorates by one or two orders of magnitude and usually is about 20' [45]. In a multistage laser system the beam divergence increases further as the light passes through the amplifier rods which act as an inhomogeneous amplifying medium.

The diameter of the laser beam is usually the same as the rod diameter of the final amplifier.

The diameter alone does not determine the energy efficiency of a laser. In fact, the reflected signal depends on the luminous energy density at the target which in turn is proportional to W_{tr}/α_T^2, where W_{tr} is the output energy and α_T is the divergence of the light beam after the telescope. The divergence α_T is given by [51]

$$\alpha_T = \alpha_L \frac{d_L}{D_T},$$

where d_L is the diameter of the laser beam, D_T is the diameter of the telescope, and α_L is the laser beam divergence (which depends weakly on its diameter).

If we use the fact that $W_{tr} \sim d_L^2$ as noted above, then $W_{tr}/\alpha_T^2 \cong$ const. Nevertheless, it is convenient to have the final-stage amplifier rod diameter as large as possible since when the energy density is conserved this leads to an increase in the beam divergence after the telescope, α_T, which makes it easier to aim the laser beam at the target.

The laser pulse duration mainly determines the precision of a single measurement of the distance to the object.

In a conventional-mode ruby laser the pulse duration depends on the duration of the flash-lamp pulse and is about 1 msec. Q-switching [52-55] makes it possible to obtain pulses lasting $(1-5) \cdot 10^{-8}$ sec. There are several methods of shortening the pulse to $\geq 10^{-9}$ sec. In [56, 57] a method of producing a pulse lasting several nanoseconds based on nonlinear amplification of a pulse with a leading edge shortened to about 10^{-9} seconds was studied. The method proposed and studied in [58-61] makes it possible to obtain pulses lasting 2-5 nsec. Pulses lasting about a nanosecond can also be obtained by shortening the pulses from a Q-switched laser with an extra electrooptical gate [62, 63]. Still shorter pulses ($\sim 10^{-10}$-10^{-11} sec) can be obtained by mode locking [64, 65].

The pulse-repetition rate of the laser determines the rate at which the measurement data can be accumulated and is limited by the time it takes to cool the rods and bring them into thermal equilibrium in order to obtain minimal divergence. For these reasons the repetition rate of a ruby laser cannot exceed a few Hertz.

2. The Receiver

The basic characteristics determining the quality of the receiver are its sensitivity, time resolution, spectral selectivity, angular field of view, and intrinsic noise.

As shown above, the reflected signal consists of a few photons. The problem is to detect this signal with a high temporal resolution. The most appropriate apparatus for this purpose is the photomultiplier [66-69].

High spectral selectivity of the receiver can be ensured by using a narrow-band filter.

Thus, the main parts of the receiver have to be a photomultiplier and a narrow-band interference filter.

The sensitivity of the photomultiplier is determined by the efficiency with which the optical radiation is converted into a photocurrent by the photocathode, i.e., the quantum efficiency. The quantum efficiency of the best multialkali photocathodes [70] is about 5% for $\lambda = 0.7$ μm with a low noise level. Recently, still more sensitive photocathodes based on GaAs[Cs_2O] and InAs$_x$P$_{1-x}$[Cs_2O] have been developed [71]. The quantum efficiency of the first of these can reach 25% for $\lambda = 0.7$ μm [72]. However, such photocathodes are not often made because of technical difficulties. The quantum efficiency of a photomultiplier can be increased by three to five times with a special prism which makes the light pass repeatedly through the photocathode [73, 74]. Especially good results have been obtained using a prism with simultaneous choice of an optimum photocathode thickness [75].

The temporal resolution of the receiver, which determines the accuracy with which time intervals are measured, is characterized by the scatter in the times at which the photomultiplier anode pulses due to single photons occur. These fluctuations are determined by the scatter in the time of flight of electrons along the dynode system of the tube, which is in turn related to the design and operating conditions of the tube. Ensuring the required response time is thus a matter of choosing a photomultiplier type and the operating regime for it.

The spectral selectivity of the receiver is determined by the narrow band interference filter [76, 77] whose spectral characteristics obey

$$k(\lambda) = \frac{k_{ph}}{1 + 4\left(\dfrac{\lambda - \lambda_{max}}{\delta\lambda_{ph}}\right)^2}, \tag{49}$$

where $k(\lambda)$ is the transmission of the filter at wavelength λ, k_{ph} is the peak transmission at λ_{max}, and $\delta\lambda_{ph}$ is the transmission bandwidth. The spectral curve is characterized by the widths $\Delta\lambda_{01}$ at the level $k(\lambda) = 0.1k_{ph}$ and $\Delta\lambda_{001}$ at the level $k(\lambda) = 0.01k_{ph}$. From Eq. (49) it follows that $\Delta\lambda_{01} \cong 3\delta\lambda_{ph}$; $\Delta\lambda_{001} \cong 10\delta\lambda_{ph}$. This shows that an ordinary interference filter has gentle fall-offs in its transmission profile. The transmission outside its passband is 10^{-3}-10^{-4}. Much better background suppression is obtained in the case of a high-contrast narrow-band dielectric interference filter whose transmission curve is characterized by $\Delta\lambda_{01} \cong 1.7\delta\lambda_{ph}$ and $\Delta\lambda_{001} \cong 3\delta\lambda_{ph}$. This means that for equal halfwidths $\delta\lambda_{ph}$ the area of the transmission curve of such a filter is roughly half that of an ordinary filter. The transmission of a high-contrast filter outside its passband is 10^{-5}-10^{-6}. The bandwidth of the filter is limited below by the spectral width and frequency stability of the laser. Under certain conditions the laser spectrum may be compressed to 0.005 Å [47]. If efficient measures are taken to stabilize the wavelength of the laser and the wavelength of the filter, it seems possible to reduce the bandpass of the filter to about 0.1 Å. There are serious technical difficulties in the way of producing such filters, but the first successes have already been achieved [46].

The angular field of view is limited by the diaphragm (D, of Fig. 11), determines the level of background illumination, and, for this reason, must be chosen as small as possible. However, its size is limited below by inaccuracies in aiming the detector at the slit and in the optical matching of the receiver with the laser. A compromise is found experimentally. Usually the field of view is taken to be 6"-10".

The intrinsic noise of a photomultiplier operating in the photon counting regime is almost completely determined by thermionic emission from the photocathode. In specially selected tubes this background may be of the order of a few pulses per second. Including the other sources of background light, such as light from the bright part of the moon, night skyglow, and so on (see Chapter V), intrinsic photomultiplier noise of up to 10^2-10^3 pulses/sec can be considered acceptable and no special measures need be taken to reduce it.

3. The Telescope and Aiming System

If we neglect the effect of the atmosphere, the amount of reflected signal at the receiver point is proportional to the fourth power of the telescope diameter. Thus it is clear that using a telescope with a maximum diameter is decisive for a laser-ranging system.

The telescope must be aimed at the appropriate point on the moon with an error no larger than the radius of the light spot on the moon, that is, with an accuracy of 1"-2". From this it follows that a separate guide mechanically coupled to the main telescope is unacceptable since the misalignment of the guide and telescope due to differential deformations as it is turned greatly exceeds this value. Thus the aiming system must be mounted at the same focus of the telescope as the laser rangefinder. This aiming accuracy must be maintained whether the measured point is on the light or dark side of the moon.

4. The Apparatus for Measuring Time Intervals

The measurement and detection apparatus together must automatically observe and record an extremely weak signal with great temporal accuracy. The apparatus for measuring the time intervals must be a digital device. Pulses lasting several seconds may be measured to an accuracy of 10^{-8} sec by directly counting the reference pulses. At the present stage of development of electronics it is practically impossible to obtain better time resolution (10^{-9}-10^{-10} sec) in this direct way. This problem is solved indirectly. The measurement is made in two stages. First the entire time interval is measured on a crude scale given by a reference oscillator with a frequency of 10-20 MHz. Then an exact measurement is made of the short intervals from the "start" and from the "stop" to the nearest reference pulses. This can be done with high accuracy using analog or vernier methods.

To avoid noise pileup the system must record only those pulses which fall in a "time window" of width chosen in accordance with the previously calculated time for the signal to propagate to the target and back.

Since it is unknown in advance which of the received pulses is the signal, it is necessary to record with a certain time resolution all pulses which fall within the "time windows" so as not to lose the necessary information. When the noise level is high or during preliminary searches for the reflector, when the "time window" is largest, this may seem practically impossible because of the low capacity of the system. In this case comparatively simple methods of recording all the pulses within the "time window" but with less accuracy may be used, such as oscilloscope pictures.

This method is used at first to observe the signal, and then the "time window" is shifted so the signal pulses appear immediately after the window gate is opened and a more precise measurement of the distance can be made. Thus the possibility of changing the width of the "time window" and its timing relative to the calculated signal arrival time must be planned for in advance.

5. Timing Service

The rate of change of the distance between fixed points on the earth and moon may be several hundred meters per second. If this distance has to be measured to within a few decimeters, then to avoid additional errors the measured distance must be related to a universal time scale with an error of less than 10^{-4} sec. This accuracy may be achieved by setting local electronic clocks to a single time scale with signals transmitted by radio or television channels.

6. The Control and Information Input and Output Systems

Besides the above elements and components, a laser-ranging system must include apparatus for feeding in ephemeris data (previously computed distances) and extracting and operating on the information obtained from the experiment, as well as devices for controlling the system as a whole. To do these things it is best to include a control computer in the system. It must:

(a) do fifth- or sixth-order interpolation from reference points of the precalculated values of the propagation time of the signal and enter this interpolation value in the delay generator for the "time window";

(b) relate the laser pulse to a time scale or, conversely, fire the laser at certain times for which the propagation times have been calculated;

(c) process and store the recorded pulses in multichannel registers according to the time analyzer program and make probability calculations with a program for optimal observation of the signal; and

(d) store all the information collected during a measurement sequence (distances and times for each pulse) and deliver this information in a form convenient for further processing.

The computer program must be fairly flexible and make it possible to obtain the output in a form that is clear and convenient for the observer so he can operationally control the measurement process and change its course if necessary. The computer must also have functional control of the entire system's operation and of its basic parameters.

We have examined the basic requirements which a lunar laser rangefinder must satisfy. Constructing an apparatus in accordance with these requirements is a fairly complicated task both because many of the components must have the maximally attainable parameters (the laser, interference filter, aiming system, etc.) and because several of these requirements are contradictory. Building such a device involves seeking a large number of compromise solutions. Naturally, the best variants could only be found by accumulating experience in building and using such apparatus.

We have built four versions of laser rangefinders. Each successive variant has been the result of developing, refining, and improving the previous one. We shall not describe the first three models. We note only that a laser rangefinder built in 1963 was used to obtain reflection from the moon [78]. The second model (1965) was used to measure the distance to the moon's surface with three times better accuracy (about 200 m) [79]. The third variant, whose parameters met the requirements of that time, was used to measure the distance to a reflector mounted on Lunokhod 1 [80]. We now consider the automatic equipment currently being used for regular measurements of the distance to all the lunar reflectors.

2. An Automatic Laser-Ranging System (1973)

Figure 12 shows the block diagram of the automatic laser-ranging system and Fig. 13 shows the arrangement of its optical system. We now examine the basic parts of the system following the block diagram.

1. The Arrangement of Apparatus on the Telescope and the Aiming System

Over a period of several years before the development of the final version of the laser transmitter for ranging of corner reflectors on the moon, several modifications of a transmitter

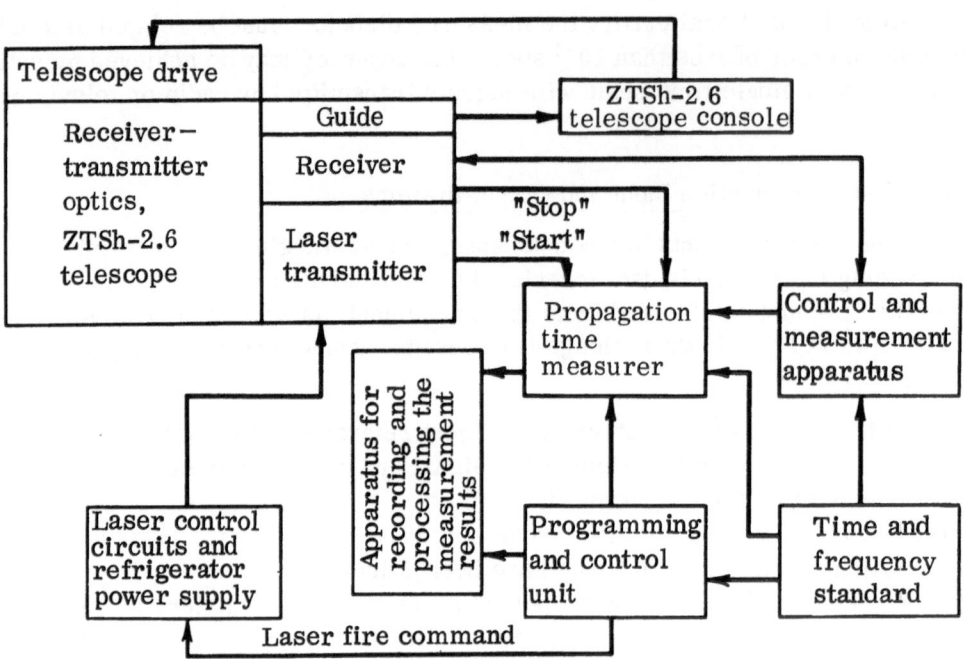

Fig. 12. A block diagram of the automatic laser-ranging system.

intended for mounting in Nasmyth, Cassegrain, and Coudé telescopes were developed and tested. Their designs were completely different from one another and from the final version. These development efforts were undertaken to find the most reasonable means of combining the laser measurements with other astrophysical work on the telescope. As opposed to earlier experiments, in which the object being probed was the surface of the moon, in developing systems for locating reflectors the problem of aiming the laser beam at the reflector becomes serious.

Our purpose was to construct the most general system possible which would enable us to aim the laser beam at a point lying on either the light or dark side of the moon. The greatest difficulty is in aiming the telescope at a point on the dark side. In this case a reference crater with known coordinates is chosen and the telescope guide is aimed at it. In order to direct the telescope, along whose axis the laser beam passes, at the required point on the moon, the telescope must be placed at an angle to the guide equal to the angular distance between the crater and this point.

The use of a separate guide in this arrangement is unacceptable because of the poorly controlled differential deformation of the guide−telescope system. Thus, the telescope must be guided by observing the crater in the field of view of the telescope in which the laser focus lies. The field of view must be sufficiently wide that the line from the crater to the observation point lies in it. In our telescope this condition is satisfied only by a direct focus and a Cassegrain focus. However, putting the apparatus in these foci is undesirable since it would then move about with the telescope tube. In one version the laser transmitter was mounted in a Cassegrain focus. Experience showed that it was very inconvenient to use this configuration. Figure 14 shows the possible ways of placing the laser-ranging equipment on a telescope.

To solve the aiming problem we have chosen an unusual place for mounting the apparatus, the center of the polar platform. To bring the focus of the telescope to this location in the optical configuration we used a hyperbolic mirror from a Nasmyth focusing system and a diagonal mirror from a Coudé focus. In this case the focal plane of the telescope is located a distance of about 1.5 m above the polar platform. According to the optical diagram (ray path) this is a contracted Coudé focus; however, it is free of two basic shortcomings of the classical Coudé

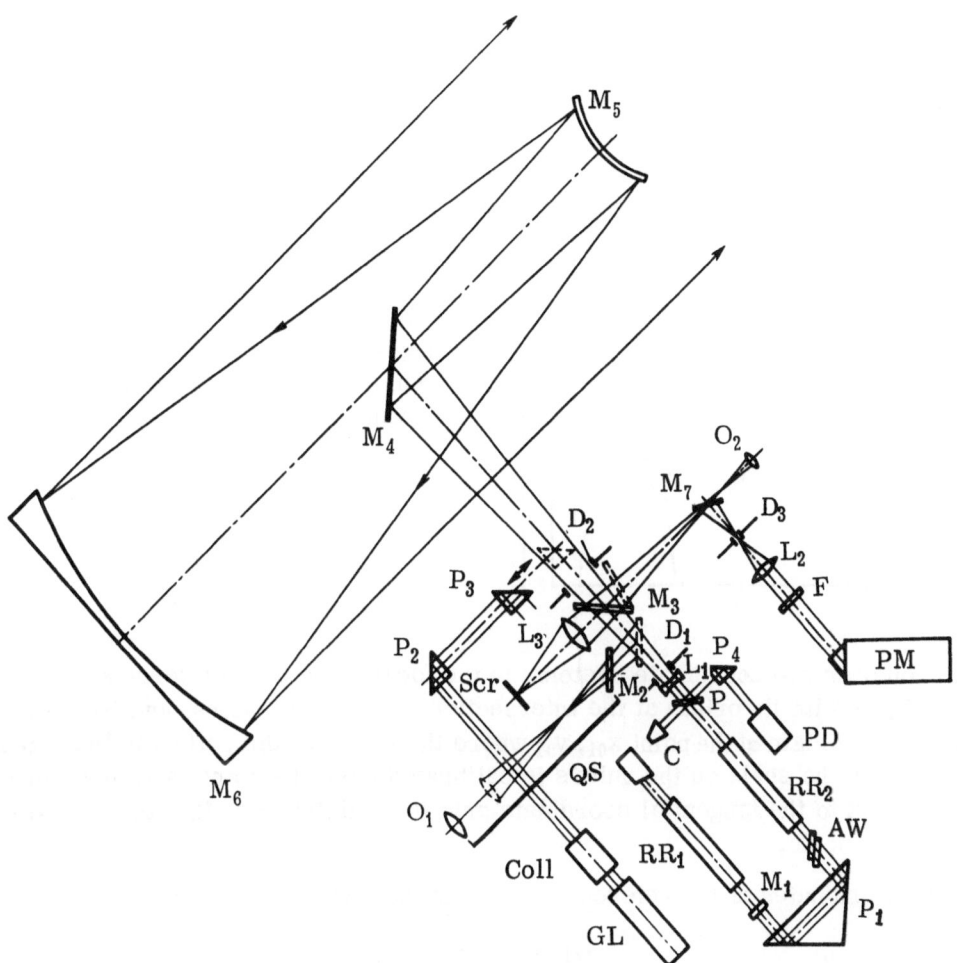

Fig. 13. The arrangement of the optical portion of the automatic laser-ranging system. The basic elements are: RR₁ and RR₂, ruby rods; QS, electrooptical Q-switch; M₁, cavity mirror; P₁, retro-prism; AW, alignment wedges; L₁, coupling lens; O₁, M₂, guide; M₃, mirror for switching in receiver or transmitter; PM, photomultiplier; F, interference filter; L₂, detector lens; D₃, detector aperture; M₄₋₆, telescope mirrors. Auxiliary elements include: GL, gas laser; Coll, collimator; P₂, P₃, prisms; Scr, screen; L₃, lens; M₇, O₂, auxiliary guide; P, beam splitter plate; C, calorimeter; P₄, prism; PD, photodiode; D₁ and D₂, diaphragms.

focus, namely, a small field of view and its rotation with respect to the geoequatorial system of coordinates when the telescope is rotated. Figure 15 shows the receiver−transmitter mounted in the center of the polar platform.

In the focal plane of the telescope there is a coordinate plane made up of two mutually perpendicular rulers along which the local guide (Fig. 16) is moved by micrometer screws. The accuracy in finding the mutual position of the foci of the guide and laser is 0.1 mm or 0".5. The guide can be moved along each of the rulers by 180 mm or 15'. In order to ensure a maximum distance between the guide and laser foci in the field of view, the laser focus is displaced to the edge of the field of view so that the maximum distance from it to the guide focus along one axis is 180 mm, and along the other, ±90 mm. And in order to be able to aim the telescope during different phases of the moon (new and old moons) the entire laser apparatus together with the coordinate rulers can be rotated 180° around the optical axis of the telescope.

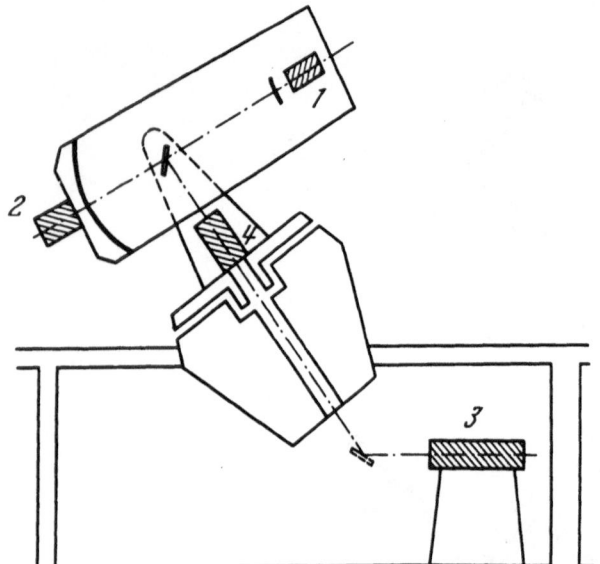

Fig. 14. Possible locations for a laser-ranging device on a telescope: 1) direct focus; 2) Cassegrain focus; 3) Coudé focus; 4) center of the polar platform.

Let us consider two coordinate systems in the focal plane of the telescope: the instrument system X_i, Y_i with its origin at the intersection point of the rulers, and the tangential system X_T, Y_T with its origin at the point x_{0i}, y_{0i} where the image of the reflector lies. Before an observation session the scale on the rulers is calibrated from the stars, and their location (angle) with respect to the tangential coordinate grid is established so that the direction of the X_i and X_T axes coincide.

Then the coordinates of the reference crater in the instrumental system are

$$x_{ci} = x_{oi} + x_{ct},$$
$$y_{ci} = y_{oi} + y_{ct}.$$

(50)

The coordinates of the reference crater (x_{ct}, y_{ct}) in the tangential system can be expressed in

Fig. 15. The optical part of the automatic laser rangefinder mounted on the polar platform of the ZTSh-2.6 telescope.

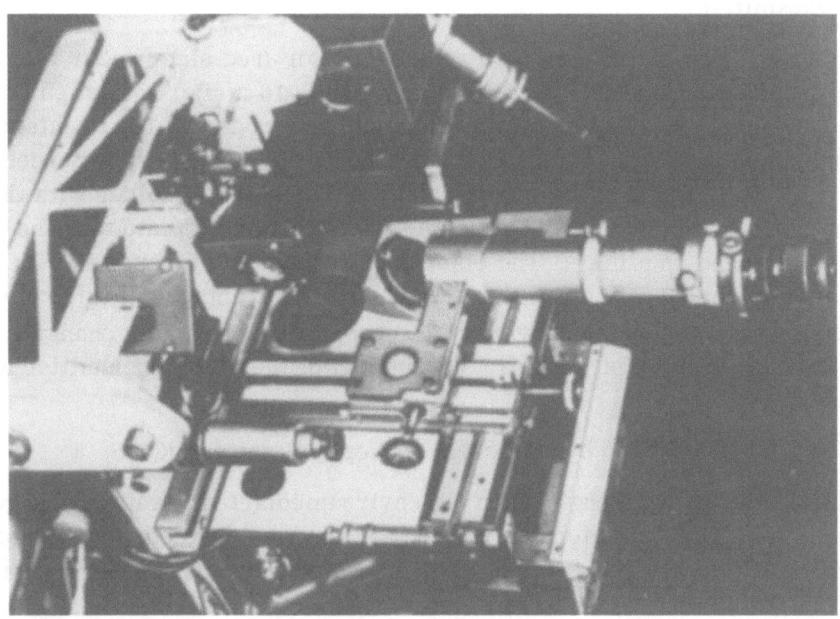

Fig. 16. The coordinate rulers and local guide.

terms of the declination and right ascension of the crater and the reflector [81] as

$$x_{ct} = \frac{1}{M} \frac{\cos \delta_c' \sin(\alpha_0' - \alpha_c')}{\sin \delta_c' \sin \delta_0' + \cos \delta_c' \cos \delta_0' \cos(\alpha_0' - \alpha_c')},$$

$$y_{ct} = \frac{1}{M} \frac{\sin \delta_c' \cos \delta_0' - \cos \delta_c' \sin \delta_0' \cos(\alpha_0' - \alpha_c')}{\sin \delta_c' \sin \delta_0' + \cos \delta_c' \cos \delta_0' \cos(\alpha_0' - \alpha_c')}, \tag{51}$$

where α_0', δ_0', α_c', and δ_c are the topocentric visual declination and right ascension of the reflector and crater and M is the instrument scale.

In aiming the laser beam at the reflector a correction ("advance") must be made for the aberration. This correction equals the angular displacement of the reflector, or equivalently, the reference crater during the time τ it takes the light to propagate to the reflector and back, that is, twice the aberration angle. Thus, in calculating the mutual location of the reflector and reference crater the coordinates $\alpha_0(T)$, $\delta_0(T)$ of the reflector must be replaced by $\alpha_0(T + \tau)$, $\delta_0(T + \tau)$.

After being reflected the laser pulse propagates the same way as rays from other features of the lunar surface including those from the reference crater. Their aberrational displacements are the same. Thus, during operation it is not necessary to shift the receiver focus with respect to the image of the reference crater.

In practice the correction for aberration is taken into account by introducing the required changes in calculating the distances between the reflector and the reference crater. To simplify optical alignment of the system the axis of the receiver was made to coincide with that of the laser while the resulting error in the direction of the receiver axis was compensated for by an expansion in its field of view (up to 5"-10").

Since the magnitude of the aberration correction does not exceed about 1".2 and the accuracy in aiming the telescope is usually about 2", this simplified approach is fully justified.

2. The Laser Transmitter

The laser transmitter is a Q-switched ruby oscillator (rod size $l = 240$ mm and $d = 10$ mm) and a ruby amplifier (rod size $l = 240$ mm and $d = 15\text{-}16$ mm). Optical pumping is by two pulsed IFP-8000 xenon lamps. The oscillator and amplifier rods and the flashlamps are mounted in a common sealed reflector case and are cooled by a water flow perpendicular to the rod axes. This reflector head is made of two parts, one of which, a holder with the ruby rods, together with the pieces of the compound reflector, is attached to a plate with the optical elements of the oscillator and amplifier. The other part of the head, a cassette with the two flashlamps, is hermetically sealed to the rod holder. The main advantages of this reflector design are rapid and uniform cooling of the ruby rods, the possibility of rapidly changing the flashlamp cassette, and conservation of the optical alignment of the laser transmitter when the cassette is changed.

The oscillator is Q-switched by a Pockels cell optical shutter (QS).

To stabilize the laser wavelength when the environmental temperature changes over a wide range ($t° = -15$ to $+20°C$), an automatic cooling system was developed and installed which consisted of a refrigerator, reservoirs with distilled water in which the refrigerator's evaporator and an electric heater were mounted, an electronic circuit for automatically controlling the temperature, and a pump to drive the water through the laser head. The accuracy of temperature regulation is $\pm 0.5°C$.

A control panel was developed which makes it possible to operationally control the main laser parameters in automatic operation and to control the laser during alignment. At the control panel there are measurement instruments, operating switches, and an intercom for communicating with the sites of the receiver detector, the laser transmitter, and the central control panel of the telescope (the ZTSh-2.6 console). Figure 17 is a block diagram of the laser oscillator-amplifier, where the names and relationships of the components are obvious.

The duration of the light pulse from the laser was measured with an FÉK-15 photodiode and an I2-7 nanosecond time interval measure (i.e., time digitizer). The accuracy of this measurement was about 10%. The amount of light needed on the photodiode was selected with the aid of neutral density filters. Time markers with a period of 10 nsec on an oscilloscope were used as a scale.

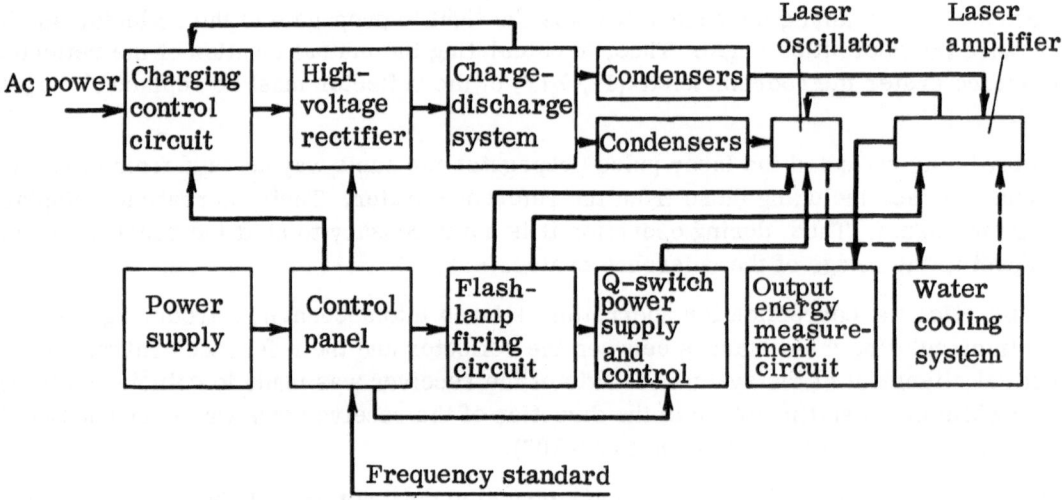

Fig. 17. Block diagram of the laser oscillator-amplifier.

The pulse energy was measured with a conical multielement calorimeter which was periodically calibrated at the working energy levels with a KIM-1 calorimeter. The accuracy of energy measurements with the latter is about 10%.

To measure the energy of each pulse in a sequence a beam splitter plate P (see Fig. 13) directed a small part (about 1%) of the output energy into a calorimeter (C). Each measurement was monitored on a chart recorder so the energy level could be held constant.

The divergence was measured in a simple way by measuring the diameter of an image of the beam focused by a lens. To do this, screens with openings of various diameters were placed in the focal plane of the lens and the energy in front of and behind the opening was measured. The beam divergence, for example, to a level of 0.1 of the energy was measured as $\alpha_{01} = d_{01}/f_L$, where d_{01} is the diameter of the aperture which transmits 0.9 of the full energy and f_L is the focal length of the lens.

The wavelength and spectral width were measured with a spectrograph and a Fabry−Perot etalon.

The parameters of the laser transmitter are: wavelength $\lambda_{las} = (6843 \pm 0.2)$ Å; pulse energy $W_{las} = 2.5-3$ J; pulse duration $\tau_{las} = 10$ nsec; beam divergence angle $\alpha_{las} = 10'$; and pulse repetition rate $f_{las} = 1/3$ Hz.

The average number of pulses before replacement of the amplifier ruby rod was $N_{las} \simeq 10^4$ pulses.

3. The Detector

In the first experiments on laser ranging of the moon's surface no special demands were placed on the temporal characteristics of the detector. In these experiments FÉU-68 and FÉU-69 photomultipliers were used.

To obtain the required time resolution in the apparatus discussed above, an FÉU-77 photomultiplier (which has been specially developed to detect laser pulses) was used. It has a multialkali photocathode and multialkali emitters to ensure high sensitivity at $\lambda \cong 7000$ Å and a high gain ($\sim 10^8$).

One feature of the FÉU-77 design is the small diameter of the working part of its photocathode (about 4 mm). Because of this it has low intrinsic noise.

In accordance with the photodetector specifications discussed in Chapter IV, Section 1, the characteristics of many photomultiplier tubes were studied and a selection was made [82].

In making the selection the sensitivities of the tubes (which had been chosen provisionally according to their test ticket data) were compared.

The photomultipliers were compared in a single-electron counting regime produced by a weak light flux from a light source that passed through a high-contrast interference filter (with $\lambda_{max} = 6943$ Å, $\delta\lambda_F = 10$ Å at $k_F = 0.45$). As a result of the comparison, a tube with a quantum efficiency of about 4.5% was chosen.

The quantum efficiency of the photocathode was increased by fastening a special prism onto the tube with immersion contact to ensure multiple passage of the light through the photocathode [75]. For this device to work, a tube with a prism has to be adjusted with respect to the light beam so that the beam will fall on the sensitive part of the photocathode at the total internal reflection angle of the cathode−vacuum interface. Since within the prism the path length of the beam before the light absorbed is about 10 cm, the angular divergence of the beam defined by the diameter of the working part of the photocathode of the FÉU-77 must not exceed a few degrees. This required rigid fixing of the tube with the prism with respect to the light beam. When these conditions were met the quantum efficiency at λ = 0.7 μm could be increased by about 2.3 times for a selected FÉU-77. Thus, the quantum efficiency of a photomultiplier with a prism is about 0.1. In laser-ranging measurements the output pulses of the photomultiplier are recorded at their leading edge.

Because of the amplitude scatter in single-electron pulses the error in the time measurement may lie within the risetime of the pulse. In order that this error not affect the measurement result it must be at least one order of magnitude less than the expected measurement error, i.e., ≤ 1 nsec. In addition, the time of flight of the electrons along the dynode system of the photomultiplier will have an effect on the time measurements. This delay between the time the light pulse arrives at the photocathode and the time an electrical pulse appears at the anode can be of the order of tens of nanoseconds. The fluctuations in this delay also must not exceed 1 nsec.

A spark gap with a light pulse lasting about 1 nsec was used to determine the temporal characteristics of the tubes. For a spark gap pulse half-width of 0.6 nsec the risetime of the photomultiplier anode pulse was 2.4 nsec and the anode pulse duration was 8-9 nsec.

The detector optical arrangement is shown in Fig. 18. The light signal collected by the telescope is directed by the receive–transmit switching mirror SM and a dielectric mirror DM onto the acceptance aperture D of the photodetector. The dielectric mirror DM has a reflectivity of about 95% for 45° incidence at a wavelength of 7000 Å and transmits the blue-green part of the spectrum. The light which goes through mirror DM is sent by mirror M into the auxiliary guide CH–EP with which the direction of the axis and the location of the photodetector focus can be monitored visually. The guide also aids in aligning the detector with the laser and in monitoring the accuracy with which the image of a star falls in the diaphragm D when calibrating the detector with the stars.

The entrance aperture D lies in the focal plane of the telescope. The scale of an image in the focus of the telescope is 0.2 mm per angular second. For this work a set of diaphragms with apertures of from 0.6 to 3 mm were made and used both in alignment and during the ob-

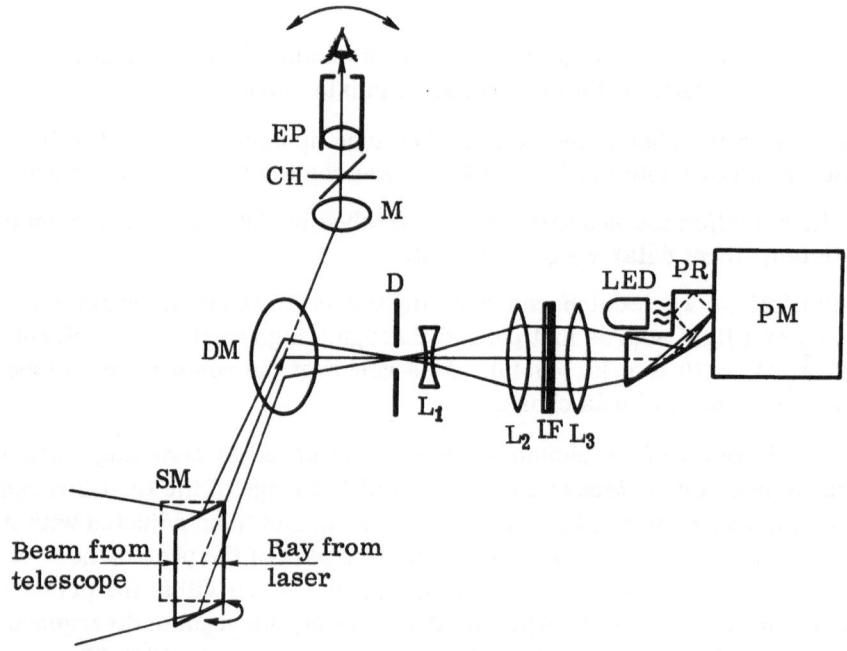

Fig. 18. The optical arrangement of the receiver.

servation sequences. The range of angular fields determined by the diaphragms, 3"-15", covers practically all possible operating conditions. After the entrance diaphragm the light beam (with an angular divergence of about 3°) is incident on a triple telescope system L_1L_2 (Galilean tube) which reduces the divergence to 1° so as to diminish its effect on the passband of the narrow-band interference filter IF.

After the telescope system the light beam passes through the interference filter IF and lens L_3 and then on through the prism PR to the photocathode of the FÉU-77.

Next to the prism attached to the photomultiplier there is a light-emitting diode LED made of silicon carbide which simulates the laser pulse to monitor the measurement-recording system.

In the receiver there is a high-contrast interference filter which provides much better attenuation of the background than an ordinary filter. This filter has the following parameters: $\lambda_{max} = 6947$ Å, $\delta\lambda_F = 5$ Å, $k_F = 0.45$, and a temperature coefficient of about 0.3 Å/deg C. Exact tuning to the working wavelength $\lambda_{las} = 6943$ Å is done by rotating the filter with respect to the direction of the incident light. The shift in the wavelength as a function of the rotation angle is given by

$$\Delta\lambda \cong -\frac{\alpha^2}{2}\lambda_{max},$$

where α is the angle between the normal to the filter surface and the beam direction. As the rotation angle of the filter is increased, $\delta\lambda_F$ increases and there is a reduction in k_F and the contrast; α must not exceed 20° for narrow-band filters.

With time there is a shift in λ_{max} to shorter wavelengths (aging of the filter). Thus it is necessary to periodically check its main characteristics. In practice a check is made to see that λ_{max} coincides exactly with the laser wavelength, and the transmission k_F is verified. To do this, some attenuators, the interference filter to be tested, and a sensitive calorimeter are put in the laser beam. The filter is put in a holder that can be rotated. In a series of successive rotations the maximum transmission of the laser light is obtained. Figure 19 shows an overall view of the detector mounted on an optical bench.

Fig. 19. The detector.

4. The Measurement and Recording Apparatus

The measurement and recording apparatus is shown in Fig. 20. After the second French reflector L-2 was placed on the moon in the beginning of 1973, this apparatus was used for regular laser ranging of observations.

Although the measurement process was automated, the apparatus was not suited to on-line observation during a measurement sequence because the preliminary estimate of the propagation time was read from punched tape. The measurement results were punched out on tape in unprocessed form, and to isolate the reflected signal the tapes had to be processed further. However, in principle the apparatus could be augmented without great alteration by a controlling computer for processing the measurement results so the reflected signal could be observed during the measurement sequence. In addition, the computer could be used for further processing of the accumulated information by different programs.

To automate all the most important and cumbersome operations involved in processing the measurement results and isolating the reflected signal, a 1001-TPA-1 computer was included in the measurement and recording apparatus [83].

The automated detection and measurement system does the following: (1) temporally select the signal according to the previously calculated ephemerides; (2) detect, store, and isolate the reflected signal from the noise; (3) measure the propagation time of the signal to an accuracy of $\pm 10^{-8}$ sec; (4) relate the measurement times (laser pulses) to the atomic time scale; (5) process and put out information on each measurement cycle; and (6) display the measurement results on a screen during the course of the experiment.

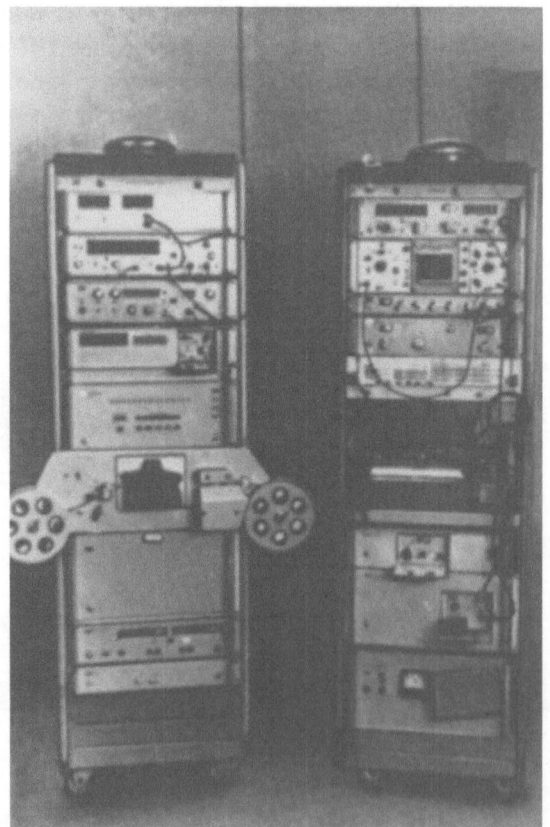

Fig. 20. The measurement and recording apparatus.

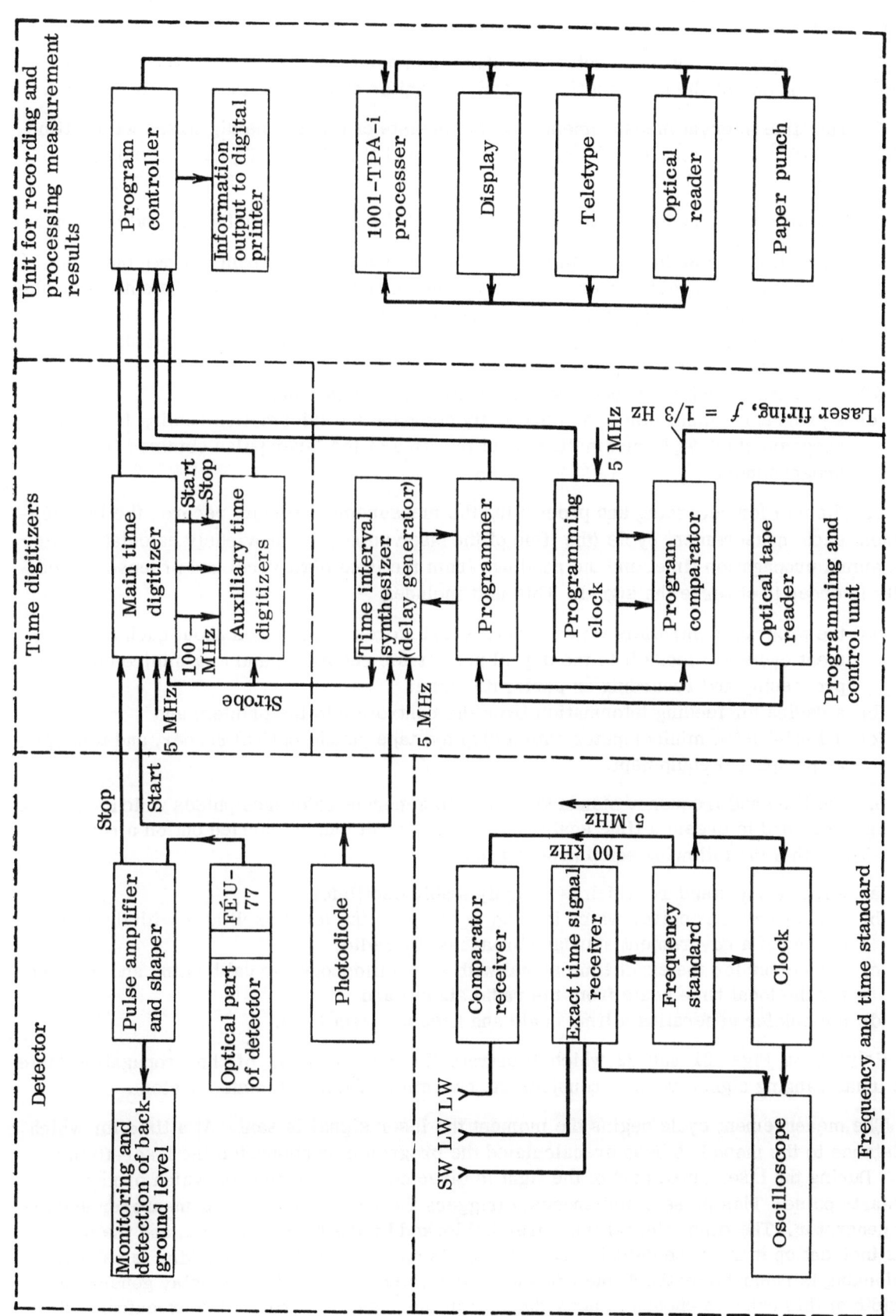

Fig. 21. Block diagram of the automatic measurement and recording system.

Figure 21 shows a block diagram of the measurement and recording apparatus which includes the following operational components.

1. The detector is meant to detect the signal and noise pulses, to form pulses which control the operation of the time digitizers, and to monitor and record the background level.

2. The time interval measurement system consists of three time digitizers with a time resolution of 10^{-8} sec to measure the propagation time of the signal. The main system measures intervals of from 10^{-8} to 10 sec, and the supplementary systems, 10^{-8} to 10^{-4} sec.

3. The programming and control unit consists of the following equipment and components:

(a) an optical reader for the data for each measurement cycle from a paper tape;
(b) a two-address programmer to store the information obtained from the optical reader and send it to the proper address in the appropriate apparatus;
(c) a delay synthesizer (delay generator) to temporally select the reflected signal with a discreteness of $\pm 10^{-7}$ sec;
(d) a program clock for producing the working time scale; and
(e) a program comparator to automatically compare the information coming from the program clock with information about the time of the laser pulse obtained from the programmer.

4. The unit for recording and processing the measurement results records the information from each measurement cycle (the time of the laser pulse, the previously calculated value of the signal propagation time, and the readings from the time digitizers) and processes these results in order to observe the signal. This unit includes

(a) the program controller, which receives and stores information about each measurement cycle and feeds it to the digital printer for recording and to a minicomputer for processing and recording on perforated tape;
(b) a device for feeding information from the controller to the printer; and
(c) a 1001-TPA-i minicomputer with a display, tape punch, optical reader, and teletype as peripheral equipment.

5. The time and frequency standard is used to generate reference pulses which scale the time digitizers and to create a time scale so the experiment can be carried out on a real time scale. To do this the following equipment is used:

(a) a frequency standard which is a highly stable oscillator;
(b) a comparator receiver which is used to compare the local oscillator with the frequency of a government standard broadcast by radio;
(c) a receiver for the exact time signal with an oscilloscope for calibrating and correcting the local time scale from the radio signal; and
(d) a clock for generating a time scale and keeping exact time.

Relying on Figs. 21 and 22 which illustrate the measurement of the propagation time of the lunar ranging signal, we now consider the operation of this automated system.

The measurement cycle begins the moment the laser signal is sent. At a time for which the distance to the moon has been precalculated the programmer generates the laser-firing pulse. During the laser pulse part of the light is directed onto a photodiode which produces the "start" pulse. This pulse simultaneously triggers the main time interval measurer and the delay generator. The time interval measurer is blocked by the "stop" channel, and the noise pulses incident on it from the detector cannot stop its operation until the blocking is removed. The blocking is removed by the "time window" pulse which comes from the delay generator. The width of this pulse is chosen in accordance with the possible error in the precalculated ephemerides. In the first experiments it was taken equal to 100 μsec. Later on, as experimen-

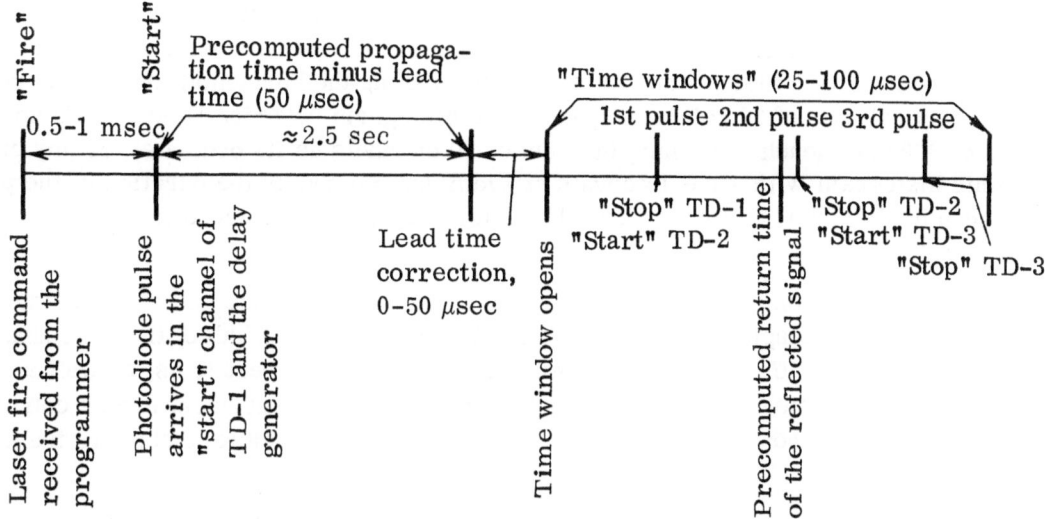

Fig. 22. Time diagram illustrating the operation of the measurement and detection system. (TD denotes a time digitizer.)

tal data were accumulated and they were used to improve the ephemerides, the errors in the precalculations were reduced. Thus, the width of the "time windows" was reduced to 20 μsec in later experiments. The delay generator included an additional manual control of the delay in the "time window" pulse with a step size of 1 μsec. This makes it possible to correct the delay as the measurements are being made.

The first falling in the "time windows" stops the main time interval measurer. If two or more pulses fall in the "time window," then they sequentially stop the auxiliary time digitizers which measure the intervals between the first and second and second and third pulses, respectively, in the "time windows."

The results of the measurements are transferred to the buffer registers of the program controller.

Along with the "time window" pulse the delay generator produces a trigger pulse for the optical reader which operates in a start−stop regime. The times of the laser pulse and the ephemeris are recorded on eight-channel paper tape. These data are read from the paper tape and transferred to the two-address programmer which transmits the information on the precalculated propagation time to the delay generator and to the buffer register of the program controller, while the information on the laser pulse time is sent to the program controller.

The instantaneous time from the program clocks is also sent to the program controller. When the code for the instantaneous time is the same as the time from the programmer, the program controller sends out a pulse to fire the laser, and a new measurement cycle begins. The information obtained in each measurement cycle (the time of the laser pulse, the readings from the main time interval measurer, the ephemeris, and readings from the two auxiliary time interval measurers) and contained in the input registers of the program controller is fed to the digital printer and also goes to the 1001-TPA-i minicomputer processor through the interface. A system based on this machine operates in real time and makes a preliminary analysis of the information which arrives from the receiver−measurement part of the system and stores it.

To observe the reflected signal during a measurement sequence, the computer operates as a time analyzer, sorting the preliminary calculated differences $\tau_0 - \tau_c$ (where τ_0 is the measured propagation time and τ_c is its precalculated value) over eight thousand channels. The width of each time channel is 10 nsec. The contents of the channels are shown on the display screen in the form of a histogram with a resolution along the horizontal axis of 100 points and along the vertical axis of 80 points. A panoramic histogram covering 80 μsec with horizontal intervals of 800 nsec (each containing the sum of 80 channels as its ordinate), or any part of this panoramic histogram with time intervals of width determined by the duration of the portion chosen, can be displayed on the screen. The information is transferred to a paper tape for later analysis.

The program allows the operator to observe the accumulated noise and signal pulses channel by channel on the display screen. Digital information on the width of the time channels and the exact value of the differences $\tau_0 - \tau_c$ for the channels of interest to us, as well as the number of pulses accumulated in the channel and the number of measurement cycles completed (i.e., laser shots), can be shown on the screen. All this makes it possible to observe the signal operationally.

The system includes a comparison of the frequency standard with a broadcast standard frequency by means of a receiver–comparator and a receiver for exact time signals. After inputting the correction (also transmitted by radio) for the separation between the atomic time scale UTC and the universal time scale UT-1 and the correction for the time for the radio signal to propagate from the radio station to the receiver point, the accuracy in matching the second markers of the clocks to the exact time signals is roughly 1 msec. The longwave radio station RBU [66.6(6) kHz] is used as a reference radiostation. A capability for operation with shortwave stations is also included.

The frequency standard is a highly stable rubidium oscillator with a daily instability of $\pm 5 \cdot 10^{-11}$. The frequency standard synchronizes all the measurement devices in the system. The frequency standard and the clock (which keeps the exact time) have an emergency battery power supply.

Figure 23 shows an overall view of the automatic measurement and recording system.

Fig. 23. The automatic measurement and recording system.

The parameters of the system are checked before each measurement sequence in order to ensure normal operation. A system for adjusting and testing the time-interval measurement unit, and the information input, output, and processing devices was developed for this purpose. With a special test it is possible to check the system under conditions identical to the working conditions at any time in a sequence without disrupting the time scale of the clock system.

V. MONITORING THE EFFECT OF THE ATMOSPHERE ON THE RANGING

SIGNAL. ISOLATING THE SIGNAL FROM THE NOISE

1. Monitoring the Composition of the Atmosphere

On the way to the moon and back the laser radiation passes through the entire atmosphere twice. This causes a considerable fluctuating attenuation of the signal. In Eq. (47) this attenuation is reflected in the coefficient k_{atm} which characterizes the attenuation under certain average atmospheric conditions and the coefficient α_L which is the total divergence of the laser beam after it passes through the atmosphere.

The attenuation in the atmosphere is due to the following processes: molecular absorption, molecular scattering, scattering on aerosol particles (dust, fog, etc.), and multiple scattering on atmospheric inhomogeneities due to turbulence.

An examination of these processes shows that whereas the first two of them can be quantitatively estimated, scattering on aerosols and, especially, scattering on atmospheric inhomogeneities are not subject to a priori quantitative evaluation.

On the other hand, to correctly plan laser-ranging measurements, one needs preliminary quantitative estimates of the state of the atmosphere. Thus, we set out to develop a system of atmospheric monitoring which would allow us to evaluate the possible attenuation of the laser beam before each measuring session.

This objective monitor helps avoid unproductive operation of the laser-ranging equipment and searches for possible causes of unsuccessful operation in those special cases when visual estimates of the atmosphere's transmission are too high.

The transmission k_{atm} of the atmosphere was determined by absolute stellar photometry. Stars with a well-known energy distribution in their spectra were used. The star to be examined was usually chosen to be in that part of the sky where the moon would be during the laser-ranging measurements. Using the catalog [84], nine fairly bright stars were chosen for which the energy distribution is given in absolute units with an interval of 50 Å. The intermediate energies are found by interpolation. The same detector that is used in the laser-ranging measurements is used to measure the energy flux from the stars. The detector efficiency is defined as $\eta_{det} = k_{PM} k_{det}$, where k_{PM} is the quantum efficiency of the photomultiplier and k_{det} is the transmission of the detector optics. Thus, η_{det} is the ratio of the number of pulses at the amplifier output to the number of photons incident on the detector aperture; $k_{PM} \cong 0.1$ (see Chapter IV). The detector optics consists of the aluminized telescope mirrors M_4, M_5, and M_6 (see Fig. 13), the aluminized mirror M_3, the dielectric mirror M_7, the three antireflection-coated lenses L_1, L_2, and L_3, and the interference filter IF (see Fig. 18). Taking the reflectivity of the aluminized mirrors to be 0.8, that of the dielectric mirror to be 0.95, the transmission of the lenses to be 0.98, and the transmission of the interference filter to be 0.45, we find the efficiency of the receiver system to be $\eta_{det} \cong 0.017$.

At the output of the receiver the signal from a given star is

$$N = \Phi S_\tau \delta \lambda_F \eta_{det} \text{ pulses/sec,} \qquad (52)$$

where Φ is the light flux from the star in photons/cm$^2 \cdot$ sec $\cdot$ Å, S_T is the area of the detector aperture in square centimeters, and $\delta\lambda_F$ is the transmission bandwidth of the receiver (filter) in angstroms. According to the catalog [84], the luminous flux in the neighborhood of $\lambda = 6950$ Å from the star Vega (α Lyrae), which is taken as a primary standard in stellar photometry, is $174 \cdot 10^{-3}$ erg/cm$^2 \cdot$ sec $\cdot$ cm or $6.1 \cdot 10^2$ photons/cm$^2 \cdot$ sec $\cdot$ Å. Taking $S_T = 4.4 \cdot 10^4$ cm^2 and $\delta\lambda_F = 5$ Å, we obtain $N_{\alpha\,Lyrae} = 1.56 \cdot 10^6$ pulses/sec neglecting atmospheric attenuation.

Measuring the luminous fluxes from the chosen stars and knowing the scaling factors of these fluxes relative to the flux from α Lyrae, the transmission coefficient of the atmosphere can be found as

$$k_{atm} = \frac{N}{N_{\alpha\,Lyrae}} k_{\alpha\,Lyrae}, \qquad (53)$$

where N is the measured light flux and $k_{\alpha\,Lyrae}$ is the scale factor relative to α Lyrae.

Evidently these measurements are not highly accurate since the quantities in Eq. (52) may have errors due, for example, to the fact that the passband of the interference filter does not overlap the laser spectrum and thus selective absorption by water vapor of the laser and star light will be somewhat different.

The monitor system must be calibrated under atmospheric conditions characterized as "extremely clear" or "very clear." In these cases the luminous flux from the stars is close to its maximum, i.e., actual, value. This calibration was done repeatedly. For example, on June 24, 1973, the measured value of the atmospheric transmission was 0.84 for $z = 62°$ and 0.86 for $z = 36°$, which is typical of "very clear" conditions.

The second parameter which characterizes the state of the atmosphere is the quantity α_L in Eq. (47). It is evaluated by measuring the diameter of the image of a star in the focal plane of the telescope. Another detector, developed for measuring the scale of the field of view, is used for this. The image size is estimated from the time for the light flux to fall to a level of 0.01 from its initial value when the star goes across the edge of a screen located in the focal plane of the detector while the telescope is held still.

The considerable inadequacy of the methods for evaluating k_{atm} and α_L due to the non-simultaneity of the monitor and laser measurements should be noted.

2. Sources of Background Light and Determining Its Level

The main sources of the background light are light from the lunar crescent (when the observed point is on the dark side of the moon, the source is light from the crescent scattered in the atmosphere and on parts of the receiver telescope), earthlight on the moon, night airglow, and the intrinsic photomultiplier noise.

Much work has been devoted to measuring the luminous flux from the moon. Of particular interest to us are the measurements made with apparatus for laser ranging of the moon since there the measurement conditions are closest to the conditions of real laser-ranging experiments. Here we present data on the background light obtained by various authors and scaled to our receiver system for our laser-ranging measurements. The parameters of this system are $\eta_{det} = 0.017$, angular field of view $\Theta_{det} = 10''$, and transmission bandwidth $\delta\lambda_F = 5$ Å. In [43] the background from the central part of the lunar disc during the lunar night was $8 \cdot 10^2$ pulses/sec and at full moon, $8 \cdot 10^6$ pulses/sec. Measurements taken at the Pic du Midi observatory [85] yielded the following values for the background light: at full moon, 10^6 pulses/sec; in the region of the terminator at a phase of 90°, $3.2 \cdot 10^3$ to $9.5 \cdot 10^3$ pulses/sec; and during the lunar night, 10^2 pulses/sec. In [86] $2.6 \cdot 10^5$ pulses/sec were obtained at a phase of 90° for the center of the lunar disc at the same observatory. The following data on the background light were obtained at the MacDonald observatory [87]: maximum background at full moon $6.2 \cdot 10^5$ pulses/sec. During the laser-ranging measurements the maximum background level was $2 \cdot 10^6$ pulses/sec. Thus, for our receiver the background may vary within the following limits: at full moon, $6.2 \cdot 10^5$ to $8 \cdot 10^6$ pulses/sec; at a phase of 90° near the terminator, $3.2 \cdot 10^2$ to $2.6 \cdot 10^5$ pulses/sec; and during the lunar night, 10^2 to $8 \cdot 10^2$ pulses/sec.

The background measurements made regularly during laser-ranging sequences with our apparatus yielded 10^6 to $6 \cdot 10^6$ pulses/sec at full moon, $2 \cdot 10^4$ to $5 \cdot 10^5$ pulses/sec in the terminator region, and $1.5 \cdot 10^3$ to $1.5 \cdot 10^4$ pulses/sec during the lunar night.

These data are in good agreement with the results on other rangefinders. The wide scatter in the background values for the same measurement condition is noteworthy. This is explained primarily by the dependence of the background components that are scattered in the atmosphere on the weather. During observations of the dark portion of the lunar disc the part of the background due to light from the crescent also depends strongly on the weather. The appearance of haze, for example, often causes the background to increase by ten times or more.

We note that the night airglow is roughly an order of magnitude weaker than the earthlight and thus may be neglected.

The intrinsic noise of the photomultiplier is about $5 \cdot 10^2$ pulses/sec and only makes a small contribution to the total background.

3. Isolating the Rangefinder Signal from the Noise

At the detector input the reflected rangefinder signal is combined with the luminous background. Spectral and spatial-temporal filtering, which are intended to reduce the background level without reducing the signal level, are used to isolate the signal. Spectral selection is achieved by using an interference filter with the narrowest possible transmission bandwidth. Spatial filtering with respect to solid angle is done by limiting the field of view of the detector, the magnitude of which is chosen on the basis of the considerations in Chapter IV, Section 1. These measures make it possible to reduce the background to a value of the order of 10^4 pulses/sec for the dark side of the moon.

The background is further limited by distance selection. This is based on the fact that the distance to the observed point on the moon can be precalculated with an error which, depending on the computational technique used, may be 1-2 km. This makes it possible to accept the reflected signal only during the "time windows" whose centers are matched to the precalculated propagation time of the laser beam to the target and back and whose durations are chosen to cover the errors in the calculation. The "time windows" were chosen to be $T_w = 100$ μsec in our initial experiments on finding the distance to the reflectors. Later, as the precalculations were refined, their width was reduced to about 20 μsec. The average background level for $T_w = 20$ μsec is $n_w = 0.2$ pulses. The magnitude of the reflected signal, as calculations and numerous measurements show, may fluctuate within the range $n_s = 0.02-0.2$ photoelectrons per laser shot. With these signal and background levels the resolving time of the detector is enough to count individual pulses, corresponding to photons, independently. The signal and background are indistinguishable in this case and obey a Poisson distribution, as shown in [82]. A signal of this magnitude can be detected only by statistical accumulation by repeated laser shorts. Any "marking" of the signal or use of coincidence schemes is impractical since this ultimately leads to a reduction in the signal level and, therefore, to greater difficulty in detecting it. Direct statistical accumulation without additional temporal selection is also ineffective. The required number of laser pulses, n_L, is determined in this situation by $\overline{N}_s \geq m_n(\overline{N}_{bg})^{1/2}$, where $\overline{N}_s = \overline{n}_s n_L$, $\overline{N}_{bg} = \overline{n}_{bg} n_L$, and m_n is the excess of signal over background fluctuations which is taken to be 4-5 from the attainable probability of a "false alarm" ($P_{fa} \sim 10^{-3}$). Then, for an average signal level of $n_s \cong 0.05$ the number of laser shots required to isolate the signal is $n_L \cong 2000$, which is unacceptable. In addition, this means of isolating the signal only allows us to establish its presence or absence within the "time windows" but does not make it possible to identify specific signal pulses.

Thus, for additional time filtering we use the fact that the error in the precalculated propagation time may be made almost constant over times corresponding to the duration of a series of measurements. As was shown in Chapter II, Section 3, the errors in the precalculations are mainly determined by inaccuracies in the accepted values of parameters of the earth—moon system. The change in this error over a given time interval may be estimated by comparing the solutions of systems of equations such as (31) which correspond to the beginning and end of the given interval. The quantity $\tau_0 - \tau_c$, where τ_0 is the measured and τ_c is the computed propagation time, will be constant to within an accuracy determined by the degree of constancy of the error in calculating τ_c and the error in measuring τ_0.

Estimates based on Eq. (31) showed that the change in $\tau_0 - \tau_c$ with time due only to errors in calculating τ_c may reach about 200 nsec over 1 hour. The experimental data obtained from measurements of the distance to the L-1, L-2, A-11, and A-15 reflectors (see Chapter VI, Sections 1 and 2) also indicate a change in $\tau_0 - \tau_c$ by 100-300 nsec over 1 hour.

The present accuracy in a single measurement of the propagation time of the laser signal, τ_0, is about ± 6 nsec. This implies that the main contribution to the change in $\tau_0 - \tau_c$ over a series of measurements (which usually lasts 5-20 min) is from errors in the precalculated value τ_c.

The maximum change in $\tau_0 - \tau_c$ over a single series of measurements is about 100 nsec. This value must then be taken as the interval width for further temporal filtering of the reflected signal.

If we divide the "time window" into intervals of this width, which we refer to as channels, then for repeated laser pulses the reflected signal will fall into a previously unknown channel but always the same one (at least in the two nearest). Thus, observing the signal pulses reduces to storing them over repeated shots and finding the channel that contains both signal and background among the remaining channels which contain only background.

We denote the channel width by t_0. The number of channels in a "time window" of duration T_w is $m = T_w/t_0$. The average background level in each channel over n_L laser shots is

$$\bar{n}_{bg} = \bar{N}_{bg}\, t_0 n_L.$$

Using the additivity property of a Poisson distribution, we may write

$$P_m(n) = \frac{m\bar{n}_{bg}^{\,n}}{n!}\, e^{-\bar{n}_{bg}} = \frac{T_w(\bar{N}_{bg}\, t_0 n_L)^n}{t_0 n!}\, e^{-\bar{N}_{bg}\, t_0 n_L} \tag{54}$$

for the probability of obtaining n background pulses in any of m channels. The probability of n or more background pulses in any of m channels is

$$P_m(\geqslant n) = \sum_n^\infty P_m(n) = \frac{T_w(\bar{N}_{bg} t_0 n_L)^n}{t_0 n!}\, e^{-\bar{N}_{bg}\, t_0 n_L} \frac{1}{1 - \dfrac{\bar{N}_{bg} t_0 n_L}{n+1}}. \tag{55}$$

Furthermore, specifying some allowable probability of a false alarm, P_{fa}, we obtain an equation for estimating the minimum observable signal:

$$P_m(\geqslant n) \leqslant P_{fa}. \tag{56}$$

TABLE 4

n_L	100	200	500	1000	2000	5000	10,000
n	4	5	7	9	12	18	32
n_s	0.04	0.025	0.014	0.009	0.006	0.0036	0.0032

As a specific example, Table 4 lists values of n for different numbers of laser shots calculated for the following parameter values in Eq. (56): $P_{fa} = 10^{-3}$; $\bar{N}_{bg} = 10^4$ pulses/sec; $T_w = 20\ \mu$sec; $t_0 = 100$ nsec.

The third row of the table shows the magnitude of the signal per laser shot. The quantity n_s thus characterizes the signal which requires at least n_L shots to be observed.

Equation (55) may serve as a basis for designing a detector−analyzer which, when one is given data on T_w, t_0, and N_{bg} after each laser shot, will analyze the data in each channel in accordance with this equation and yield information on the presence or absence of a signal, depending on whether $P_m(n)$ is smaller or greater than some previously specified P_{fa}.

The signal is accumulated both to establish the fact of its existence and, primarily, to measure the distance to the object. However, this requires accumulation of the maximum possible amount of statistical data. Thus, in a real experiment the acquisition process continues until the presence of a signal is clearly established and the probability determinations become unnecessary. In this case the relative number of background pulses which might be taken for signal pulses is greatly reduced and the accuracy of the measurement is enhanced.

A detection criterion, however, is necessary when the signal is lower than the expected level or completely absent as, for example, when the target is missed. If after n_L laser shots corresponding to an expected signal n_s no channels with a number of pulses n such that $P_m(\geq n) \leq P_{fa}$ are observed, then it is concluded that there is no signal, further acquisition is stopped, and the reasons for the failure are analyzed.

First the state of the atmosphere is checked by measuring the atmospheric transmission k_{atm} and the diameter of the scattering circle which determines α_L [see Eq. (47)]. When the state of the atmosphere is satisfactory, a search for the reflector is begun by scanning.

Finally, we have seen that even when we are observing reflectors lying on the dark side of the moon it is fairly difficult to isolate the reflected signal. Going to the bright part of the moon involves a large increase in the luminous background and requires a substantial improvement in the filtering and in the noise stability of the atmosphere.

VI. MEASUREMENTS OF THE DISTANCE TO REFLECTORS ON THE MOON AND THE FIRST DETERMINATIONS OF SEVERAL PARAMETERS OF THE EARTH−MOON SYSTEM

1. First Measurements of the Distances to the Lunokhod 1 Reflector

In November 1970 the first French reflector, mounted on Lunokhod 1 (Fig. 24), was placed on the moon. To ensure that the reflector could be oriented toward the earth it was mounted so the angle between its axis (normal to the corner element panel) and the vertical axis of the lunokhod (lunar vehicle) was equal to the angle between the selenocentric radius-vectors of the lunokhod and the point on the moon's surface with selenographic coordinates $l = 0$, $b = 0$. Orientation with respect to the local azimuth was realized by turning the entire

Fig. 24. Lunokhod-1 and the French reflector mounted on it
(to the upper left).

lunokhod about the vertical axis. This reflector could be used only during the lunar night. Thus
the vehicle was turned at the end of the daylight program just before the lunar night set in.
The accuracy of the orientation was determined by the list and trim of the lunokhod in its
nocturnal position (approximately 1-3°) and the moon's libration (roughly ± 8).

In terms of astronomical conditions, observations of this reflector during the first lunar
night after its landing on the moon were possible on December 5-8. During this period it was
possible to make two series of measurements, on December 5 and 6. No observations were
made on December 7 and 8 because of poor meteorological conditions. In these measurements
the next to the last version of the laser rangefinder was used with a time resolution of about
$2 \cdot 10^{-8}$ sec and a repetition rate of 1/15 Hz.

Figure 25 shows a sketch of the part of the moon's surface near the Lunokhod 1. The
dashed curve shows the preliminary shape and size of the laser spot.

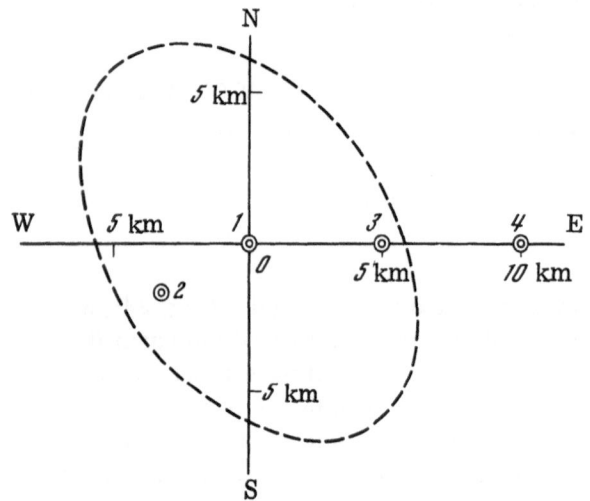

Fig. 25. A diagram of the scan for the
L-1 reflector.

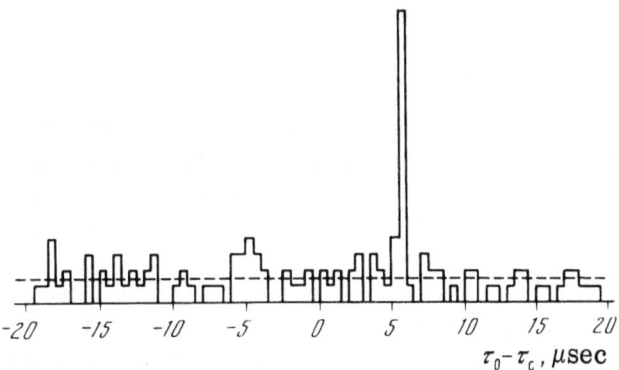

Fig. 26. A histogram with the reflected signal
from L-1 on December 6, 1970; N = 250 pulses.

Fig. 27. Dependence of $\tau_0 - \tau_c$ on time (December 6, 1970, L-1): 1) noise pulses; 2) reflected pulses recorded with an oscilloscope detector with an accuracy of $\pm 10^{-7}$ sec; 3) pulses detected with a time interval with an accuracy of $\pm 10^{-8}$ sec.

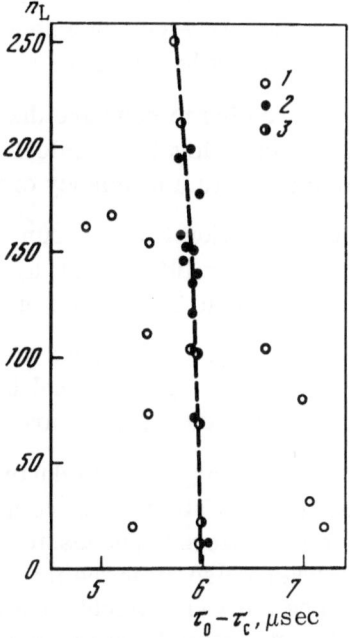

A program for searching for the reflector by successively "shooting at" 25 points uniformly distributed in a square of size 25×25 km was developed. The search was begun at the central point of this square, 34°53' west longitude, 38°17' south latitude (point 1 in the figure), where rangefinding was attempted during the entire sequence of December 5. As a result of these observations, lasting roughly 40 min (170 pulses), a fairly distinct reflected signal was observed although it was smaller than the calculated value.

During the sequence of December 6 the laser beam was successively aimed at points 2, 3, and 4, each of which was measured for 1 hour (250 shots per point). With the system aimed at point 2 the detected reflected signal was roughly the same size as for point 1. At point 3 the signal was very weak, and at point 4 it was almost absent. No survey of the area to the north, west, and south of point 2 was made because of the limited duration of the sequences.

For preliminary isolation of the reflected signal, histograms of the distribution of $\tau_0 - \tau_c$ (Fig. 26) were made.†

---

† The τ_c ephemerides were calculated at the Institute of Theoretical Astrophysics of the Academy of Sciences of the USSR.

Only part of the "time window" (an interval ± 20 μsec from the center) is shown in this histogram since it seemed that the reflected signal lay near the center and, therefore, the "time window" had been chosen with a wide margin.

For final isolation of the signal points and studies of the change in the measured distance with time, the parts of the "time window" corresponding to the reflected signal were examined separately. In Fig. 27 all the recorded photomultiplier pulses, including noise pulses, are plotted as a function of the number of the laser pulse (or, equivalently, the time). It can be seen that the signal points are grouped in an interval of width roughly 0.3 μsec and have a smooth time variation during the sequence. If we take the width of the signal interval to be 0.3 μsec, then the signal-to-noise ratio for the sequence of December 5 is of order 27 and that of December 6 (point 2) is of order 21. The average background is shown in Fig. 26 as a horizontal dotted line.

Only the signal points obtained with the time digitizer with an accuracy of $\pm 10^{-8}$ sec were used to determine the distances to the reflector. If we approximate the variation of these points during the sequences by smooth curves, then the deviation of the experimental points from these curves is less than $\pm 2 \cdot 10^{-8}$ sec, which corresponds to an error in the distance of ± 3 m.

It is interesting to compare the experimentally obtained values with the expected signals. The reflected signal level was calculated according to Eq. (47) with an additional coefficient to account for the quantum efficiency of the photomultiplier, k_{PM}.

In the calculation it was assumed that: $W_{tr} = 4$ J; $S_{refl} = 640$ cm^2; $\gamma = 0.8$; $S_T = 4.4$ m^2; $R = 380,000$ km; $\alpha_L = 10"$; $\Theta = 6"$; $k_{tr} = 0.6$; $k_D = 0.2$; $k_{PM} = 0.1$; $k_{refl} = 0.5$; $k_{atm} = 0.8$. Then the expected signal should be $n_c \sim 0.3$ electrons per laser pulse.

Experimentally it was found with the laser aimed at point 1 that 1 $n_c = 0.065$ and at point 2 that $n_c = 0.075$; that is, the signal is considerably less than the calculated value. The possible reasons for this discrepancy are discussed in Section 4 of this chapter.

In later lunar nights (both during the active operational period of Lunokhod 1 after it finally came to rest) attempts were made to observe the L-1 reflector. However, for a long time these efforts were unsuccessful. The situation was similar at the Pic du Midi observatory in France. There also after the first successful observations in December 1970 it was not possible to see the L-1 reflector again. Several unsuccessful attempts were also made to locate the L-1 reflector from the MacDonald observatory in the U.S.A.

All these failures led to the conclusion that the efficiency of the reflector had been substantially reduced due either to its being covered with dust or to mechanical damage during the active period of the lunokhod.

Nevertheless, it was not true that all possibility of observing the L-1 reflector were exhausted. In the Soviet and French observatories there still was not a sufficiently good range-finder in 1971-1972, so no systematic search was made. The failure of the American efforts might be attributable to disparities between the Soviet and American lunar maps. Therefore, in 1974 we made another attempt to find the L-1 reflector with an improved rangefinder. These attempts were successful and since May 1974 regular observations of the reflector have been made.

It should be noted that the same apparatus that was used for the first measurements of the distance to the L-1 reflector was used in 1971 for several observations of the American A-15 reflector.

2. Measurements of the Distances from Reflectors on the Moon Using the Automated System

In the beginning of 1973 the automatic system described in Chapter IV was put into use. Many of its major parameters were substantially improved over the previous model. The basic advantage of this system was its high speed (1/3 Hz) and the possibility of operator control during the measurements.

Despite its obvious advantages this apparatus, like its predecessor, was intended for operation with a rather low background light level. Suitable conditions in this sense occur only at nighttime and when points that lie on the dark side of the moon are being observed. A further limit on the possible observation time is due to the strong absorption of the laser light in the atmosphere at large zenith angles. From this standpoint zenith angles < 60° are accessible. In accordance with these limitations, favorable astronomical conditions exist in separate seasons of the year and are determined ultimately by the location of the observed point on the moon.

In January, 1973, the second French reflector, mounted on Lunokhod 2, was placed on the moon. At the time it was put on the moon the astronomical observation conditions were extremely unfavorable and continued to deteriorate until April, 1973. Adding to this the poor weather in the winter-spring period of that year, it was impossible to seriously count on any success during the first months of operation.

Nevertheless, despite the very unfavorable weather and astronomical conditions, attempts were made to observe the Lunokhod 2 reflector beginning right after the lunar night set in. Fourteen measurement sequences were planned up to June, 1973. Of these ten did not take place because of continuous cloud cover and four did take place under unfavorable conditions (large zenith distances, strong illumination due to unfavorable lunar phase, poor visibility, and strong atmospheric turbulence accompanied these observations). The efforts of the first months, as might be expected, were fruitless. Only with the improvement in astronomical and meteorological conditions in June, 1973, were the first results obtained.

On June 22, 23, and 24 the first successful observations were made of the L-2 reflector, and on the last of these three days the A-15 reflector was observed as well.[†] At the time these observations began the TPA-i minicomputer was added to the rangefinder system. This greatly eased observation and preliminary processing of the signal.

Figure 28 shows the display of a histogram with the first signal from L-2 recorded on June 22, 1973. Each point corresponds to a photoelectron whose time of arrival was recorded by the time interval measurer. The numbers on the photograph are, at the left, the time interval between the beginning of the time scale (τ axis) of the histogram the moment the "time window" is opened (the unit of time is 10 nsec); in the middle, the number of laser pulses; at the upper right, the width of a time channel (the unit of measurement is 1 nsec); and at the lower right, the number of signal points accumulated during the measurement series.

The reference point for aiming the laser at the L-2 reflector was the Archimedes-A crater.

To test the correctness of the aiming system a search was made during the sequence of June 23 for the optimum direction of the laser beam. To do this the telescope was successively shifted to the north, south, east, and west of the calculated point by a distance equal to the diameter of the light spot on the moon. On moving away from the calculated point the reflected signal fell. This test was repeated on June 24. After a definite reflection was obtained from L-2 the telescope was shifted to the side while the ranging system operated for 10 min (about 200 shots). There was no reflected signal during this period.

---

†We note that the astronomical conditions for the A-15 reflector were more favorable at this time of year and that regular measurements were begun with it somewhat earlier, in March, 1973.

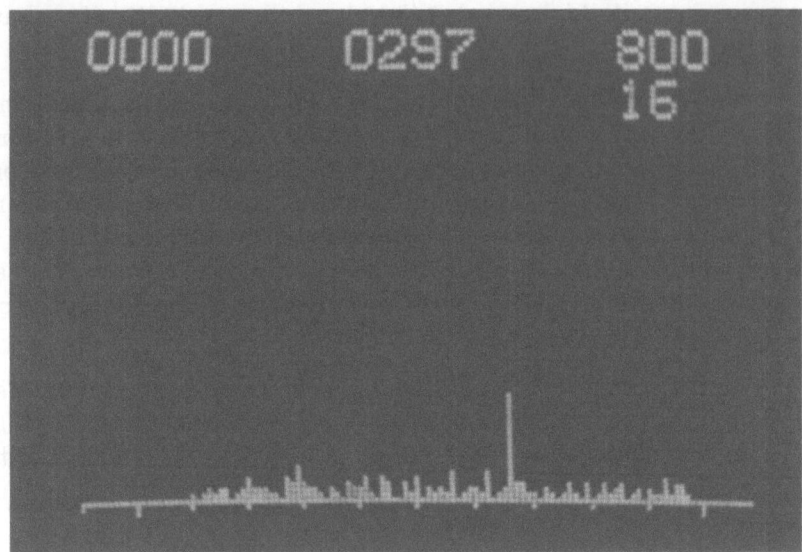

Fig. 28. A histogram with the reflected signal from L-2 (June 22, 1973).

With these sequences the regular measurement of the distances to the L-2 and A-15 reflectors was begun. Later, after August, 1973, observations of the A-11 reflector (which was less efficient) were added to these, and after May, 1974, observations of the L-1 reflector were renewed. During 1973-1974 about 900 distance measurements were made, and of these roughly 500 were made in 1974. In many cases it was possible to observe two or three reflectors during a single sequence with repeated transfer from one reflector to another.

From all the data accumulated up to now we may characterize the average levels of the reflected signals by $n_{L2} = 0.045$, $n_{A15} = 0.083$, and $n_{A11} = 0.037$. These quantities are more than an order of magnitude less than the expected reflected signal levels given by Eq. (47). This discrepancy requires a specific analysis. For now we can only point out some possible reasons for discrepancy.

Equation (47) does not take into account the structure of the laser beam after it passes through the atmosphere. It was derived assuming a uniform energy distribution over the light spot on the moon. In passing through an inhomogeneous atmosphere the laser beam acquires a complicated spatial structure. Over its cross section there will be narrow maxima whose intensity exceeds the average by several times. The bulk of the cross section will have an intensity much less than the average. Thus, if the reflector has equal probability of being at an arbitrary point of the spot and the reflected signal from the bright maxima is much greater than 1 photoelectron (and is recorded as a single-electron pulse), then there will be a drop in the average reflected signal over many shots. No quantitative estimate of this effect can be made without knowledge of the structure of the beam. One fact in favor of this explanation is that no discrepancy is observed between the computed and experimental magnitudes of the reflected signal for ranging of the lunar surface (without a reflector), when the signal level is determined by the average intensity over the spot on the moon. Another confirmation of this explanation may be that the signal level varies over wide limits both during a single sequence and from one sequence to another. Thus, the average signal per shot during a sequence in the case of the L-2 reflector, for example, varies by more than an order of magnitude (from $0.9 \cdot 10^{-2}$ to $11 \cdot 10^{-2}$).

Some reduction in the reflected signal may be due to inaccurate aiming of the telescope at the target. Inaccurate aiming may be due to errors in the calculations of the relative an-

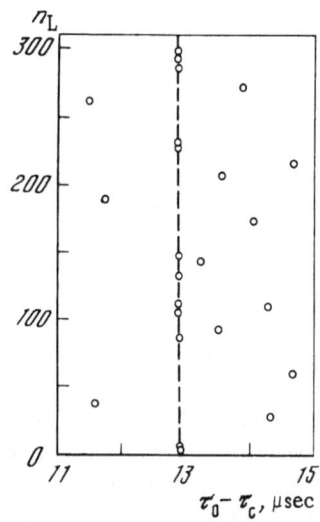

Fig. 29. The time dependence of $\tau_0 - \tau_c$
(August 20, 1973, A-11 reflector).

gular coordinates of the reflectors and the reference craters both because of inaccurate knowledge of the selenographic coordinates of the reflectors and because of inaccuracies in the
theories used in the calculations. Errors in aiming the telescope may also be a result of atmospheric instabilities and, finally, insufficient experimental skill on the part of the observer.

Figure 29 shows a typical plot of the variation in the experimentally obtained differences
$\tau_0 - \tau_c$ with the laser shot number (or, equivalently, with the time).

3. Measurement Errors and Calibration of the System

The mathematical scheme discussed in Chapter II presupposes the use of distance measurements between fixed points on the earth and moon. On the moon these points are the corner reflectors. On the earth one such fixed point is the intersection of the α and δ axes of the
telescope (the center of the diagonal mirror M_4 in Fig. 13). The optical paths to this point
from the laser and the detector and from this point in the telescope to the exit from the telescope are constant and independent of the location of the telescope.

It is thus necessary to relate the measurement results to this point. The correction
which must be introduced in the measurements consists of two parts, geometrical and electronic.

The geometrical correction, which is easily determined by measuring the corresponding
segments of the optical paths, is 37.5 m in our apparatus (or 125 nsec in time units).

The electronic part of the correction is determined by the difference in the electrical
path lengths of the "start" and "stop" circuits.

The principal components of this correction are the delay in the photomultiplier, $\tau_{PM} = $
45 nsec, in the amplifer-pulse generator, $\tau_{amp} = 28$ nsec, and in the connecting cables, $\tau_{cab} = $
26 nsec.

The combined total correction due to delay in the electronic circuits is $\tau_{PM} + \tau_{amp} + \tau_{cab} = $
89 ± 2 nsec. This quantity was also measured as a whole. The "start" of the time interval
measurer was initiated by the laser pulse incident on the photodiode and the "stop," by the
same pulse received by the photomultiplier. The result was 90 ± 4 nsec.

It should also be noted that the propagation time is measured by a time interval measurer
with a step size of $\varepsilon = 10$ nsec. Time is reckoned from the earliest reference oscillator (100
MHz) pulse after the laser is fired. Thus, the delay in the timing origin with respect to the

laser pulse may with equal probability be any amount in the interval 0 to ε with an average of 0.5ε. The timing scale is ended at the reference oscillator pulse before the reflected laser signal (the pulse after it is not counted). Thus, the time count stops in advance of the time the reflected pulse arrives. The average amount of this advance is also 0.5ε.

Therefore, on the average the time measured with the time digitizer is always less than the actual time by ε = 10 nsec. This amount must be added to the time inverval measurement.

The total instrumental correction to the measured distance is thus 204 ± 2 nsec.

The errors in the measurement of the propagation time of the light pulse are made up of several components. The error due to the finite duration of the laser pulse is statistical. For a pulse of width 15 nsec at the 0.1 level the root-mean-square value of this error is roughly σ_L = 4 nsec. The root-mean-square error associated with the discreteness in counting the beginning of the measured time interval is σ_ε = 0.29ε $\cong$ 3 nsec [88]. The same error arises due to discreteness in counting the end of the measurement interval. The error due to the finite risetime of the photomultiplier is σ_{PM} $\cong$ 1.3 nsec (see Chapter IV, Section 3). Besides these, the corrections and errors due to the propagation of the laser light through the atmosphere must be evaluated. Here we must distinguish two effects: the bending of the beam trajectory in the atmosphere due to the gradient in the refractive index with altitude (astronomical refraction) and the lengthening of the optical path due to the difference of the refractive index from unity.

The increase in the length of the trajectory due to bending of the beam in the atmosphere is negligibly small (of the order of millimeters [89]) for zenith angles of up to 70° and may be neglected.

The lengthening of the optical path due to the difference of the refractive index from unity may be several meters for large zenith angles. This lengthening may be taken into account by calculating the integral of the refractive index with respect to altitude.

In [90, 91] it was shown that the lengthening may be calculated approximately using only data on the pressure, temperature, and humidity of the air near the earth's surface. If the pressure and temperature are measured with an accuracy of about 0.1%, then the lengthening may be determined with an accuracy of 1-2 cm. The lengthening of the optical path along the vertical for the Crimean observatory (altitude 550 m above sea level), calculated by the method of [90, 91], is 225 cm at standard temperature (t_0 = 20°C) and pressure (p = 1013 mb). This quantity cannot change by more than about ±10 cm for the actual changes in the meteorological conditions in the Crimea. With the presently achieved accuracy of about ±1 m in the laser measurements there is evidently no particular sense in taking weather conditions into precise account. Thus, the atmospheric correction is introduced by the formula D_{atm} = 2.25 sec z (m), where z is the zenith angle. Up to zenith angles z ~ 70° this formula describes the change in the thickness of the atmosphere with an accuracy of about 1%. This correction expressed in time units and including the double passage of the laser signal through the atmosphere is τ_{atm} = 1.5 · 10^{-8} sec z sec, while the error in this is σ_{atm} $\cong$ 0.6 nsec (for z = 0°).

A small additional scatter in the measurement results given in time units by σ_T $\cong$ 0.5-2.0 nsec is due to imprecision in coupling the measurement times to the time scale (~ 10^{-3} sec).

Therefore, the overall root-mean-square error in the measurement of the time intervals is $\sigma_0 = (\sigma_L^2 + 2\sigma_\varepsilon^2 + \sigma_{PM}^2 + \sigma_{atm}^2 + \sigma_T^2)^{1/2} \cong$ 6.0 nsec, which corresponds to an error in the distance measurement of ±0.9 m.

For an experimental estimate of the errors in the measurements we turn again to Fig. 29 which shows the difference $\tau_0 - \tau_c$ as a function of the laser shot number. Since to find these differences the values of τ_c are fed into the computer with a step size ε = 10 nsec (although τ_c

is precalculated with an accuracy two orders of magnitude greater), the root-mean-square error in the difference $\tau_0 - \tau_c$ will be $\sigma_{0c} = (\sigma_0^2 + \sigma_c^2)^{1/2}$, where $\sigma_0 \cong 6$ nsec is the root-mean-square error in the measurement of τ_0 and $\sigma_c = \sigma_\varepsilon \cong 3$ nsec is the error due to the discreteness in the τ_c data. Thus, $\sigma_{0c} \cong 7$ nsec.

The root-mean-square deviations in the experimental points from the regression curves oscillate from one sequence to the next within limits of 8-9 nsec, which is in good agreement with our estimate.

4. Coupling the Measurements to the Time Scale

The process of measuring the distance between the rangefinder and the reflector takes a finite time during which there is a relative displacement of the earth and moon. Thus it is necessary to examine the relationship of the time at which the distance measurement was made to some time scale. The measured time interval between the time T_1 the laser is fired and the time T_2 the reflected signal arrives is broken down into two unequal intervals τ_1 and τ_2 which are the times the signal takes to propagate to the target and back. The measured interval $T_2 - T_1 = \tau_1 + \tau_2$ must be compared with the sum of the precalculated times, $\tau_{c1} + \tau_{c2}$, the first of which is determined by the location of the earth at time T_1 and the moon at time $T_1 + \tau_{c1}$, and the second of which is determined by the location of the earth at time T_2 and the moon at time $T_1 + \tau_{c1}$.

In practice the time of the measurement is easily matched to the time scale by referring the measured distance to some average time between T_1 and T_2. Let us introduce a topocentric rectangular coordinate system rigidly attached to the geoequatorial system. In this system the laser transmitter is fixed and only the reflector moves. The light signal sent at time T reaches the reflector at time $T + \tau$ and returns in the same interval τ. Thus, τ is an expression in units of the speed of light of the distance to the reflector at time $T + \tau$. Precalculating τ at time $T + \tau$ already requires a priori knowledge of τ. This difficulty may be avoided by the method of successive approximations. The value $\tau_0 = D/c$ is calculated from the distance D at time T. Then D_1 at time $T_1 = T + \tau_0$ is found along with the corresponding value $\tau_1 = D_1/c$, and so on.

Thus, the distance D corresponding to the measured value τ can be referred to time $T + \tau$ with an accuracy determined by the degree of approximation which is found from the condition that $T_n - T_{n-1}$ must be less than the error in measuring τ.

We now consider the problem of coupling the measurements to the time scales. Evidently, the times $T + \tau$ must be defined on a single time scale for all the measurements. There are several time scales which are coupled to one another with a certain accuracy. They include [92] the following: ET, ephemeris time in which the geocentric distances to the reflectors are calculated; UT, universal time to which the times of firing the laser are matched and in which the location of the earth is calculated at these times; and UTC, the coordinated atomic time scale.

We shall first consider with what accuracy the times of the laser pulses must be matched to the time scale from the standpoint of observing the reflected signal. To observe the signal in the presence of noise it is necessary that the difference $\tau_0 - \tau_c$ be minimal and, more importantly, constant, since this determines the required width of the "time window" and, consequently, the noise level and the permissible channel width in the time analyzer (see Chapter IV). Here it is convenient to take the working scale to be UT1 in which the angular motion of the earth, that is, the fastest component in the mutual motion of the earth and moon, is determined. The timing service gives the time with an accuracy of about 10^{-2} sec on this time scale. Besides this fast component, the topocentric ephemeris of the reflector contains components of the motion with longer periods which describe the geocentric motion of the reflector, which is calculated in the ET scale. On going to the UT scale with the formula ET = UT + ΔT the ephemeris will be matched to this scale with an error determined by the inaccuracy in the correction ΔT. Since, however, this inaccuracy enters in a slowly varying orbital component of the reflector motion with greater constancy in its velocity than the components related to the earth's rotation, it leads to a relatively slow variation in $\tau_0 - \tau_c$. In short series of observations this is expressed in a shift in the reflected signal with respect to its calculated time and, to a much smaller extent, by its being blurred.

An analysis of the conditional equations for the corrections dT and ds, where s is the local sidereal time, shows that the change in $\tau_0 - \tau_c$ due to the change from the ET to the UT1 system is an order of magnitude less than that due to the inaccuracy of coupling to the UT1 scale itself. Thus, we can tolerate an inaccuracy in the accepted value of the correction of $\Delta T \sim 10^{-1}$ sec.

The times of the measurements which have already been made must be coupled to one another with an accuracy of at least about 10^{-3} sec. In this case the error in the distance due to the inaccuracy in the coupling would not be more than about $2 \cdot 10^{-9}$ sec; that is, it does not exceed the measurement error. If we are speaking of the mutual coupling of observations separated by intervals of several years, then the stability of the clocks which give the time scale must be at least 10^{-12}. Such stability can be ensured by atomic frequency standards. Thus only the UTC scale can meet these requirements.

The timing service for the laser rangefinder system couples the times of the measurements to the UTC scale with an accuracy of about 10^{-3} sec with the aid of signals transmitted by radio. For the accuracy (about 10^{-8} sec) attained up to now in ranging measurements, this coupling accuracy seems adequate. To couple the measurement times to the working UT1 scale, the UT1–UTC correction broadcast on the radio is used. The accuracy of this correction (about 10^{-2} sec) is, as noted previously, sufficient for observation of the reflected signal.

5. A Determination of the Geocentric Coordinates of the Observation Point and the Selenographic Coordinates of the Lunar Reflectors

It was shown earlier that long-term laser rangefinding measurements are needed for the most accurate determination of the parameters of the earth−moon system (up to 18.6 years). However, it is of interest to examine the prospects for determining several parameters without waiting to accumulate data over such a long time. This interest is dictated both by a desire to obtain results over a short time and, mainly, by the fact that such geodynamic problems as the study of the variations in the speed of rotation of the earth, the motion of the pole, the continental drift, and tidal effects require determination of the "instantaneous" coordinates of the observation points over short series of measurements. To evaluate the prospects for using the results of such short time measurements we turn to Eq. (31). Using the expressions for the coefficients of this equation [given after Eq. (31)] we made a crude estimate of the contributions to $D_0 - D_c$ from inaccuracies in the values of the parameters used in the precalculation. These contributions are estimated by multiplying the accepted probable errors in the parameters by the corresponding coefficients in the conditional equation. In determining these coefficients we have taken the following values of the quantities contained in them:

$$\mathscr{E}', \mathscr{E}'', \mathscr{E}''', \mathscr{G}', \mathscr{G}'', \mathscr{G}''', \mathscr{H}', \mathscr{H}''' \leqslant 1;$$

$$A', A'', B', B'', C', C'' \leqslant 1; \qquad k_i \cong 1 \, (i = 1, 2, \ldots, 7);$$

$$Q \cong 1; \quad \frac{R_{\mathbb{C}}}{\Delta'} \sim \frac{1}{240}; \quad \frac{w}{\Delta'} \sim \frac{1}{100}; \quad \frac{h}{\Delta'} \sim \frac{1}{100}.$$

TABLE 5

Parameter	Δ_0	e	e'	γ	α	l_0
Probable error in the parameter	200—500 m	$0''.02$—$0''.1$	$(5$—$75)\cdot10^{-8}$	$0''.015$	$1\cdot10^7$	$0.''04$
Contribution to the difference $D_0 - D_c$	200—500 m	40—200 m	10 m	1 m	5 m	1.3 m

Parameter	l_0'	F_0	D_0	s	w	h	ε
Probable error in the parameter	$9''.0$	$1''.0$	$0''.11$	$0''.1$—$10''.0$	100—200 m	100—200 m	$0.''1$
Contribution to the difference $D_0 - D_c$	10 m	3 m	9 m	3—300 m	100—200 m	100—200 m	3 m

Parameter	$R_{\mathbb{C}}$	l	b	f	g
Probable error in the parameter	100—1000 m	$1'$—$5'$	$1'$—$5'$	0.01—0.1	$1''$—$10''$
Contribution to the difference $D_0 - D_c$	Several hundred meters	Several hundred meters	Several hundred meters	4—40 m	0.2—4.0 m

In the coefficients D_s and D_w the value of $\cos\delta_{\mathbb{C}}$ is assumed to be unity. In addition the relation $\left|\Delta\frac{\partial Q}{\partial e}\right|\gg\left|w\frac{\partial\lambda_{\mathbb{C}}}{de}\right|>\left|w\frac{\partial\beta_{\mathbb{C}}}{\partial e}\right|$ has been used. The probable errors in the parameters are taken from [36, 93-95].

Table 5 lists the probable errors of all 18 parameters in the conditional equation (31) and their contributions to $D_0 - D_c$.

From this quantity it follows that all terms of the conditional equation can be divided into two groups depending on their size. The main errors in $D_0 - D_c$ are due to the errors in Δ_0, e, s, w, h, $R_{\mathbb{C}}$, l, and b. If we neglect the errors in the remaining parameters, the order of the system of conditional equations is reduced from 18 to 8 and the conditional equation takes the form

$$\frac{1}{Q}d\Delta_0 - \frac{1}{Q^2}\Delta_0 de + w\sin t\,d\lambda - \cos t\,dw + \sin\delta dh + \cos\omega' dR_{\mathbb{C}} -$$
$$- \cos b\sin(l-l_0)R_{\mathbb{C}}dl + [\cos b\sin b_0' - \sin b\cos(l-l_0')]R_{\mathbb{C}}db = D_0 - D_c. \qquad (57)$$

We note that of the eight quantities to be determined in this equation the local sidereal time, s, is not a parameter. It is proportional to the time. However, from Eqs. (3) and (17) we may conclude that s is expressed in terms of a number of parameters such as the geocentric longitude of the observer λ, the coefficients of the polynomial representing the right ascension of the mean equatorial sun (the longitude at the initial epoch L_0', the tropical motion of the mean sun μ), and the constants entering in the expression for the nutation in the right ascension N_α. In this discussion we can identify the correction to the local sidereal time ds with the correction to the accepted value of the observer's longitude $d\lambda$ since the values of L_0', μ, etc. are known with much greater accuracy than the observer's longitude λ [94, 95]. However, in a strict discussion we must take ds in Eq. (57) to mean the correction to the value of the local sidereal time taken at the time of the measurements. Finding this correction by laser ranging makes it possible to measure the difference in longitude of points on the earth's surface and is therefore just as important as determining the correction to the longitude $d\lambda$.

A further simplification in the conditional equation (57) is based on an analysis of the time variation of its coefficients. Evidently, the coefficients in the corrections to the observer's coordinates λ, w, h change fastest (with a period of about a day) while the coefficient for dh generally changes little. Since the coefficients in the remaining corrections change little over daily intervals, it is possible to divide the corrections in Eq. (57) into two groups to the first of which $d\lambda$ and dw belong, while all the others belong to the second. The terms in the equation belonging to the second group may, on the basis of one general feature, the small variation in their correction coefficients over a day, be combined into one term, and the system of conditional equations may therefore be reduced to a third-order system. Formally, the conditional equation may be simplified this way if we leave only terms containing $d\lambda$, dw, and (for example) $d\Delta_0$ in its left-hand side. The equation then takes the form

$$\frac{1}{Q} d\Delta_0 + w \sin t\, d\lambda - \cos t\, dw = D_0 - D_c. \tag{58}$$

Transforming Eq. (57) to the form (58) by discarding part of the terms introduces additional errors in the solutions $d\Delta_0$, $d\lambda$, and dw of the latter. These errors can be estimated by solving the system of equations whose left-hand sides are the rejected terms of Eq. (57) for three specific measurement times.

A system of conditional equations of the form (58) can be used to analyze the results of laser-ranging measurements made over intervals of up to several days.

One indication of the value of a series of measurement data taken to improve the geodesic parameters λ and w is the time interval during the day in which the measurements were taken. The best interval would be one located symmetrically with respect to the time the moon crosses the meridian and encompassing as many hours as possible.

Making sequences of measurements that are long enough at present encounters a series of limitations due to the impossibility of daytime (earth's) observations or of observations at times when the reflector is illuminated by the sun (lunar day), as well as to the difficulty of working at large zenith angles (strong atmospheric absorption, reduced visibility). In terms of length the best series of observations during a single night were those of the L-2 reflector taken on September 18, 1973 and September 7, 1974.

These series were analyzed by solving a simplified system of equations of the form (58) by the method described in [35].

The precalculated values of τ_c were derived on the basis of the theory of lunar motion with an IAU nomenclature index j = 2. The components of the moon's libration, τ, ρ, σ, were taken in the form of trigonometric series [25] and the parameters of the physical libration were taken to be $f = 0.73$ and $\mathcal{J} = 1'32'20''$.

The results of this analysis are given in Table 6.

To refine the parameters of the lunar orbital and rotational motion, the series of measurements must be long enough to include as large a part as possible of the periods of the fundamental arguments of the theory of lunar motion. Table 7 shows the distribution of our

TABLE 6

Corrections to the coordinates of the observation point	Date; hour angle interval	
	September 18, 1973, -5^h01^m to -0^h58^m	September 7, 1974, -4^h08^m to 0^h26^m
$d\lambda''$,	3.3±0.1	3.4±0.1
dw, m	18.1±10.9	19.0±16.5

TABLE 7

Fundamental arguments	Intervals of variation of the fundamental arguments, degrees			
	December 5 and 6, 1970	November 12, 1971	March 9- September 20, 1973	May 28-September 10, 1974
$\mathbb{C}$	331—334	162	345—90	225—30
π	340	18	72—94	121—134
Ω	327	309	284—273	260—252
L'	254	230	346—178	70—171
π'	282	282	282	282

observations over the periods of the fundamental arguments $\mathbb{C}$, π, Ω (the mean longitude of the moon, the mean longitudes of the perigee, and node of the lunar orbit), and L' and π', the mean longitude of the sun and the longitude of the perigee of the sun's orbit. The table includes isolated measurements made in 1970-1971.

It was shown above that along with the geocentric coordinates of the observer the greatest contribution to $D_0 - D_c$ is from inaccuracies in the selenographic coordinates of the reflector, the average distance to the moon, and the eccentricity of the moon's orbit. As can be seen from Table 7, the observations cover only small portions of the periods of the fundamental arguments and are of insufficient duration to determine the orbital parameters. Thus it seemed possible to determine those corrections to the selenographic coordinates of the reflectors for which the coefficients in the conditional equations (57) vary with an overall period of about a month.

The most measurements were made using the L-2 reflector. This series was taken for the basis of the corrections. The experimental values of $\tau_0 - \tau_c$ for this series lie within the interval $(2-4) \cdot 10^{-6}$ sec.

Taking the corrections dλ and dw to the geocentric coordinates of the observer obtained here, the solution of a system of conditional equations of the form (57), constructed for the entire set of measurements during 1973-1974, was obtained for the selenographic coordinates l_{L2} and b_{L2}, for the L-2 reflector and for its selenographic radius-vector, R_{L2}. As a result, the following corrections to these coordinates were found: $dl_{L2} = 2.8'$, $db_{L2} = -3.0'$, and $dR_{L2} = -580$ m.

A recalculation of τ_c for the same series of measurements using the refined coordinates of the observation point, λ, w, and of the reflector, l_{L2}, b_{L2}, R_{L2}, reduces the root-mean-square value of the difference $\tau_0 - \tau_c$ to $\sigma_{\tau_0-\tau_c} = 36 \cdot 10^{-8}$ sec.

Along with analyzing the distance measurements to reflector L-2, we analyzed a series of measurements of the distance to the A-15 reflector. This series was much shorter than that for L-2 and also covers small parts of the periods of the fundamental arguments. Thus, we can speak of attempts to match the measured quantities with the precalculated values.

The values of τ_c for the A-15 reflector were precalculated taking the above corrections in the coordinates of the observation point into account as well. Then the selenographic coordinates of the A-15 reflector were refined by solving the corresponding conditional equations.

As a result we obtained the following corrections to the coordinates of the A-15 reflector: $dl_{A15} = -4.2'$, $db_{A15} = -4.8'$, and $dR_{A15} = -1130$ m.

After these corrections were included the root-mean-square residual difference was found to be $\sigma_{\tau_0-\tau_c} = 9 \cdot 10^{-8}$ sec.

These refinements must be regarded as a first attempt to determine a consistent set of parameters corresponding to the chosen mathematical model of the earth—moon system from the results of laser-ranging measurements.

LITERATURE CITED

1. T. H. Maiman, Nature, 187:493 (1960).
2. Yu. L. Kokurin, V. V. Kurbasov, V. F. Lobanov, V. M. Mozhzherin, A. N. Sukhanovskii, and N. S. Chernykh, Kosm. Issled., 4:414 (1966).
3. A. A. Yakovkin, in: The Moon [in Russian], Fizmatgiz, Moscow (1960), p. 39.
4. R. Baldwin, What Do We Know about the Moon? [Russian translation], Mir, Moscow (1967).
5. G. MacDonald, Rotation of the Earth, Cambridge University Press (1975).
6. N. N. Pavlov, in: The Earth's Rotation and the Determination of Time [in Russian], Nauka, Moscow (1969), p. 5.
7. A. Wegener, Origin of Continents and Oceans, Peter Smith Publishing, New York.
8. Continental Drift [Russian translation], Mir, Moscow (1966).
9. H. Takeuchi, S. Ueda, and H. Kanamori, Do the Continents Move? [Russian translation], Mir, Moscow (1970).
10. M. Tarling and D. Tarling, Continental Drift, Doubleday, New York (1971).
11. V. K. Abalakin and Yu. L. Kokurin, in: Some Problems in the Physics of Outer Space [in Russian], VAGO, Moscow (1974), p. 63.
12. K. Froome and L. Essen, Velocity of Light, Academic Press, New York (1969).
13. Yu. L. Kokurin, V. V. Kurbasov, V. F. Lobanov, A. N. Sukhanovskii, and N. S. Chernykh, Kosm. Issled., 5:219 (1967).
14. Yu. L. Kokurin and V. F. Lobanov, Kosm. Issled., 6:247 (1968).
15. E. W. Brown, Mem. Roy. Astron. Soc., 53:39 (1900).
16. E. W. Brown, Mem. Roy. Astron. Soc. 54:1 (1904).
17. E. W. Brown, Mem. Roy. Astron. Soc., 57:51 (1905).
18. E. W. Brown, Mem. Roy. Astron. Soc., 59:1 (1909).
19. E. W. Brown, Tables of the Motion of the Moon, Yale University Press, New Haven; Oxford Univ. Press, London (1919).
20. W. J. Eckert, R. Jones, and H. K. Clark, Improved Lunar Ephemerides 1952-1959, U.S. Government Printing Office, Washington (1954).
21. M. A. Fursenko, Byull. Inst. Teor. Astron., 10:4 (1965).
22. V. K. Abalakin and M. A. Fursenko, Byull. Inst. Teor. Astron., 11:8 (1968).
23. D. V. Zagrebin, Introduction to Astrometry [in Russian], Nauka, Moscow—Leningrad (1966).
24. K. A. Kulikov, The Variability of Latitudes and Longitudes [in Russian], Fizmatgiz, Moscow (1962).
25. Yu. A. Chikanov, Tr. Kazan. Gos. Astron. Obs., 35:116 (1968).
26. F. Hayn, Die Rotation des Mondes, Encycl. Math. Wiss., 6:1020 (1923).
27. K. Kosiel, Acta Astron., a4:153 (1949).
28. Astronomical Yearbook of the USSR for 1973 [in Russian], Nauka, Moscow (1970).
29. V. K. Abalakin, E. P. Aksenov, E. A. Grebenikov, and Yu. A. Ryabov, Handbook on Celestial Mechanics and Astrodynamics [in Russian], Nauka, Moscow (1971).
30. S. Newcomb, Astron. Pap., 6:1 (1895).
31. E. W. Woolard, Astron. Pap., 15:1 (1953).
32. P. Escobal, Methods of Orbit Determination, Wiley, New York (1965).
33. D. Brouwer and G. Clemence, Methods of Celestial Mechanics, Academic Press, New York (1971).

34. Yu. L. Kokurin, V. F. Lobanov, A. N. Sukhanovsky, V. K. Abalakin, M. A. Fursenko, and A. A. Gurstein, Phys. Earth Planet. Inter. 7:491 (1973).

35. V. K. Abalkin, V. N. Boiko, Yu. L. Kokurin, V. F. Lobanov, and M. A. Fursenko, Astron. Zh., 52:387 (1975).

36. T. C. Van Flandern, Cel. Mech., 1:163 (1969).

37. R. H. Dicke, The Earth–Moon System, Plenum Press, New York (1966), p. 98.

38. Z. Kopal, ed., The Physics and Astronomy of the Moon [Russian translation], Mir, Moscow (1973).

39. A. F. Bogomolov, The Elements of Radar [in Russian], Sovetskoe Radio, Moscow (1954).

40. A. I. Tudorovskii, Tr. Gos. Opt. Inst., 14:137 (1941).

41. P. R. Yoder, J. Opt. Soc. Am., 48:496 (1958).

42. K. N. Chandler, J. Opt. Soc. Am., 50:203 (1960).

43. R. S. Jullian, Measure of the Moon, D. Reidel Publishing Company, Dordrecht, Holland (1967), p. 181.

44. R. K. Long, Proc. IEEE, 51:859 (1963).

45. B. I. Stepanov, ed., Computational Techniques for Lasers [in Russian], Nauka i Tekhnika, Minsk (1968).

46. Electro Optics, No. 7, p. 2 (1971), Information.

47. A. L. Mikaélyan, V. G. Savel'ev, and Yu. G. Turkov, Pis'ma Zh. Éksp. Teor. Fiz., 6:6 (1967).

48. V. Ya. Anaton'yants, N. S. Ivanova, A. L. Mikaélyan, V. P. Minaev, and Yu. G. Turkov, Kvant. Élektron., No. 5(11) p. 106 (1972).

49. V. E. Zuev, The Laser Meteorologist [in Russian], Gidrometeoizdat, Leningrad (1974).

50. F. P. Burns, IEEE Spectrum, 3:115 (1967).

51. E. F. Ishchenko and Yu. M. Klimkov, Lasers [in Russian], Sovetskoe Radio, Moscow (1968).

52. R. W. Hellwarth, Advances in Quantum Electronics, Columbia University Press, New York (1961), p. 334.

53. N. G. Basov, V. S. Zuev, and P. G. Kryukov, Zh. Éksp. Teor. Fiz., 43:353 (1962).

54. F. J. McClung and R. W. Hellwarth, J. Appl. Phys. 33:828 (1962).

55. R. J. Coolins and P. Kisliuk, J. Appl. Phys. 33:2009 (1962).

56. N. G. Basov and V. S. Letokhov, Dokl. Akad. Nauk SSSR, 167:73 (1966).

57. N. G. Basov, V. S. Zuev, P. G. Kryukov, V. S. Letokhov, Yu. V. Senatskii, and S. V. Chekalin, Zh. Éksp. Teor. Fiz., 54:767 (1968).

58. A. A. Vuylsteke, J. Appl. Phys., 34:1615 (1963).

59. W. R. Hook, R. H. Dishington, and R. P. Hilberg, Appl. Phys. Lett., 9:125 (1966).

60. W. R. Hook, R. P. Holberg, and R. H. Dishington, Proc. IEEE, 54:1954 (1966).

61. J. Ernest, M. Michon, and J. Debrie, Phys. Lett., 22:147 (1966).

62. N. G. Basov, P. G. Kriukov, V. S. Letokhov, and Yu. V. Senatskii, IEEE J. Quant. Electron., 4:606 (1968).

63. G. L. Kashen, L. Steinmetz, and J. Kysilka, Appl. Phys. Lett., 13:229 (1968).

64. A. J. DeMaria, R. Gagosz, H. A. Heynan, A. W. Penney, and G. Wisner, J. Appl. Phys., 38:2693 (1967).

65. N. G. Basov, P. G. Kriukov, and V. S. Letokhov, Laser Magazine, 4:39 (1969).

66. N. O. Chechik, S. M. Fainshtein, and T. M. Lifshits, Electron Multipliers [in Russian], Gostekhizdat, Moscow (1957).

67. A. N. Pertsev and A. N. Pisarevskii, Single-Electron Characteristics of Photomultipliers and Their Applications [in Russian], Atomizdat, Moscow (1971).

68. G. A. Norton, Appl. Opt., 7:1 (1968).

69. S. K. Poultney, Adv. Electron. Electron. Phys., 31:39 (1972).

70. A. H. Sommer, Rev. Sci. Instrum., 26:725 (1955).

71.   J. J. Scheer and J. van Laar, Solid State Commun., 3:189 (1965).

72.   B. F. Williams and J. J. Tietjen, Proc. IEEE, 59:1489 (1971).

73.   W. D. Gunter, G. R. Grant, and S. A. Shaw, Appl. Opt., 9:251 (1970).

74.   W. D. Gunter, E. F. Ericson, and G. R. Grant, Appl. Opt., 4:512 (1965).

75.   V. P. Vasil'ev, G. É. Kufal', L. F. Pliev, and M. F. Fok, Zh. Prikl. Spektrosk., 14:2 (1971).

76.   Sh. A. Furman, Opt. Mekh. Promst., 9:50 (1968).

77.   Sh. A. Furman, Opt. Spektrosk., 33:553 (1972).

78.   A. Z. Grasyuk, V. S. Zuev, Yu. L. Kokurin, P. G. Kryukov, V. V. Kurbasov, V. F. Lobanov, V. M. Mozhzherin, A. N. Sukhanovskii, N. S. Chernykh, and K. K. Chuvaev, Dokl. Akad. Nauk SSSR, 154:1303 (1964).

79.   Yu. L. Kokurin, V. V. Kurbasov, V. F. Lobanov, V. M. Mozhzherin, A. N. Sukhanovskii, and N. S. Chernykh, Pis'ma Zh. Éksp. Teor. Fiz., 3:219 (1966).

80.   Yu. L. Kokurin, V. V. Kurbasov, V. F. Lobanov, A. N. Sukhanovskii, and N. S. Chernykh, Kosm. Issled., 9:912 (1971).

81.   A. Konig, Astronomical Techniques [Russian translation], Mir, Moscow (1967).

82.   V. V. Kurbasov, Candidate's Dissertation, Phys. Inst., Acad. Sci. USSR, Moscow (1974).

83.   V. N. Zaitsev, V. V. Kurbasov, A. V. Kutsenko, N. M. Lypkan', and B. A. Polos'yants, FIAN Preprint No. 47 (1974).

84.   V. N. Tereshchenko and A. V. Kharitonov, Zone Spectrometric Standards. Studies of the Energy Distribution in the Spectra of 109 Stars in Absolute Units [in Russian], Nauka, Alma-Ata (1972).

85.   Rapport d'étude théorique, Phase 1-Lot, TL-2, Sud Aviation, Paris (1969).

86.   A. Orszag, Thèse de Doctorat en Sciences, Paris (1972).

87.   S. K. Poultney, The Detector Package, Technical Report 957, University of Maryland (1969).

88.   D. Hudson, Statistics for Physicists [Russian translation], Mir, Moscow (1967).

89.   B. R. Bean and G. D. Theyer, J. Res. Nat. Bur. Stand. (Radio Propagation), 67D:273 (1963).

90.   H. S. Hopfield, Radio Sci., 6:357 (1969).

91.   H. S. Hopfield, J. Geophys. Res., 74:4487 (1969).

92.   P. I. Bakulin and N. S. Blinov, Exact Time Services [in Russian], Nauka, Moscow (1968).

93.   C. O. Alley, P. L. Bender, R. F. Chang, D. G. Currie, et al., in: Apollo-11 Preliminary Science Report, NASA SP-214, Washington, D.C. (1969), p. 163.

94.   R. L. Duncombe, Astron. J., 61:226 (1956).

95.   Sur le système de constantes astronomiques, Bull. Astron., p. 25, Paris (1965).